50% OFF Online GED Pı

Dear Customer,

We consider it an honor and a privilege that you chose our GED Study Guide. As a way of showing our appreciation and to help us better serve you, we have partnered with Mometrix Test Preparation to offer **50% off their online GED Prep Course.** Many GED courses are needlessly expensive and don't deliver enough value. With their course, you get access to the best GED prep material, and you only pay half price.

Mometrix has structured their online course to perfectly complement your printed study guide. The GED Prep Course contains **over 140 lessons** that cover all the most important topics, **180+ video reviews** that explain difficult concepts, **over 700 practice questions** to ensure you feel prepared, and **digital flashcards,** so you can fit some studying in while you're on the go.

Online GED Prep Course

Topics Covered:
- Reasoning Through Language Arts
  - *Reading Comprehension*
  - *Critical Thinking*
  - *Writing*
- Mathematical Reasoning
  - *Number Operations*
  - *Algebra, Functions, and Patterns*
  - *Measurement and Geometry*
- Science
  - *Physical Science*
  - *Earth and Space Science*
- Social Studies
  - *History and Government*
  - *Economics*

Course Features:
- GED Study Guide
  - Get content that complements our best-selling study guide.
- 3 Full-Length Practice Tests
  - With over 700 practice questions, you can test yourself again and again.
- Mobile Friendly
  - If you need to study on-the-go, the course is easily accessible from your mobile device.
- GED Flashcards
  - The course includes a flashcard mode consisting of over 400 content cards to help you study.

To receive this discount, simply head to their website: mometrix.com/university/ged and add the course to your cart. At the checkout page, enter the discount code: **APEXGED50**

If you have any questions or concerns, please don't hesitate to contact Mometrix at universityhelp@mometrix.com.

Sincerely,

 in partnership with

Free Study Tips DVD

In addition to the tips and content in this guide, we have created a FREE DVD with helpful study tips to further assist your exam preparation. **This FREE Study Tips DVD provides you with top-notch tips to conquer your exam and reach your goals.**

Our simple request in exchange for the strategy-packed DVD is that you email us your feedback about our study guide. We would love to hear what you thought about the guide, and we welcome any and all feedback—positive, negative, or neutral. It is our #1 goal to provide you with top quality products and customer service.

To receive your **FREE Study Tips DVD**, email freedvd@apexprep.com. Please put "FREE DVD" in the subject line and put the following in the email:

> a. The name of the study guide you purchased.
>
> b. Your rating of the study guide on a scale of 1-5, with 5 being the highest score.
>
> c. Any thoughts or feedback about your study guide.
>
> d. Your first and last name and your mailing address, so we know where to send your free DVD!

Thank you!

The GED Tutor

Study Guide 2021 & 2022 All Subjects

GED Test Prep with 3 Full-Length Practice Exams [4th Edition Review]

Matthew Lanni

Written and edited by APEX Publishing.

ISBN 13: 9781637755068
ISBN 10: 1637755066

APEX Publishing is not connected with or endorsed by any official testing organization. APEX Publishing creates and publishes unofficial educational products. All test and organization names are trademarks of their respective owners.

The material in this publication is included for utilitarian purposes only and does not constitute an endorsement by APEX Publishing of any particular point of view.

For additional information or for bulk orders, contact info@apexprep.com.

Table of Contents

Test Taking Strategies

1. Reading the Whole Question

A popular assumption in Western culture is the idea that we don't have enough time for anything. We speed while driving to work, we want to read an assignment for class as quickly as possible, or we want the line in the supermarket to dwindle faster. However, speeding through such events robs us from being able to thoroughly appreciate and understand what's happening around us. While taking a timed test, the feeling one might have while reading a question is to find the correct answer as quickly as possible. Although pace is important, don't let it deter you from reading the whole question. Test writers know how to subtly change a test question toward the end in various ways, such as adding a negative or changing focus. If the question has a passage, carefully read the whole passage as well before moving on to the questions. This will help you process the information in the passage rather than worrying about the questions you've just read and where to find them. A thorough understanding of the passage or question is an important way for test takers to be able to succeed on an exam.

2. Examining Every Answer Choice

Let's say we're at the market buying apples. The first apple we see on top of the heap may *look* like the best apple, but if we turn it over, we can see bruising on the skin. We must examine several apples before deciding which apple is the best. Finding the correct answer choice is like finding the best apple. Although it's tempting to choose an answer that seems correct at first without reading the others, it's important to read each answer choice thoroughly before making a final decision on the answer. The aim of a test writer might be to get as close as possible to the correct answer, so watch out for subtle words that may indicate an answer is incorrect. Once the correct answer choice is selected, read the question again and the answer in response to make sure all your bases are covered.

3. Eliminating Wrong Answer Choices

Sometimes we become paralyzed when we are confronted with too many choices. Which frozen yogurt flavor is the tastiest? Which pair of shoes look the best with this outfit? What type of car will fill my needs as a consumer? If you are unsure of which answer would be the best to choose, it may help to use process of elimination. We use "filtering" all the time on sites such as eBay® or Craigslist® to eliminate the ads that are not right for us. We can do the same thing on an exam. Process of elimination is crossing out the answer choices we know for sure are wrong and leaving the ones that might be correct. It may help to cover up the incorrect answer choice. Covering incorrect choices is a psychological act that alleviates stress due to the brain being exposed to a smaller amount of information. Choosing between two answer choices is much easier than choosing between all of them, and you have a better chance of selecting the correct answer if you have less to focus on.

4. Sticking to the World of the Question

When we are attempting to answer questions, our minds will often wander away from the question and what it is asking. We begin to see answer choices that are true in the real world instead of true in the world of the question. It may be helpful to think of each test question as its own little world. This world may be different from ours. This world may know as a truth that the chicken came before the egg or may assert that two plus two equals five. Remember that, no matter what hypothetical nonsense may be in the question, assume it to be true. If the question states that the chicken came before the egg, then choose your answer based on that truth. Sticking to the world of the question means placing all of our biases and assumptions aside and relying on the question to guide us to the correct answer. If we are simply looking for answers that are correct based on our own judgment, then we may choose incorrectly. Remember an answer that is true does not necessarily answer the question.

5. Key Words

If you come across a complex test question that you have to read over and over again, try pulling out some key words from the question in order to understand what exactly it is asking. Key words may be words that surround the question, such as *main idea, analogous, parallel, resembles, structured,* or *defines*. The question may be asking for the main idea, or it may be asking you to define something. Deconstructing the sentence may also be helpful in making the question simpler before trying to answer it. This means taking the sentence apart and obtaining meaning in pieces, or separating the question from the foundation of the question. For example, let's look at this question:

> Given the author's description of the content of paleontology in the first paragraph, which of the following is most parallel to what it taught?

The question asks which one of the answers most *parallels* the following information: The *description* of paleontology in the first paragraph. The first step would be to see *how* paleontology is described in the first paragraph. Then, we would find an answer choice that parallels that description. The question seems complex at first, but after we deconstruct it, the answer becomes much more attainable.

6. Subtle Negatives

Negative words in question stems will be words such as *not, but, neither,* or *except*. Test writers often use these words in order to trick unsuspecting test takers into selecting the wrong answer—or, at least, to test their reading comprehension of the question. Many exams will feature the negative words in all caps (*which of the following is NOT an example*), but some questions will add the negative word seamlessly into the sentence. The following is an example of a subtle negative used in a question stem:

> According to the passage, which of the following is *not* considered to be an example of paleontology?

If we rush through the exam, we might skip that tiny word, *not*, inside the question, and choose an answer that is opposite of the correct choice. Again, it's important to read the question fully, and double check for any words that may negate the statement in any way.

7. Spotting the Hedges

The word "hedging" refers to language that remains vague or avoids absolute terminology. Absolute terminology consists of words like *always, never, all, every, just, only, none,* and *must*. Hedging refers to words like *seem, tend, might, most, some, sometimes, perhaps, possibly, probability,* and *often*. In some cases, we want to choose answer choices that use hedging and avoid answer choices that use absolute terminology. It's important to pay attention to what subject you are on and adjust your response accordingly.

8. Restating to Understand

Every now and then we come across questions that we don't understand. The language may be too complex, or the question is structured in a way that is meant to confuse the test taker. When you come across a question like this, it may be worth your time to rewrite or restate the question in your own words in order to understand it better. For example, let's look at the following complicated question:

> Which of the following words, if substituted for the word *parochial* in the first paragraph, would LEAST change the meaning of the sentence?

Let's restate the question in order to understand it better. We know that they want the word *parochial* replaced. We also know that this new word would "least" or "not" change the meaning of the sentence. Now let's try the sentence again:

> Which word could we replace with *parochial,* and it would not change the meaning?

Restating it this way, we see that the question is asking for a synonym. Now, let's restate the question so we can answer it better:

> Which word is a synonym for the word *parochial?*

Before we even look at the answer choices, we have a simpler, restated version of a complicated question.

9. Predicting the Answer

After you read the question, try predicting the answer *before* reading the answer choices. By formulating an answer in your mind, you will be less likely to be distracted by any wrong answer choices. Using predictions will also help you feel more confident in the answer choice you select. Once you've chosen your answer, go back and reread the question and answer choices to make sure you have the best fit. If you have no idea what the answer may be for a particular question, forego using this strategy.

10. Avoiding Patterns

One popular myth in grade school relating to standardized testing is that test writers will often put multiple-choice answers in patterns. A runoff example of this kind of thinking is that the most common answer choice is "C," with "B" following close behind. Or, some will advocate certain made-up word patterns that simply do not exist. Test writers do not arrange their correct answer choices in any kind of pattern; their choices are randomized. There may even be times where the correct answer choice will be the same letter for two or three questions in a row, but we have no way of knowing when or if this might happen. Instead of trying to figure out what choice the test writer probably set as being correct, focus on what the *best answer choice* would be out of the answers you are presented with. Use the tips above, general knowledge, and reading comprehension skills in order to best answer the question, rather than looking for patterns that do not exist.

FREE DVD OFFER

Achieving a high score on your exam depends not only on understanding the content, but also on understanding how to apply your knowledge and your command of test taking strategies. **Because your success is our primary goal, we offer a FREE Study Tips DVD, which provides top-notch test taking strategies to help you optimize your testing experience.**

Our simple request in exchange for the strategy-packed DVD is that you email us your feedback about our study guide.

To receive your **FREE Study Tips DVD**, email freedvd@apexprep.com. Please put "FREE DVD" in the subject line and put the following in the email:

   a. The name of the study guide you purchased.

   b. Your rating of the study guide on a scale of 1-5, with 5 being the highest score.

   c. Any thoughts or feedback about your study guide.

   d. Your first and last name and your mailing address, so we know where to send your free DVD!

Introduction to the GED

Function of the Test

The General Education Development (GED) test is designed to offer individuals without a high school diploma the opportunity to earn a high school equivalency credential. Developed and administered by the GED Testing Service, a joint venture of the American Council on Education and Pearson VUE, the GED exam evaluates the test taker's knowledge of core high school subjects in alignment with current high school graduation standards in the United States.

The GED, which is now only offered via computer administration, is the only nationally-recognized high school equivalency credential. According to MyGED, the GED is accepted as the equivalent of a high school diploma at approximately 98% of colleges and universities in the United States. The majority of test takers are those who did not graduate from high school, but are interested in furthering their education or careers. Since the GED was revamped in 2014, the passing rate has dropped to around 63% from 76% in 2013.

Test Administration

The GED may be taken at over 3,000 sites around the United States and Canada; other international options are available as well. Community colleges, adult education centers, and local school boards serve as many of the testing sites. The GED Testing Service offers a comprehensive search of test centers by geographic location. Testing is typically offered year round, though the individual site may have a designated testing schedule.

Although the administration rules and policies may vary by region, all GED tests are taken in person on a computer. Test takers may opt to take one or multiple subjects on a given date of administration. There are no official requirements as to when or how quickly all four subjects must be passed. Generally, test takers are able to take any subject subtest three times without any restrictions on retesting. However, after three failed attempts, test takers are required to wait at least 60 days to retake the subtest. Testing accommodations are available for those with documented disabilities, in accordance with the Americans with Disabilities Act.

Test Format

The GED consists of four sections, or modules: Mathematical Reasoning, Reasoning Through Language Arts, Science, and Social Studies. More than just assessing rote memorization, each module has questions that assess the test taker's ability to apply reasoning skills, comprehension skills, and multidisciplinary thinking. If multiple modules are attempted in one day, a 10-minute scheduled break is offered between sections. Unscheduled breaks (leaving the testing room during a module) are not permitted.

Test modules vary in length:

Subject	Time	Topics
Mathematical Reasoning	115 minutes	Basic math, geometry, basic algebra, graphs and functions
Reasoning Through Language Arts	150 minutes (includes 45 minutes for written essay)	Reading for meaning, identifying and creating arguments, grammar and language
Science	90 minutes	Reading for meaning in science, designing and interpreting science experiments, using numbers and graphics in science
Social Studies	70 minutes	Reading for meaning in social studies, analyzing historical events and arguments in social studies, using numbers and graphics in social studies

Test questions are in a variety of formats including multiple choice, fill-in-the-blank, drop-down, select an area, short response, written essay, and drag and drop.

Scoring

Scores become available on MyGED within 24 hours after the test taker completes the test. Each of the four modules are scored on a scale of 100–200. A test taker must achieve at least 145 points on each of the four modules in order to earn high school equivalency. Scores cannot be combined between sections, meaning that a high score on one subject cannot be used to provide additional points to make up for an insufficient score in another module. Test scores are divided into four ranges:

1. Scores lower than 145 points earn a "Not Passing" designation. Test takers must retake the module to earn high school equivalency.

2. Score of 145 points or higher earn the "GED Passing Score/High School Equivalency" designation.

3. Test takers who achieve scores of 165-175 points are considered "GED College Ready." This alerts colleges and universities that the test taker may be exempt from taking placement tests or engaging in remedial coursework; policies vary by institution.

4. Test takers with scores greater than 175 points achieve "GED College Ready + Credit" status. Again, while policies vary between institutions, test takers may be afforded college credit for certain courses.

Study Prep Plan for the GED

 **Breathe**

Reducing stress is key when preparing for your test.

 Build

Create a study plan to help you stay on track.

 **Begin**

Stick with your study plan. You've got this!

1 Week Study Plan

Day 1	Day 2	Day 3	Day 4	Day 5	Day 6	Day 7
Mathematical Reasoning	Reasoning Through Language Arts	Science	Practice Test #1	Practice Test #2	Practice Test #3	Take your exam!

2 Week Study Plan

Day 1	Day 2	Day 3	Day 4	Day 5	Day 6	Day 7
Basic Math	Geometry	Mean, Median, Mode, and Range	Basic Algebra	Graphs and Functions	Reading for Meaning	Identifying and Creating Arguments

Day 8	Day 9	Day 10	Day 11	Day 12	Day 13	Day 14
Reading for Meaning in Science	Reading for Meaning in Social Studies	Analyzing Historical Events and Arguments	Practice Test #1	Practice Test #2	Practice Test #3	Take your exam!

30 Day Study Plan

Day 1	Day 2	Day 3	Day 4	Day 5	Day 6	Day 7
Fractions and Decimals in Order	Operations with Decimals	Multiple-Step Problems that Use Ratios, Proportions, and Percents	Area, Circumference, Radius, and Diameter of a Circle	Graphical Data Including Graphs, Tables, and More	Probability of an Event	Adding, Subtracting, Multiplying, and Factoring Linear Expressions

Day 8	Day 9	Day 10	Day 11	Day 12	Day 13	Day 14
Factoring	Solving a System of Two Linear Equations	Features of Graphs and Tables for Linear and Nonlinear Relationships	Events, Plots, Characters, Settings, and Ideas	Tone and Figurative Language	The Relationship Between Evidence and Main Ideas and Details	Determining Whether Evidence is Relevant and Sufficient

Day 15	Day 16	Day 17	Day 18	Day 19	Day 20	Day 21
Word Usage	Editing Sentences for Parallel Structure and Correct Use of Conjunctions	Using Apostrophes with Possessive Nouns Correctly	Understanding Symbols, Terms, and Phrases in Science	Using Evidence to Draw Conclusions or Make Predictions	Using Numbers or Symbols to Display Science Information	Social Studies Vocabulary

Day 22	Day 23	Day 24	Day 25	Day 26	Day 27	Day 28
Connections Between Different Social Studies Elements	Using Data Presented in Visual Form, Including Maps, Charts, Graphs, and Tables	Practice Test #1	Review Answer Explanations	Practice Test #2	Review Answer Explanations	Practice Test #3

Day 29	Day 30
Review Answer Explanations	Take your exam!

Mathematical Reasoning

Basic Math

Fractions and Decimals in Order

Ordering rational numbers is a way to compare two or more different numerical values. Determining whether two amounts are equal, less than, or greater than is the basis for comparing both positive and negative numbers. Also, a group of numbers can be compared by ordering them from the smallest amount to the largest amount. A few symbols are necessary to use when ordering rational numbers. The equals sign, $=$, shows that the two quantities on either side of the symbol have the same value.

For example, $\frac{12}{3} = 4$ because both values are equivalent. Another symbol that is used to compare numbers is $<$, which represents "less than." With this symbol, the smaller number is placed on the left and the larger number is placed on the right. Always remember that the symbol's "mouth" opens up to the larger number. When comparing negative and positive numbers, it is important to remember that the number occurring to the left on the number line is always smaller and is placed to the left of the symbol. This idea might seem confusing because some values could appear at first glance to be larger, even though they are not. For example, $-5 < 4$ is read "negative 5 is less than 4." Here is an image of a number line for help:

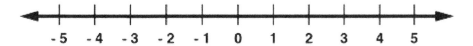

The symbol $\leq$ represents "less than or equal to," and it joins $<$ with equality. Therefore, both $-5 \leq 4$ and $-5 \leq -5$ are true statements and "-5 is less than or equal to both 4 and -5." Other symbols are $>$ and $\geq$, which represent "greater than" and "greater than or equal to." Both $4 \geq -1$ and $-1 \geq -1$ are correct ways to use these symbols.

Here is a chart of these four inequality symbols:

Symbol	Definition
$<$	less than
$\leq$	less than or equal to
$>$	greater than
$\geq$	greater than or equal to

Comparing integers is a straightforward process, especially when using the number line, but the comparison of decimals and fractions is not as obvious. When comparing two non-negative decimals, compare digit by digit, starting from the left. The larger value contains the first larger digit. For example, 0.1456 is larger than 0.1234 because the value 4 in the hundredths place in the first decimal is larger than the value 2 in the hundredths place in the second decimal. When comparing a fraction with a decimal, convert the fraction to a decimal and then compare in the same manner. Finally, there are a few options when comparing fractions. If two non-negative fractions have the same denominator, the fraction with the larger numerator is the larger value. If they have different denominators, they can be converted to equivalent fractions with a common denominator to be compared, or they can be converted to decimals to be compared. When comparing two negative decimals or fractions, a different approach must be used.

It is important to remember that the smaller number exists to the left on the number line. Therefore, when comparing two negative decimals by place value, the number with the larger first place value is smaller due to the negative sign. Whichever value is closer to 0 is larger. For instance, -0.456 is larger than -0.498 because of the values in the hundredth places. If two negative fractions have the same denominator, the fraction with the larger numerator is smaller because of the negative sign.

Multiples and Factors

Factors and multiples are different types of quantities, but they both involve the operation of multiplication. Every number has at least one pair of **factors**, which are defined to be quantities that are multiplied together to obtain the given number. For instance, the positive integer 6 has two sets of factors: 3 and 2 and 6 and 1 because both $3 \times 2 = 6$ and $6 \times 1 = 6$. Every number is its own factor because it can always be multiplied by 1 to obtain itself. An integer that has a single pair of factors, itself and 1, is known as a prime number. For example, 7 is a prime number because its only factors are 7 and 1.

Multiples are results of the multiplication of any number times an integer. For instance, when we multiply 3 times 2, we obtain 6. Therefore, 6 is a multiple of 3. Other multiples of 3 are 9, 12, 15, 18, 21, etc. We can also obtain negative multiples by multiplying by negative integers. For instance, 3 times -1 is -3. Therefore, -3 is another multiple of 3. Other negative multiples of 3 are -6, -9, -12, -15, etc.

Factorization is the process of breaking up a mathematical quantity, such as a number or polynomial, into a product of two or more factors. For example, a factorization of the number 16 is $16 = 8 \times 2$. If multiplied out, the factorization results in the original number. A **prime factorization** is a specific factorization when the number is factored completely using prime numbers only. For example, the prime factorization of 16 is:

$$16 = 2 \times 2 \times 2 \times 2$$

A factor tree can be used to find the prime factorization of any number. Within a factor tree, pairs of factors are found until no other factors can be used, as in the following factor tree of the number 72:

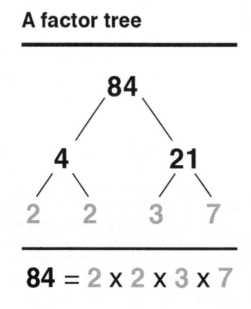

A factor tree

$84 = 2 \times 2 \times 3 \times 7$

It first breaks 84 into 21×4, which is not a prime factorization. Then, both 4 and 21 are factored into their primes. The final numbers on each branch consist of the numbers within the prime factorization. Therefore:

$$84 = 2 \times 2 \times 3 \times 7$$

Factorization can be helpful in finding greatest common divisors and least common denominators.

Simplifying Exponents

Another arithmetic concept is the use of repeated multiplication, which can be written in a more compact notation using exponents. The guide previously mentioned the following examples:

$$(-1)(-1)(-1)(-1) = 1 \text{ and } (-1)(-1)(-1)(-1)(-1) = -1$$

The first example can be written as $(-1)^4 = 1$, and the second example can be written as:

$$(-1)^5 = -1$$

Both are exponential expressions; -1 is the base in both instances, and 4 and 5 are the respective exponents. Note that a negative number raised to an odd power is always negative, and a negative number raised to an even power is always positive. Also, $(-1)^4$ is not the same as -1^4. In the first expression, the negative is included in the parentheses, but it is not in the second expression. The second expression is found by evaluating 1^4 first to get 1 and then by applying the negative sign to obtain -1.

Numbers can also be written using exponents. The number 7,000 can be written as $7 \times 1,000$ because 7 is in the thousands place. It can also be written as 7×10^3 because $1,000 = 10^3$. Another number that can use this notation is 500. It can be written as 5×100, or 5×10^2, because $100 = 10^2$. The number 30 can be written as 3×10, or 3×10^1, because $10 = 10^1$. Notice that each one of the exponents of 10 is equal to the number of zeros in the number. Seven is in the thousands place, with three zeros, and the exponent on ten is 3. The five is in the hundreds place, with two zeros, and the exponent on the ten is 2. A question may give the number 40,000 and ask for it to be rewritten using exponents with a base of ten. Because the number has a four in the ten-thousands place and four zeros, it can be written using an exponent of four: 4×10^4.

When performing operations with exponents, one must make sure that the order of operations is followed. Therefore, once all operations within any parentheses or grouping symbols are performed, all exponents must be evaluated before any other operation is completed. For example, to evaluate $(3 - 5)^2 + 4 - 5^3$, the subtraction in parentheses is done first. Then the exponents are evaluated to obtain $(-2)^2 + 4 - 5^3$, or $4 + 4 - 125$, which is equivalent to -117. A common mistake involves negative signs combined with exponents. Note that -4^2 is not equal to $(-4)^2$. The negative sign is in front of the first exponential expression, so the result is:

$$-(4 \times 4) = -16$$

However, the negative sign is inside the parentheses in the second expression, so the result is:

$$(-4 \times -4) = 16.$$

Laws of exponents can also help when performing operations with exponents. If two exponential expressions have the same base and are being multiplied, just add the exponents.

For example:

$$2^5 \times 2^7 = 2^{5+7} = 2^{12}$$

If two exponential expressions have the same base and are being divided, subtract the exponents. For example:

$$\frac{4^{30}}{4^2} = 4^{30-2} = 4^{28}$$

If an exponential expression is being raised to another exponent, multiply the exponents together. For example:

$$(3^2)^5 = 3^{2 \times 5} = 3^{10}$$

Rational exponents are used to express the root of a number raised to a specific power. For example, $3^{\frac{1}{2}}$ has a base of 3 and rational exponent of $\frac{1}{2}$. The square root of 3 raised to the first power can be written as $\sqrt[2]{3^1}$. Any number with a rational exponent can be written this way. The **numerator**, or number on top of the fraction, becomes the whole number exponent and the **denominator**, or bottom number of the fraction, becomes the root. Another example is $4^{\frac{3}{2}}$. It can be rewritten as the square root of four to the third power, or $\sqrt[2]{4^3}$. This can be simplified by performing the operations 4 to the third power, $4^3 = 4 \times 4 \times 4 = 64$, and then taking the square root of 64, $\sqrt[2]{64}$, which yields an answer of 8. Another way of stating the answer would be 4 to power of $\frac{3}{2}$ is eight, or that 4 to the power of $\frac{3}{2}$ is the square root of 4 cubed:

$$\sqrt[2]{4}^3 = 2^3 = 2 \times 2 \times 2 = 8$$

Distance Between Numbers on a Number Line

A number line is a great way to visualize the integers. Integers are labeled on the following number line:

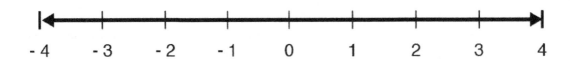

The arrows on the right- and left-hand sides of the number line show that the line continues indefinitely in both directions.

The number line contains all real numbers. To graph a number other than an integer on a number line, it needs to be plotted between two integers. For example, 3.5 would be plotted halfway between 3 and 4.

The **absolute value** of any real number is the distance from that number to 0 on the number line. The absolute value of a number can never be negative. For example, the absolute value of both 8 and −8 is 8 because they are both 8 units away from 0 on the number line. This is written as:

$$|8| = |-8| = 8$$

12

Whole Numbers, Fractions, and Decimal Problems

Operations with Whole Numbers

The four basic operations include addition, subtraction, multiplication, and division. The result of addition is a **sum**, the result of subtraction is a **difference**, the result of multiplication is a **product,** and the result of division is a **quotient**. Each type of operation can be used when working with rational numbers; however, the basic operations need to be understood first while using simpler numbers before working with fractions and decimals.

These operations should first be learned using whole numbers. Addition needs to be done column by column. To add two whole numbers, add the ones column first, then the tens columns, then the hundreds, etc. If the sum of any column is greater than 9, a one must be carried over to the next column. For example, the following is the result of 482+924:

$$
\begin{array}{r}
1 \\
482 \\
+924 \\
\hline
1406
\end{array}
$$

Notice that the sum of the tens column was 10, so a one was carried over to the hundreds column. Subtraction is also performed column by column. Subtraction is performed in the ones column first, then the tens, etc. If the number on top is less than the number below, a one must be borrowed from the column to the left.

For example, the following is the result of 5,424 – 756:

$$
\begin{array}{r}
4\ 13\ 11\ 14 \\
\cancel{5}\ \cancel{4}\ \cancel{2}\ \cancel{4} \\
-\quad 7\ \ 5\ \ 6 \\
\hline
4\ \ 6\ \ 6\ \ 8
\end{array}
$$

Notice that a one is borrowed from the tens, hundreds, and thousands place. After subtraction, the answer can be checked through addition. A check of this problem would be to show that 756+4,668 = 5,424.

In multiplication, the number on top is known as the multiplicand, and the number below is the multiplier. Complete the problem by multiplying the multiplicand by each digit of the multiplier. Make sure to place the ones value of each result under the multiplying digit in the multiplier. Each value to the right is then a

0. The product is found by adding each product. The following example shows the process of multiplying 46 times 37:

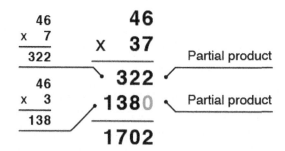

Finally, division can be performed using long division. When dividing, the first number is known as the **dividend**, and the second is the **divisor.** For example, with $a \div b = c$, *a* is the dividend, *b* is the divisor, and *c* is the quotient. For long division, place the dividend within the division bar and the divisor on the outside. For example, with $8,764 \div 4$, refer to the first problem in the diagram below. First, there are two 4's in the first digit, 8. This number 2 gets written above the 8. Then, multiply 4 times 2 to get 8, and that product goes below the 8. Subtract to get 0, and then carry down the second digit, 7. Continue the same steps. $7 \div 4 = 1$ R3, so 1 is written above the 7. Multiply 4 times 1 to get 4 and write it below the 7. Subtract to get 3 and carry the 6 down next to the 3. The resulting steps give a 9 and a 1. The final subtraction results in a 0, which means that 8,764 is evenly divisible by 4 with no remaining numbers.

The second example shows that:

$$4,536 \div 216 = 21$$

The steps are a little different because 216 cannot be contained in 4 or 5, so the first step is placing a 2 above the 3 because there are two 216's in 453.

Finally, the third example shows that:

$$546 \div 31 = 17 \, \text{R}19$$

The 19 is a remainder. Notice that the final subtraction does not result in a 0, which means that 546 is not divisible by 31. The remainder can also be written as a fraction over the divisor to say that $546 \div 31 = 17\frac{19}{31}$.

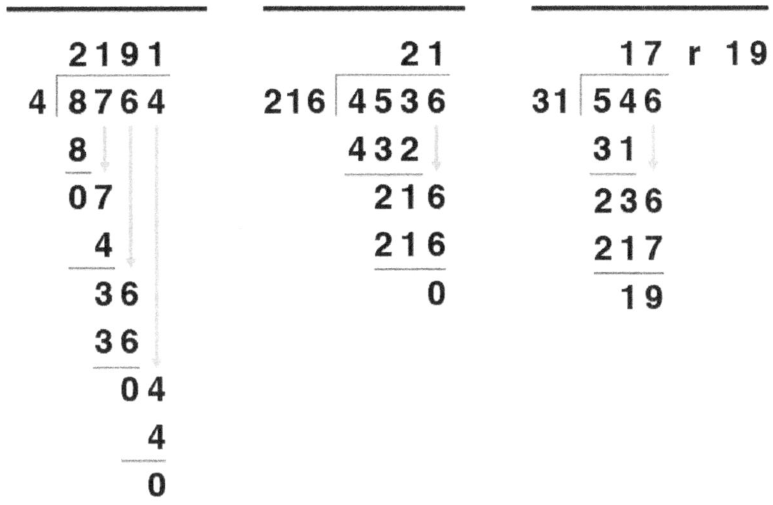

```
    2191            21                17 r 19
  4│8764        216│4536          31│546
    8              432               31
    07             216               236
    4              216               217
    36               0                19
    36
      04
       4
       0
```

A remainder can have meaning in a division problem with real-world application. For example, consider the third example:

$$546 \div 31 = 17 \, \text{R}19$$

Let's say that we had $546 to spend on calculators that cost $31 each, and we wanted to know how many we could buy. The division problem would answer this question. The result states that 17 calculators could be purchased, with $19 left over. Notice that the remainder will never be greater than or equal to the divisor.

Once the operations are understood with whole numbers, they can be used with negative numbers. There are many rules surrounding operations with negative numbers. First, consider addition with integers. The sum of two numbers can first be shown using a number line. For example, to add $-5 + (-6)$, plot the point -5 on the number line. Adding a negative number is the same as subtracting, so move 6 units to the left. This process results in landing on -11 on the number line, which is the sum of -5 and -6. If adding a positive number, move to the right. While visualizing this process using a number line is useful for understanding, it is more efficient to learn the rules of operations. When adding two numbers with the same sign, add the absolute values of both numbers, and use the common sign of both numbers as the sign of the sum. For example, to add $-5 + (-6)$, add their absolute values:

$$5 + 6 = 11$$

Then, introduce a negative number because both addends are negative. The result is -11. To add two integers with unlike signs, subtract the lesser absolute value from the greater absolute value, and apply the sign of the number with the greater absolute value to the result. For example, the sum $-7 + 4$ can be computed by finding the difference $7 - 4 = 3$ and then applying a negative because the value with the larger absolute value is negative. The result is -3. Similarly, the sum $-4 + 7$ can be found by computing the same difference but leaving it as a positive result because the addend with the larger absolute value is positive. Also, recall that any number plus 0 equals that number. This is known as the **Addition Property of 0.**

Subtracting two integers with opposite signs can be computed by changing to addition to avoid confusion. The rule is to add the first number to the opposite of the second number. The opposite of a number is the number with the same value on the other side of 0 on the number line. For example, -2 and 2 are opposites. Consider $4 - 8$. Change this to adding the opposite as follows: $4 + (-8)$. Then, follow the rules of addition of integers to obtain -4. Secondly, consider $-8 - (-2)$. Change this problem to adding the opposite as $-8 + 2$, which equals -6. Notice that subtracting a negative number functions the same as adding a positive number.

Multiplication and division of integers are actually less confusing than addition and subtraction because the rules are simpler to understand. If two factors in a multiplication problem have the same sign, the result is positive. If one factor is positive and one factor is negative, the result, known as the *product,* is negative. For example:

$$(-9)(-3) = 27 \ and \ 9(-3) = -27$$

Also, any number times 0 always results in 0. If a problem consists of several multipliers, the result is negative if it contains an odd number of negative factors, and the result is positive if it contains an even number of negative factors. For example:

$$(-1)(-1)(-1)(-1) = 1 \ and \ (-1)(-1)(-1)(-1)(-1) = -1$$

Similar rules apply within division. First, consider some vocabulary. When dividing 14 by 2, it can be written in the following ways: $14 \div 2 = 7$ or $\frac{14}{2} = 7$. 14 is the dividend, 2 is the divisor, and 7 is the quotient. If two numbers in a division problem have the same sign, the quotient is positive. If two numbers in a division problem have different signs, the quotient is negative. For example:

$$14 \div (-2) = -7 \ and \ -14 \div (-2) = 7$$

To check division, multiply the quotient times the divisor to obtain the dividend. Also, remember that 0 divided by any number is equal to 0. However, any number divided by 0 is undefined. It just does not make sense to divide a number by 0 parts.

If more than one operation is to be completed in a problem, follow the Order of Operations. The mnemonic device, PEMDAS, states the order in which addition, subtraction, multiplication, and division need to be done. It also includes when to evaluate operations within grouping symbols and when to incorporate exponents. PEMDAS, which some remember by thinking "please excuse my dear Aunt Sally," refers to parentheses, exponents, multiplication, division, addition, and subtraction.

First, complete any operation within parentheses or any other grouping symbol like brackets, braces, or absolute value symbols. Note that this does not refer to when parentheses are used to represent multiplication like $(2)(5)$. An operation is not within parentheses like it is in (2×5). Then, any exponents must be computed. Next, multiplication and division are performed from left to right. Finally, addition and subtraction are performed from left to right. The following is an example in which the operations within the parentheses need to be performed first, so the order of operations must be applied to the exponent, subtraction, addition, and multiplication within the grouping symbol:

$$9 - 3(3^2 - 3 + 4 \cdot 3)$$

$$9 - 3(3^2 - 3 + 4 \cdot 3) \quad \text{Work within the parentheses first}$$

$$= 9 - 3(9 - 3 + 12)$$

$$= 9 - 3(18)$$

$$= 9 - 54$$

$$= -45$$

Interpreting Remainders in Division Problems

Understanding remainders begins with understanding the division problem. The problem $24 \div 7$ can be read as "twenty-four divided by seven." The problem is asking how many groups of 7 will fit into 24. Counting by seven, the multiples are 7, 14, 21, 28. Twenty-one, which is three groups of 7, is the closest to 24. The difference between 21 and 24 is 3, which is called the remainder. This is a remainder because it is the number that is left out after the three groups of seven are taken from 24. The answer to this division problem can be written as 3 with a remainder 3, or $3\frac{3}{7}$.

The fraction $\frac{3}{7}$ can be used because it shows the part of the whole left when the division is complete. Another division problem may have the following numbers: $36 \div 5$. This problem is asking how many groups of 5 will fit evenly into 36. When counting by multiples of 5, the following list is generated: 5, 10, 15, 20, 25, 30, 35, 40. As seen in the list, there are seven groups of five that make 35. To get to the total of 36, there needs to be one additional number. The answer to the division problem would be $36 \div 5 = 7$ R1, or $7\frac{1}{5}$. The fractional part represents the number that cannot make up a whole group of five.

Operations with Decimals

Operations can be performed on rational numbers in decimal form. Recall that to write a fraction as an equivalent decimal expression, divide the numerator by the denominator. For example:

$$\frac{1}{8} = 1 \div 8 = 0.125$$

With the case of decimals, it is important to keep track of place value. To add decimals, make sure the decimal places are in alignment and add vertically. If the numbers do not line up because there are extra or missing place values in one of the numbers, then zeros may be used as placeholders. For example, $0.123 + 0.23$ becomes:

$$
\begin{array}{r}
0.123 \\
\underline{0.230} \\
0.353
\end{array}
$$

Subtraction is done the same way. Multiplication and division are more complicated. To multiply two decimals, place one on top of the other as in a regular multiplication process and do not worry about lining up the decimal points. Then, multiply as with whole numbers, ignoring the decimals. Finally, in the solution, insert the decimal point as many places to the left as there are total decimal values in the original problem. Here is an example of a decimal multiplication problem:

$$\begin{array}{r} 0.52 \\ \times \quad 0.2 \\ \hline 0.104 \end{array}$$ *2 decimal places*
1 decimal place
3 decimal places

The answer to 52 times 2 is 104, and because there are three decimal values in the problem, the decimal point is positioned three units to the left in the answer.

The decimal point plays an integral role throughout the whole problem when dividing with decimals. First, set up the problem in a long division format. If the divisor is not an integer, move the decimal to the right as many units as needed to make it an integer. The decimal in the dividend must be moved to the right the same number of places to maintain equality.

Then, complete division normally. Here is an example of long division with decimals:

Long division with decimals

$$\begin{array}{r} 212 \\ 6 \overline{\smash)1272} \\ \underline{12} \\ 07 \\ \underline{6} \\ 12 \end{array}$$

The decimal point in 0.06 needed to move two units to the right to turn it into an integer (6), so it also needed to move two units to the right in 12.72 to make it 1,272. The quotient is 212. To check a division problem, multiply the answer by the divisor to see if the result is equal to the dividend.

Sometimes it is helpful to round answers that are in decimal form. First, find the place to which the rounding needs to be done. Then, look at the digit to the right of it. If that digit is 4 or less, the number in the place value to its left stays the same, and everything to its right becomes a 0. This process is known as *rounding down*. If that digit is 5 or higher, the number in the place value to its left increases by 1, and every number to its right becomes a 0. This is called rounding up. Excess 0s at the end of a decimal can be dropped. For example, 0.145 rounded to the nearest hundredth place would be rounded up to 0.15, and 0.145 rounded to the nearest tenth place would be rounded down to 0.1.

Operations with Fractions

Recall that a rational number can be written as a fraction and can be converted to a decimal through division. If a rational number is negative, the rules for adding, subtracting, multiplying, and dividing integers must be used. If a rational number is in fraction form, performing addition, subtraction, multiplication, and division is more complicated than when working with integers. First, consider addition. To add two fractions having the same denominator, add the numerators and then reduce the fraction. When an answer is a fraction, it should always be in lowest terms. *Lowest terms* means that every common factor, other than 1, between the numerator and denominator is divided out. For example:

$$\frac{2}{8} + \frac{4}{8} = \frac{6}{8} = \frac{6 \div 2}{8 \div 2} = \frac{3}{4}$$

Both the numerator and denominator of $\frac{6}{8}$ have a common factor of 2, so 2 is divided out of each number to put the fraction in lowest terms. If denominators are different in an addition problem, the fractions must be converted to have common denominators. The **least common denominator (LCD)** of all the given denominators must be found, and this value is equal to the **least common multiple (LCM)** of the denominators. This non-zero value is the smallest number that is a multiple of both denominators. Then, rewrite each original fraction as an equivalent fraction using the new denominator. Once in this form, apply the process of adding with like denominators. For example, consider $\frac{1}{3} + \frac{4}{9}$. The LCD is 9 because it is the smallest multiple of both 3 and 9. The fraction $\frac{1}{3}$ must be rewritten with 9 as its denominator. Therefore, multiply both the numerator and denominator times 3. Multiplying times $\frac{3}{3}$ is the same as multiplying times 1, which does not change the value of the fraction. Therefore, an equivalent fraction is $\frac{3}{9}$, and $\frac{1}{3} + \frac{4}{9} = \frac{3}{9} + \frac{4}{9} = \frac{7}{9}$, which is in lowest terms. Subtraction is performed in a similar manner; once the denominators are equal, the numerators are then subtracted. The following is an example of addition of a positive and a negative fraction:

$$-\frac{5}{12} + \frac{5}{9} = -\frac{5 \times 3}{12 \times 3} + \frac{5 \times 4}{9 \times 4}$$

$$-\frac{15}{36} + \frac{20}{36} = \frac{5}{36}$$

Common denominators are not used in multiplication and division. To multiply two fractions, multiply the numerators together and the denominators together. Then, write the result in lowest terms. For example:

$$\frac{2}{3} \times \frac{9}{4} = \frac{18}{12} = \frac{3}{2}$$

Alternatively, the fractions could be factored first to cancel out any common factors before performing the multiplication. For example:

$$\frac{2}{3} \times \frac{9}{4}$$

$$\frac{2}{3} \times \frac{3 \times 3}{2 \times 2} = \frac{3}{2}$$

This second approach is helpful when working with larger numbers, as common factors might not be obvious. Multiplication and division of fractions are related because the division of two fractions is

19

changed into a multiplication problem. This means that dividing a fraction by another fraction is the same as multiplying the first fraction by the *reciprocal* of the second fraction, so that second fraction must be inverted, or "flipped," to be in reciprocal form. For example:

$$\frac{11}{15} \div \frac{3}{5}$$

$$\frac{11}{15} \times \frac{5}{3} = \frac{55}{45} = \frac{11}{9}$$

The fraction $\frac{5}{3}$ is the reciprocal of $\frac{3}{5}$. It is possible to multiply and divide numbers containing a mix of integers and fractions. In this case, convert the integer to a fraction by placing it over a denominator of 1. For example, a division problem involving an integer and a fraction is:

$$3 \div \frac{1}{2}$$

$$\frac{3}{1} \times \frac{2}{1} = \frac{6}{1} = 6$$

Finally, when performing operations with rational numbers that are negative, the same rules apply as when performing operations with integers. For example, a negative fraction times a negative fraction results in a positive value, and a negative fraction subtracted from a negative fraction results in a negative value.

Operations with Rational Numbers

As mentioned, **rational numbers** are any numbers that can be written as a fraction of integers. Operations to be performed on rational numbers include adding, subtracting, multiplying, and dividing. Essentially, this refers to performing these operations on fractions. Adding and subtracting fractions must be completed by first finding the least common denominator. For example, the problem $\frac{3}{5} + \frac{6}{7}$ requires that the common multiple be found between 5 and 7. The smallest number that divides evenly by 5 and 7 is 35. For the denominators to become 35, they must be multiplied by 7 and 5 respectively. The fraction $\frac{3}{5}$ can be multiplied by 7 on the top and bottom to yield the fraction $\frac{21}{35}$. The fraction $\frac{6}{7}$ can be multiplied by 5 to yield the fraction $\frac{30}{35}$. Now that the fractions have the same denominator, the numerators can be added. The answer to the addition problem becomes:

$$\frac{3}{5} + \frac{6}{7} = \frac{21}{35} + \frac{30}{35} = \frac{51}{35}$$

The same technique can be used for subtraction of rational numbers. The operations multiplication and division may seem easier to perform because finding common denominators is unnecessary. If the problems reads $\frac{1}{3} \times \frac{4}{5}$, then the numerators and denominators are multiplied by each other and the answer is found to be $\frac{4}{15}$. For division, the problem must be changed to multiplication before performing operations. The following words can be used to remember to leave, change, and flip before multiplying. If the problems reads $\frac{3}{7} \div \frac{3}{4}$, then the first fraction is *left* alone, the operation is *changed* to multiplication, and then the last fraction is *flipped*.

The problem becomes:

$$\frac{3}{7} \times \frac{4}{3} = \frac{12}{21}$$

Rational numbers can also be negative. When two negative numbers are added, the result is a negative number with an even greater magnitude. When a negative number is added to a positive number, the result depends on the value of each addend. For example, $-4 + 8 = 4$ because the positive number is larger than the negative number. For multiplying two negative numbers, the result is positive. For example, $-4 \times -3 = 12$, where the negatives cancel out and yield a positive answer.

Squares, Square Roots, Cubes, and Cube Roots

Another operation that can be performed on rational numbers is the square root. Dealing with real numbers only, the positive square root of a number is equal to one of the two repeated positive factors of that number. For example:

$$\sqrt{49} = \sqrt{7 \times 7} = 7$$

A **perfect square** is a number that has a whole number as its square root. Examples of perfect squares are 1, 4, 9, 16, 25, etc. If a number is not a perfect square, an approximation can be used with a calculator. For example, $\sqrt{67} = 8.185$, rounded to the nearest thousandth place. Taking the square root of a fraction that includes perfect squares involves breaking up the problem into the square root of the numerator separate from the square root of the denominator. For example:

$$\sqrt{\frac{16}{25}} = \frac{\sqrt{16}}{\sqrt{25}} = \frac{4}{5}$$

If the fraction does not contain perfect squares, a calculator can be used. Therefore, $\sqrt{\frac{2}{5}} = 0.632$, rounded to the nearest thousandth place.

In addition to the square root, the cube root is another operation. If a number is a *perfect cube*, the cube root of that number is equal to one of the three repeated factors. For example:

$$\sqrt[3]{27} = \sqrt[3]{3 \times 3 \times 3} = 3$$

A negative number has a cube root, which will also be a negative number. For example:

$$\sqrt[3]{-27} = \sqrt[3]{(-3)(-3)(-3)} = -3$$

Similar to square roots, if the number is not a perfect cube, a calculator can be used to find an approximation. Therefore, $\sqrt[3]{\frac{2}{3}} = 0.873$, rounded to the nearest thousandth place.

Higher-order roots also exist. The number relating to the root is known as the **index.** Given the following root, $\sqrt[3]{64}$, 3 is the index, and 64 is the **radicand.** The entire expression is known as the **radical.** Higher-order roots exist when the index is larger than 3. They can be broken up into two groups: even and odd roots. Even roots, when the index is an even number, follow the properties of square roots. They are found by finding the number that, when multiplied by itself the number of times indicated by the index, results in the radicand. For example, the fifth root of 32 is equal to 2 because:

$$\sqrt[5]{32} = \sqrt[5]{2 \times 2 \times 2 \times 2 \times 2} = 2$$

Odd roots, when the index is an odd number, follow the properties of cube roots. A negative number has an odd root. Similarly, an odd root is found by finding the single factor that is repeated that many times to obtain the radicand. For example, the 4th root of 81 is equal to 3 because $3^4 = 81$. This radical is written as $\sqrt[4]{81} = 3$. Higher-order roots can also be evaluated on fractions and decimals, for example, because:

$$\left(\frac{2}{7}\right)^4 = \frac{16}{2,401}$$

$$\sqrt[4]{\frac{16}{2,401}} = \frac{2}{7}$$

And because:

$$(0.1)^5 = 0.00001$$

$$\sqrt[5]{0.00001} = 0.1$$

Undefined Expressions

To check a solution to any division problem, multiply the quotient times the divisor to obtain the dividend. Also, remember that 0 divided by any number is equal to 0. However, any number divided by 0 is undefined. It just does not make sense to divide a number by 0 parts. Square roots of negative numbers are also undefined.

Unit Rates

A **unit rate** is a rate with a denominator of one. It is a comparison of two values with different units where one value is equal to one. Examples of unit rates include 60 miles per hour and 200 words per minute. Problems involving unit rates may require some work to find the unit rate. For example, if Mary travels 360 miles in 5 hours, what is her speed, expressed as a unit rate? The rate can be expressed as the following fraction:

$$\frac{360 \; miles}{5 \; hours}$$

The denominator can be changed to one by dividing by five. The numerator will also need to be divided by five to follow the rules of equality. This division turns the fraction into $\frac{72 \; miles}{1 \; hour}$, which can now be labeled as a unit rate because one unit has a value of one. Another type question involves the use of unit rates to solve problems. For example, if Trey needs to read 300 pages and his average speed is 75 pages per hour, will he be able to finish the reading in 5 hours? The unit rate is 75 pages per hour, so the total of 300 pages can be divided by 75 to find the time. After the division, the time it takes to read is four hours. The answer to the question is yes, Trey will finish the reading within 5 hours.

Objects at Scale

Another application of proportions involves objects at scale. This concept can be demonstrated by examining similar triangles. If two triangles have the same measurement as two triangles in another triangle, the triangles are said to be **similar**. If two are the same, the third pair of angles are equal as well because the sum of all angles in a triangle is equal to 180 degrees. Each pair of equivalent angles are known as **corresponding angles. Corresponding sides** face the corresponding angles, and it is true that corresponding sides are in proportion. For example, consider the following set of similar triangles:

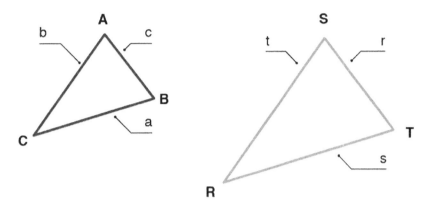

Angles A and S have the same measurement, angles C and R have the same measurement, and angles B and T have the same measurement. Therefore, the following proportion can be set up from the sides:

$$\frac{c}{r} = \frac{a}{s} = \frac{b}{t}$$

This proportion can be helpful in finding missing lengths in pairs of similar triangles.

For example, if the following triangles are similar, a proportion can be used to find the missing side lengths, a and b.

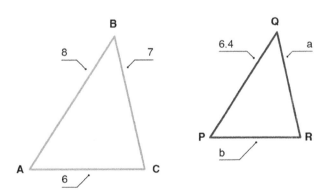

The proportions $\frac{8}{6.4} = \frac{6}{b}$ and $\frac{8}{6.4} = \frac{7}{a}$ can both be cross-multiplied and solved to obtain $a = 5.6$ and $b = 4.8$.

A real-life situation that uses similar triangles involves measuring shadows to find heights of unknown objects. Consider the following problem: A building casts a shadow that is 120 feet long, and at the same time, another building that is 80 feet high casts a shadow that is 60 feet long. How tall is the first building? Each building, together with the sun rays and shadows casted on the ground, forms a triangle. They are similar because each building forms a right angle with the ground, and the sun rays form equivalent angles. Therefore, these two pairs of angles are both equal. Because all angles in a triangle add up to 180 degrees, the third angles are equal as well. Both shadows form corresponding sides of the triangle, the buildings form corresponding sides, and the sun rays form corresponding sides. Therefore, the triangles are similar, and the following proportion can be used to find the missing building length:

$$\frac{120}{x} = \frac{60}{80}$$

Cross-multiply to obtain the cross products:

$$9600 = 60x$$

Then, divide both sides by 60 to obtain:

$$x = 160$$

This solution means that the other building is 160 feet high.

Multiple-Step Problems that Use Ratios, Proportions, and Percents

Ratios
Recall that a **ratio** is the comparison of two different quantities. Comparing 2 apples to 3 oranges results in the ratio 2:3, which can be expressed as the fraction $\frac{2}{5}$. Note that order is important when discussing ratios. The number mentioned first is the **antecedent**, and the number mentioned second is the **consequent**. Note that the consequent of the ratio and the denominator of the fraction are *not* the same. When there are 2 apples to 3 oranges, there are five fruit total; two fifths of the fruit are apples, while three fifths are oranges. The ratio 2:3 represents a different relationship that the ratio 3:2. Also, it is important to make sure that when discussing ratios that have units attached to them, the two quantities use the same units. For example, to think of 8 feet to 4 yards, it would make sense to convert 4 yards to feet by multiplying by 3. Therefore, the ratio would be 8 feet to 12 feet, which can be expressed as the fraction $\frac{8}{20}$. Also, note that it is proper to refer to ratios in lowest terms. Therefore, the ratio of 8 feet to 4 yards is equivalent to the fraction $\frac{2}{5}$.

Many real-world problems involve ratios. Often, problems with ratios involve proportions, as when two ratios are set equal to find the missing amount. However, some problems involve deciphering single ratios. For example, consider an amusement park that sold 345 tickets last Saturday. If 145 tickets were sold to adults and the rest of the tickets were sold to children, what would the ratio of the number of adult tickets to children's tickets be? A common mistake would be to say the ratio is 145:345. However, 345 is the total number of tickets sold. There were 345 − 145 = 200 tickets sold to children. Thus, the correct ratio of adult to children's tickets is 145:200. As a fraction, this expression is written as $\frac{145}{200}$, which can be reduced to $\frac{29}{40}$.

While a ratio compares two measurements using the same units, **rates** compare two measurements with different units. Examples of rates would be $200 for 8 hours of work, or 500 miles traveled per 20 gallons. Because the units are different, it is important to always include the units when discussing rates. Rates can be easily seen because if they are expressed in words, the two quantities are usually split up using one of the following words: *for, per, on, from, in.* Just as with ratios, it is important to write rates in lowest terms. A common rate that can be found in many real-life situations is cost per unit. This quantity describes how much one item or one unit costs. This rate allows the best buy to be determined, given a couple of different sizes of an item with different costs. For example, if 2 quarts of soup was sold for $3.50 and 3 quarts was sold for $4.60, to determine the best buy, the cost per quart should be found. $\frac{\$3.50}{2\text{ qt}} = \$1.75$ per quart, and $\frac{\$4.60}{3\text{ qt}} = \$1.53$ per quart. Therefore, the better deal would be the 3-quart option.

Rate of change problems involve calculating a quantity per some unit of measurement. Usually the unit of measurement is time. For example, meters per second is a common rate of change. To calculate this measurement, find the distance traveled in meters and divide by total time traveled. The calculation is an average of the speed over the entire time interval. Another common rate of change used in the real world is miles per hour.

Consider the following problem that involves calculating an average rate of change in temperature. Last Saturday, the temperature at 1:00 a.m. was 34 degrees Fahrenheit, and at noon, the temperature had increased to 75 degrees Fahrenheit. What was the average rate of change over that time interval? The average rate of change is calculated by finding the change in temperature and dividing it by the total hours elapsed. Therefore, the rate of change was equal to:

$$\frac{75-34}{12-1} = \frac{41}{11} \text{ degrees per hour}$$

This quantity rounded to two decimal places is equal to 3.72 degrees per hour.

A common rate of change that appears in algebra is the slope calculation. Given a linear equation in one variable, $y = mx + b$, the **slope**, m, is equal to:

$$\frac{rise}{run} \text{ or } \frac{change\ in\ y}{change\ in\ x}$$

In other words, slope is equivalent to the ratio of the vertical and horizontal changes between any two points on a line. The vertical change is known as the **rise**, and the horizontal change is known as the **run**. Given any two points on a line (x_1, y_1) and (x_2, y_2), slope can be calculated with the formula:

$$m = \frac{y_2 - y_1}{x_2 - x_1} = \frac{\Delta y}{\Delta x}$$

Common real-world applications of slope include determining how steep a staircase should be, calculating how steep a road is, and determining how to build a wheelchair ramp.

Many times, problems involving rates and ratios involve proportions. A proportion states that two ratios (or rates) are equal. The property of cross products can be used to determine if a proportion is true, meaning both ratios are equivalent. If $\frac{a}{b} = \frac{c}{d}$, then to clear the fractions, multiply both sides by the least common denominator, bd. This results in $ad = bc$, which is equal to the result of multiplying along both diagonals. For example, $\frac{4}{40} = \frac{1}{10}$ grants the cross product $4 \times 10 = 40 \times 1$, which is equivalent to $40 = 40$ and shows that this proportion is true.

Cross products are used when proportions are involved in real-world problems. Consider the following: If 3 pounds of fertilizer will cover 75 square feet of grass, how many pounds are needed for 375 square feet? To solve this problem, a proportion can be set up using two ratios. Let x equal the unknown quantity, pounds needed for 375 feet. Then, the equation found by setting the two given ratios equal to one another is $\frac{3}{75} = \frac{x}{375}$. Cross-multiplication gives:

$$3 \times 375 = 75x$$

Therefore, $1,125 = 75x$. Divide both sides by 75 to get $x = 15$. Therefore, 15 pounds of fertilizer are needed to cover 375 square feet of grass.

Proportions
Fractions appear in everyday situations, and in many scenarios, they appear in the real-world as ratios and in proportions. A **ratio** is formed when two different quantities are compared. For example, in a group of 50 people, if there are 33 females and 17 males, the ratio of females to males is 33 to 17. This expression can be written in the fraction form as $\frac{33}{50}$, where the denominator is the sum of females and males, or by using the ratio symbol, 33:17. The order of the number matters when forming ratios. In the same setting, the ratio of males to females is 17 to 33, which is equivalent to $\frac{17}{50}$ or 17:33.

A **proportion** is an equation involving two ratios. The equation $\frac{a}{b} = \frac{c}{d}$, or $a : b = c : d$ is a proportion, for real numbers a, b, c, and d. Usually, in one ratio, one of the quantities is unknown, and cross-multiplication is used to solve for the unknown. Consider $\frac{1}{4} = \frac{x}{5}$. To solve for x, cross-multiply to obtain $5 = 4x$. Divide each side by 4 to obtain the solution $x = \frac{5}{4}$. It is also true that percentages are ratios in which the second term is 100 minus the first term. For example, 65% is 65:35 or $\frac{65}{100}$. Therefore, when working with percentages, one is also working with ratios.

Real-world problems frequently involve proportions. For example, consider the following problem: If 2 out of 50 pizzas are usually delivered late from a local Italian restaurant, how many would be late out of 235 orders? The following proportion would be solved with x as the unknown quantity of late pizzas:

$$\frac{2}{50} = \frac{x}{235}$$

Cross multiplying results in $470 = 50x$. Divide both sides by 50 to obtain $x = \frac{470}{50}$, which in lowest terms is equal to $\frac{47}{5}$. In decimal form, this improper fraction is equal to 9.4. Because it does not make sense to answer this question with decimals (portions of pizzas do not get delivered) the answer must be rounded. Traditional rounding rules would say that 9 pizzas would be expected to be delivered late. However, to be safe, rounding up to 10 pizzas out of 235 would probably make more sense.

Percentages
Percentages are defined to be parts per one hundred. To convert a decimal to a percentage, move the decimal point two units to the right and place the percent sign after the number. Percentages appear in many scenarios in the real world. It is important to make sure the statement containing the percentage is translated to a correct mathematical expression. Be aware that it is extremely common to make a mistake when working with percentages within word problems.

An example of a word problem containing a percentage is the following: 35% of people speed when driving to work. In a group of 5,600 commuters, how many would be expected to speed on the way to their place of employment? The answer to this problem is found by finding 35% of 5,600. First, change the percentage to the decimal 0.35. Then compute the product: $0.35 \times 5{,}600 = 1{,}960$. Therefore, it would be expected that 1,960 of those commuters would speed on their way to work based on the data given.

In this situation, the word "of" signals to use multiplication to find the answer. Another way percentages are used is in the following problem: Teachers work 8 months out of the year. What percent of the year do they work? To answer this problem, find what percent of 12 the number 8 is, because there are 12 months in a year. Therefore, divide 8 by 12, and convert that number to a percentage:

$$\frac{8}{12} = \frac{2}{3} = 0.66\overline{6}$$

The percentage rounded to the nearest tenth place tells us that teachers work 66.7% of the year. Percentages also appear in real-world application problems involving finding missing quantities like in the following question: 60% of what number is 75? To find the missing quantity, an equation can be used. Let x be equal to the missing quantity. Therefore, $0.60x = 75$. Divide each side by 0.60 to obtain 125. Therefore, 60% of 125 is equal to 75.

Sales tax is an important application relating to percentages because tax rates are usually given as percentages. For example, a city might have an 8% sales tax rate. Therefore, when an item is purchased with that tax rate, the real cost to the customer is 1.08 times the price in the store. For example, a $25 pair of jeans costs the customer $25 \times 1.08 = \$27$. Sales tax rates can also be determined if they are unknown when an item is purchased. If a customer visits a store and purchases an item for $21.44, but the price in the store was $19, they can find the tax rate by first subtracting $21.44 - \$19$ to obtain $2.44, the sales tax amount. The sales tax is a percentage of the in-store price. Therefore, the tax rate is $\frac{2.44}{19} = 0.128$, which has been rounded to the nearest thousandths place. In this scenario, the actual sales tax rate given as a percentage is 12.8%.

Geometry

Side Lengths of Shapes When Given the Area or Perimeter

Solving for missing values in shapes requires knowledge of the shape and its characteristics. For example, a triangle has three sides and three angles that add up to 180 degrees. If two angle measurements are given, the third can be calculated. For the triangle below, the one given angle has a measure of 55 degrees. The missing angle is x. The third angle is labeled with a square, which indicates a measure of 90 degrees. Because all angles must sum to 180 degrees, the following equation can be used to find the missing x-value:

$$55° + 90° + x = 180°$$

Adding the two given angles and subtracting the total from 180, the missing angle is found to be 35 degrees.

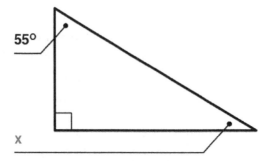

A similar problem can be solved with circles. If the radius is given but the circumference is unknown, the circumference can be calculated based on the formula $C = 2\pi r$. This example can be used in the figure below. The radius can be substituted for r in the formula. Then the circumference can be found as:

$$C = 2\pi \times 8 = 16\pi = 50.24 \text{ cm}$$

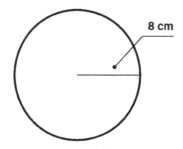

Other figures that may have missing values could be the length of a square, given the area, or the perimeter of a rectangle, given the length and width. All of the missing values can be found by first identifying all the characteristics that are known about the shape, then looking for ways to connect the missing value to the given information.

Area and Perimeter of Two-Dimensional Shapes

Perimeter and area are two commonly used geometric quantities that describe objects. **Perimeter** is the distance around an object. The perimeter of an object can be found by adding the lengths of all sides. Perimeter may be used in problems dealing with lengths around objects such as fences or borders. It may also be used in finding missing lengths, or working backwards. If the perimeter is given, but a length is missing, subtraction can be used to find the missing length. Given a square with side length s, the formula for perimeter is $P = 4s$.

Given a rectangle with length l and width w, the formula for perimeter is:

$$P = 2l + 2w$$

The perimeter of a triangle is found by adding the three side lengths, and the perimeter of a trapezoid is found by adding the four side lengths. The units for perimeter are always the original units of length, such as meters, inches, miles, etc. When discussing a circle, the distance around the object is referred to as its

circumference, not perimeter. The formula for circumference of a circle is $C = 2\pi r$, where r represents the radius of the circle. This formula can also be written as $C = d\pi$, where d represents the diameter of the circle.

Area is the two-dimensional space covered by an object. These problems may include the area of a rectangle, a yard, or a wall to be painted. Finding the area may be a simple formula, or it may require multiple formulas to be used together. The units for area are square units, such as square meters, square inches, and square miles. Given a square with side length s, the formula for its area is $A = s^2$.

Some other common shapes are shown below:

Shape	Formula	Graphic
Rectangle	$Area = length \times width$	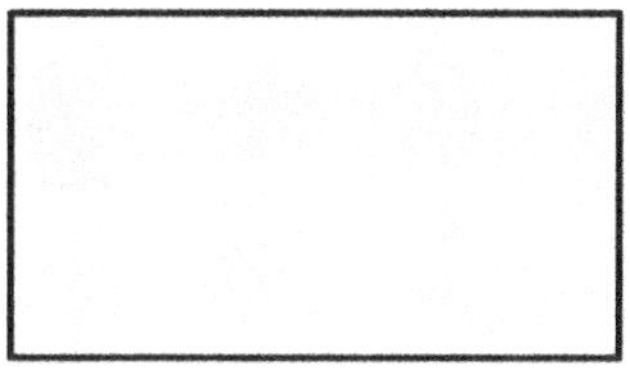
Triangle	$Area = \frac{1}{2} \times base \times height$	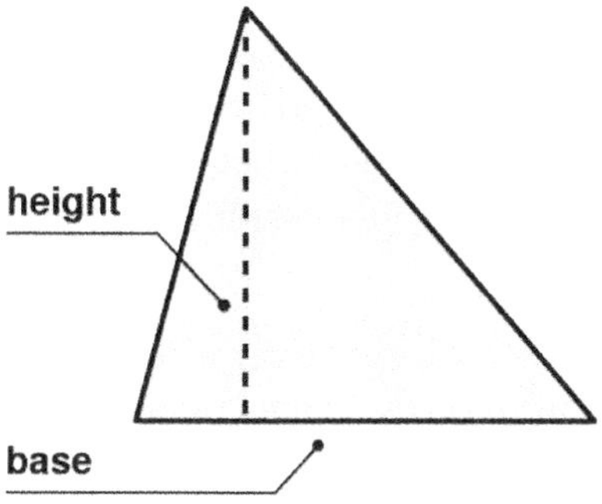
Circle	$Area = \pi \times radius^2$	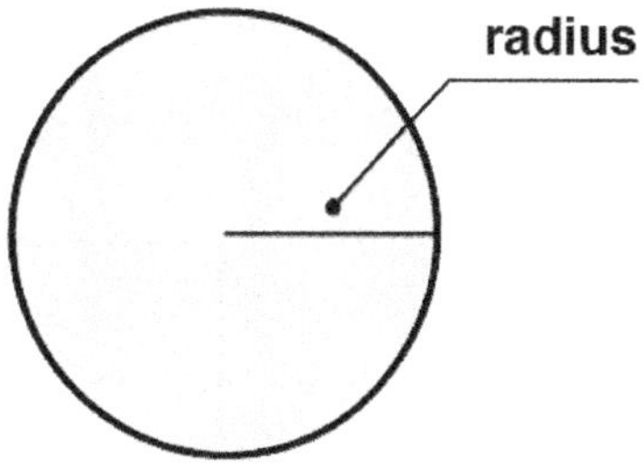

The following formula, not as widely used as those shown above, but very important, is the area of a trapezoid:

Area of a Trapezoid

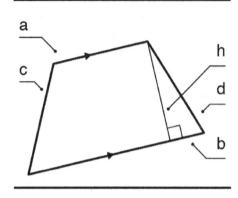

$$A = \frac{1}{2}(a + b)h$$

To find the area of the shapes above, use the given dimensions of the shape in the formula. Complex shapes might require more than one formula. To find the area of the figure below, break the figure into two shapes. The rectangle has dimensions 6 cm by 7 cm. The triangle has dimensions 6 cm by 6 cm. Plug the dimensions into the rectangle formula:

$$A = 6 \times 7$$

Multiplication yields an area of 42 cm². The triangle area can be found using the formula:

$$A = \frac{1}{2} \times 4 \times 6$$

Multiplication yields an area of 12 cm². Add the areas of the two shapes to find the total area of the figure, which is 54 cm².

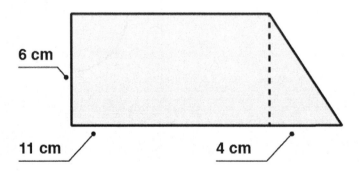

Instead of combining areas, some problems may require subtracting them, or finding the difference.

To find the area of the shaded region in the figure below, determine the area of the whole figure. Then subtract the area of the circle from the whole.

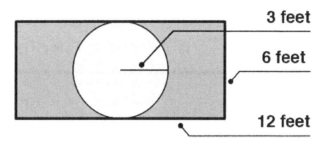

The following formula shows the area of the outside rectangle:

$$A = 12 \times 6 = 72 \text{ ft}^2$$

The area of the inside circle can be found by the following formula:

$$A = \pi(3)^2 = 9\pi = 28.3 \text{ ft}^2$$

As the shaded area is outside the circle, the area for the circle can be subtracted from the area of the rectangle to yield an area of 43.7 ft².

While some geometric figures may be given as pictures, others may be described in words. If a rectangular playing field with dimensions 95 meters long by 50 meters wide is measured for perimeter, the distance around the field must be found. The perimeter includes two lengths and two widths to measure the entire outside of the field. This quantity can be calculated using the following equation:

$$P = 2(95) + 2(50) = 290 \, m$$

The distance around the field is 290 meters.

Area, Circumference, Radius, and Diameter of a Circle

The formula for area of a circle is $A = \pi r^2$ and therefore, formula for area of a **sector** is $\pi r^2 \dfrac{A}{360}$, a fraction of the entire area of the circle. If the radius of a circle and arc length is known, the central angle measurement in degrees can be found by using the formula:

$$\frac{360 \cdot arclength}{2\pi r}$$

If the desired central angle measurement is in radians, the formula for the central angle measurement is much simpler as:

$$\frac{arc\ length}{r}$$

The Center, Radius, Central Angle, a Sector, and an Arc of a Circle

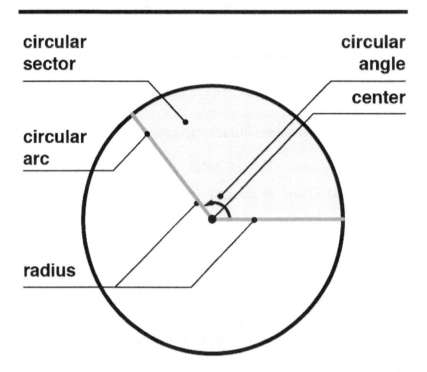

A **chord** of a circle is a straight-line segment that connects any two points on a circle. The line segment does not have to travel through the center, as the diameter does. Also, note that the chord stops at the circumference of the circle. If it did not stop and extended toward infinity, it would be known as a **secant line.** The following shows a diagram of a circle with a chord shown by the dotted line.

The radius is r and the central angle is A:

A Circle with a Chord

Chord Length $= 2\,r\sin\dfrac{A}{2}$

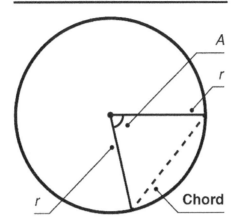

One formula for chord length can be seen in the diagram, and is equal to $2r\sin\dfrac{A}{2}$, where A is the central angle. Another formula for chord length is: chord length $= 2\sqrt{r^2 - D^2}$, where D is equal to the distance from the chord to the center of the circle. This formula is basically a version of the Pythagorean Theorem.

Formulas for chord lengths vary based on what type of information is known. If the radius and central angle are known, the first formula listed above should be used by plugging the radius and angle in directly. If the radius and the distance from the center to the chord are known, the second formula listed previously (chord length $= 2\sqrt{r^2 - D^2}$) should be used.

Many theorems exist between arc lengths, angle measures, chord lengths, and areas of sectors. For instance, when two chords intersect in a circle, the product of the lengths of the individual line segments are equal. For instance, in the following diagram, $A \times B = C \times D$.

A x B = C x D

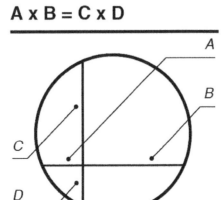

Pythagorean Theorem

The Pythagorean theorem states that given a right triangle, the sum of the squares of the two legs equals the square of the hypotenuse. For example, consider the following right triangle:

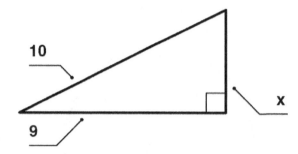

The missing side, x, can be found using the Pythagorean theorem. Since:

$$9^2 + x^2 = 10^2$$

$$81 + x^2 = 100$$

which gives:

$$x^2 = 19$$

To solve for x, take the square root of both sides. Therefore, $x = \sqrt{19} = 4.36$, which has been rounded to two decimal places.

Volume and Surface Area of Three-Dimensional Shapes

Volume is three-dimensional and describes the amount of space that an object occupies, but it's different from area because it has three dimensions instead of two. The units for volume are cubic units, such as cubic meters, cubic inches, and cubic miles. Volume can be found by using formulas for common objects such as cylinders and boxes.

The following chart shows a diagram and formula for the volume of two objects.

Shape	Formula	Diagram
Rectangular Prism (box)	$V = length \times width \times height$	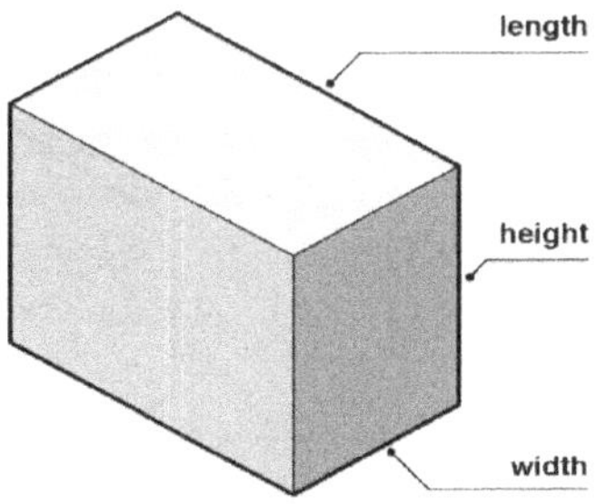
Cylinder	$V = \pi \times radius^2 \times height$	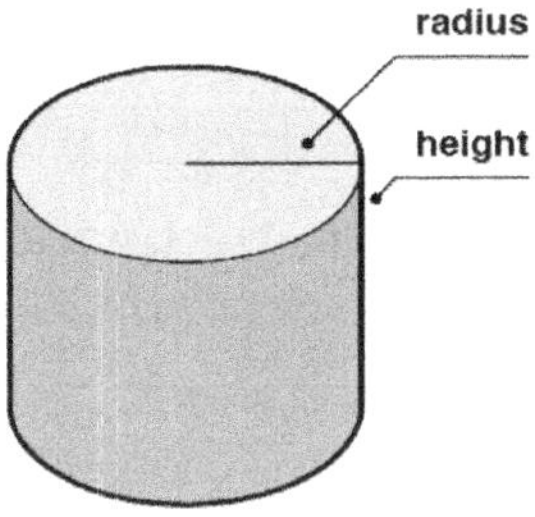

Volume formulas of these two objects are derived by finding the area of the bottom two-dimensional shape, such as the circle or rectangle, and then multiplying times the height of the three-dimensional shape. Other volume formulas include the volume of a cube with side length s: $V = s^3$; the volume of a sphere with radius r: $V = \frac{4}{3}\pi r^3$; and the volume of a cone with radius r and height h:

$$V = \frac{1}{3}\pi r^2 h$$

If a soda can has a height of 5 inches and a radius on the top of 1.5 inches, the volume can be found using one of the given formulas. A soda can is a cylinder. Knowing the given dimensions, the formula can be completed as follows:

$$V = \pi(radius)^2 \times height$$

$$\pi(1.5 \text{ in})^2 \times 5 \text{ in} = 35.325 \text{ in}^3$$

Notice that the units for volume are inches cubed because it refers to the number of cubic inches required to fill the can.

Three-Dimensional Figures

A **rectangular solid** is a six-sided figure with sides that are rectangles. All of the faces meet at right angles, and it looks like a box. Its three measurements are length *l,* width *w,* and height *h.*

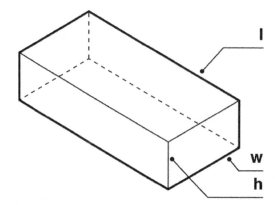

If all sides are equal in a rectangular solid, the solid is known as a *cube.* The cube has six congruent faces that meet at right angles, and each side length is the same and is labeled *s.*

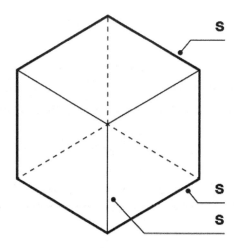

A **cylinder** is a three-dimensional geometric figure consisting of two parallel circles and two parallel lines connecting the ends. The circle has radius *r*, and the cylinder has height *h*.

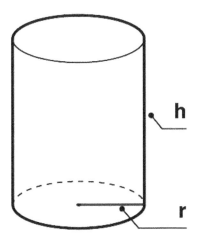

Finally, a **sphere** is a symmetrical three-dimensional shape, where every point on the surface is equal distance from its center. It has a radius *r*.

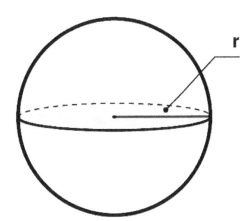

Graphical Data Including Graphs, Tables, and More

Tables, charts, and graphs can be used to convey information about different variables. They are all used to organize, categorize, and compare data, and they all come in different shapes and sizes. Each type has its own way of showing information, whether it is in a column, shape, or picture. To answer a question relating to a table, chart, or graph, some steps should be followed. First, the problem should be read thoroughly to determine what is being asked to determine what quantity is unknown. Then, the title of the table, chart, or graph should be read.

The title should clarify what actual data is being summarized in the table. Next, look at the key and both the horizontal and vertical axis labels, if they are given. These items will provide information about how the data is organized. Finally, look to see if there is any more labeling inside the table. Taking the time to get a good idea of what the table is summarizing will be helpful as it is used to interpret information.

Tables are a good way of showing a lot of information in a small space. The information in a table is organized in columns and rows. For example, a table may be used to show the number of votes each candidate received in an election. By interpreting the table, one may observe which candidate won the election and which candidates came in second and third. In using a bar chart to display monthly rainfall amounts in different countries, rainfall can be compared between countries at different times of the year. **Graphs** are also a useful way to show change in variables over time, as in a line graph, or percentages of a whole, as in a pie graph.

The table below relates the number of items to the total cost. The table shows that 1 item costs $5. By looking at the table further, 5 items cost $25, 10 items cost $50, and 50 items cost $250. This cost can be extended for any number of items. Since 1 item costs $5, then 2 items would cost $10. Though this information isn't in the table, the given price can be used to calculate unknown information.

Number of Items	1	5	10	50
Cost ($)	5	25	50	250

A **bar graph** is a graph that summarizes data using bars of different heights. It is useful when comparing two or more items or when seeing how a quantity changes over time. It has both a horizontal and vertical axis. Interpreting bar graphs includes recognizing what each bar represents and connecting that to the two variables. The bar graph below shows the scores for six people on three different games. The color of the bar shows which game each person played, and the height of the bar indicates their score for that game. William scored 25 on game 3, and Abigail scored 38 on game 3. By comparing the bars, it's obvious that Williams scored lower than Abigail.

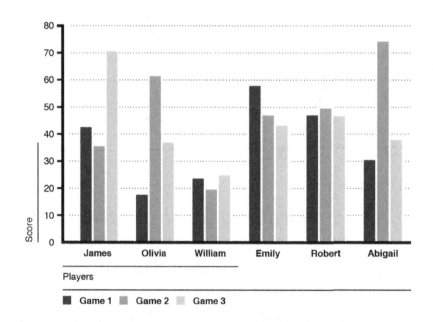

A **line graph** is a way to compare two variables. Each variable is plotted along an axis, and the graph contains both a horizontal and a vertical axis. On a line graph, the line indicates a continuous change. The change can be seen in how the line rises or falls, known as its slope, or rate of change. Often, in line graphs, the horizontal axis represents a variable of time. Audiences can quickly see if an amount has increased or decreased over time. The bottom of the graph, or the x-axis, shows the units for time, such as days, hours, months, etc. If there are multiple lines, a comparison can be made between what the two lines represent. For example, the following line graph shows the change in temperature over five days. The top line represents the high, and the bottom line represents the low for each day. Looking at the top line alone, the high decreases for a day, then increases on Wednesday. Then it decreases on Thursday and increases again on Friday. The low temperatures have a similar trend, shown in bottom line. The range in temperatures each day can also be calculated by finding the difference between the top line and bottom line on a particular day. On Wednesday, the range was 14 degrees, from 62 to 76° F.

Daily Temperatures

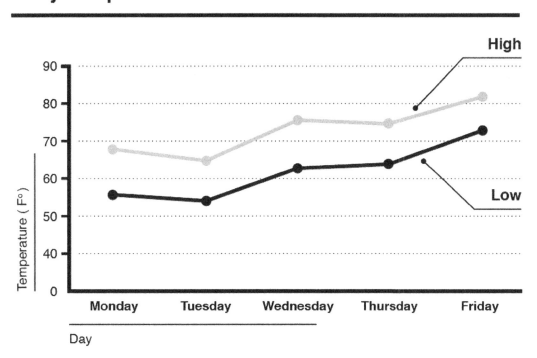

Pie charts are used to show percentages of a whole, as each category is given a piece of the pie, and together all the pieces make up a whole. They are a circular representation of data which are used to highlight numerical proportion. It is true that the arc length of each pie slice is proportional to the amount it individually represents. When a pie chart is shown, an audience can quickly make comparisons by comparing the sizes of the pieces of the pie. They can be useful for comparison between different categories. The following pie chart is a simple example of three different categories shown in comparison to each other.

Light gray represents cats, dark gray represents dogs, and the gray between those two represents other pets. As the pie is cut into three equal pieces, each value represents just more than 33 percent, or $\frac{1}{3}$ of the whole. Values 1 and 2 may be combined to represent $\frac{2}{3}$ of the whole. In an example where the total pie represents 75,000 animals, then cats would be equal to $\frac{1}{3}$ of the total, or 25,000. Dogs would equal 25,000 and other pets also equal 25,000.

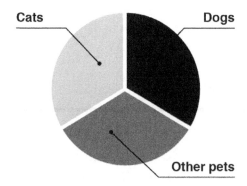

The fact that a circle is 360 degrees is used to create a pie chart. Because each piece of the pie is a percentage of a whole, that percentage is multiplied times 360 to get the number of degrees each piece represents. In the example above, each piece is $\frac{1}{3}$ of the whole, so each piece is equivalent to 120 degrees. Together, all three pieces add up to 360 degrees.

Stacked bar graphs, also used fairly frequently, are used when comparing multiple variables at one time. They combine some elements of both pie charts and bar graphs, using the organization of bar graphs and the proportionality aspect of pie charts. The following is an example of a stacked bar graph that represents the number of students in a band playing drums, flute, trombone, and clarinet. Each bar graph is broken up further into girls and boys:

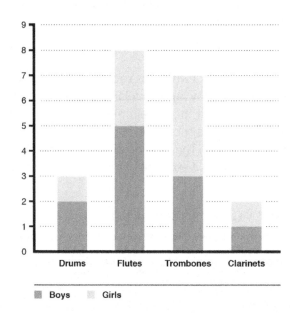

To determine how many boys play trombone, refer to the darker portion of the trombone bar, resulting in 3 students.

A **scatterplot** is another way to represent paired data. It uses **Cartesian coordinates**, like a line graph, meaning it has both a horizontal and vertical axis. Each data point is represented as a dot on the graph. The dots are never connected with a line. For example, the following is a scatterplot showing people's height versus age.

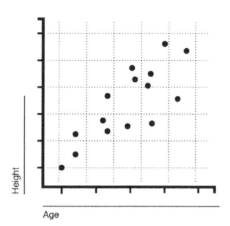

A scatterplot, also known as a **scattergram**, can be used to predict another value and to see if an association, known as a **correlation**, exists between a set of data. If the data resembles a straight line, the data is **associated.** The following is an example of a scatterplot in which the data does not seem to have an association:

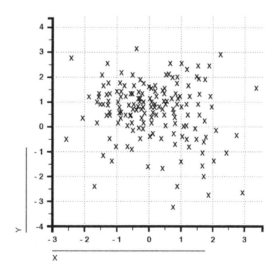

Mean, Median, Mode, and Range

A **data set** can be described by calculating the mean, median, and mode. These values, called **measures of center**, allow the data to be described with a single value that is representative of the data set.

The most common measure of center is the **mean**, also referred to as the **average.**

41

To calculate the mean,

- Add all data values together
- Divide by the sample size (the number of data points in the set)

The **median** is middle data value, so that half of the data lies below this value and half lies above it.

To calculate the median,

- Order the data from least to greatest
- The point in the middle of the set is the median
  - In the event that there is an even number of data points, add the two middle points and divide by 2

The **mode** is the data value that occurs most often.

To calculate the mode,

- Order the data from least to greatest
- Find the value that occurs most often

Example: Amelia is a leading scorer on the school's basketball team. The following data set represents the number of points that Amelia has scored in each game this season. Use the mean, median, and mode to describe the data.

16, 12, 26, 14, 28, 14, 12, 15, 25

Solution:

Mean: 16 + 12 + 26 + 14 + 28 + 14 + 12 + 15 + 25 = 162

$162 \div 9 = 18$

Amelia averages 18 points per game.

Median: 12, 12, 14, 14, **15**, 16, 25, 26, 28

Amelia's median score is 15.

Mode: 12, 12, 14, 14, 15, 16, 25, 26, 28

12 and 14 each occur twice in the data set, so this set has 2 modes: 12 and 14.

The **range** is the difference between the largest and smallest values in the set. In the example above, the range is 28 − 12 = 16.

Counting Techniques

The total number of events in the sample space must be known to solve probability problems. Different methods can be used to count the number of possible outcomes, depending on whether different arrangements of the same items are counted only once or separately. **Permutations** are arrangements in which different sequences are counted separately. Therefore, order matters in permutations. **Combinations** are arrangements in which different sequences are not counted separately. Therefore, order does not matter in combinations. For example, if 123 is considered different from 321, permutations would be discussed. However, if 123 is considered the same as 321, combinations would be considered.

If the sample space contains n different permutations of n different items and all of them must be selected, there are $n!$ different possibilities. For example, five different books can be rearranged 5! = 120 times. The probability of one person randomly ordering those five books in the same way as another person is $\frac{1}{120}$. A different calculation is necessary if a number less than n is to be selected or if order does not matter. In general, the notation $P(n,r)$ represents the number of ways to arrange r objects from a set of n if order does matter, and:

$$P(n,r) = \frac{n!}{(n-r)!}$$

Therefore, in order to calculate the number of ways five books can be arranged in three slots if order matters, plug n = 5 and r = 3 in the formula to obtain:

$$P(5,3) = \frac{5!}{(5-3)!} = \frac{5!}{2!} = 60$$

Secondly, $C(n,r)$ represents the total number of r combinations selected out of n items when order does not matter, and:

$$C(n,r) = \frac{n!}{(n-r)!\ r!}$$

Therefore, the number of ways five books can be arranged in three slots if order does not matter is:

$$C(5,3) = \frac{5!}{(5-3)!\ 3!} = 10$$

The following relationship exists between permutations and combinations:

$$C(n,r) = \frac{P(n,r)}{r!}$$

Sets of numbers and other similarly organized data can also be represented graphically. Venn diagrams are a common way to do so. A Venn diagram represents each set of data as a circle. The circles overlap, showing that each set of data is overlapping. A Venn diagram is also known as a **logic diagram** because it visualizes all possible logical combinations between two sets. Common elements of two sets are represented by the area of overlap.

The following is an example of a Venn diagram of two sets A and B:

Parts of the Venn Diagram

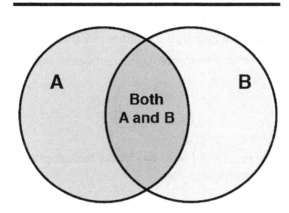

Another name for the area of overlap is the **intersection.** The intersection of A and B, $A \cap B$, contains all elements that are in both sets A and B. The **union** of A and B, $A \cup B$, contains all elements that are in either set A or set B. Finally, the **complement** of $A \cup B$ is equal to all elements that are not in either set A or set B. These elements are placed outside of the circles.

The following is an example of a Venn diagram in which 24 students were surveyed asking if they had brothers or sisters or both. Ten students only had brothers, 7 students only had sisters, and 5 had both brothers and sisters. This number 5 is the intersection and is placed where the circles overlap. Two students did not have a cat or a dog. Two is therefore the complement and is placed outside of the circles.

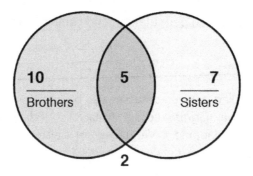

Venn diagrams can have more than two sets of data. The more circles, the more logical combinations are represented by the overlapping. The following is a Venn diagram that represents students who like the colors green, pink, or blue. There were 30 students surveyed. The innermost region represents those students that like green, pink, and blue. Therefore, 2 students like all three. In this example, all students like at least one of the colors, so no one exists in the complement.

30 students

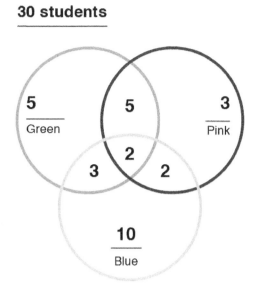

Venn diagrams are typically not drawn to scale, but if they are and their area is proportional to the amount of data it represents, it is known as an **area-proportional** Venn diagram.

Probability of an Event

A probability experiment is an action that causes specific results, such as counts or measurements. The result of such an experiment is known as an outcome, and the set of all potential outcomes is known as the sample space. An **event** consists of one or more of those outcomes. For example, consider the probability experiment of tossing a coin and rolling a six-sided die. The coin has two possible outcomes— a heads or a tails—and the die has six possible outcomes—rolling each number 1–6. Therefore, the sample space has twelve possible outcomes: a heads or a tails paired with each roll of the die.

A simple event is an event that consists of a single outcome. For instance, selecting a queen of hearts from a standard fifty-two-card deck is a simple event; however, selecting a queen is not a simple event because there are four possibilities.

Classical, or **theoretical, probability** is when each outcome in a sample space has the same chance to occur. The probability for an event is equal to the number of outcomes in that event divided by the total number of outcomes in the sample space. For example, consider rolling a six-sided die. The probability of rolling a 2 is $\frac{1}{6}$, and the probability of rolling an even number is $\frac{3}{6}$, or $\frac{1}{2}$, because there are three even numbers on the die. This type of probability is based on what should happen in theory but not what actually happens in real life.

Empirical probability is based on actual experiments or observations. For instance, if a die is rolled eight times, and a 1 is rolled two times, the empirical probability of rolling a 1 is $\frac{2}{8} = \frac{1}{4}$, which is higher than the theoretical probability. The Law of Large Numbers states that as an experiment is completed repeatedly, the empirical probability of an event should get closer to the theoretical probability of an event.

Probabilities range from 0 to 1. The closer the probability of an event occurring is to 0, the less likely it will occur. The closer it is to 1, the more likely it is to occur.

The **addition rule** is necessary to find the probability of event A or event B occurring or both occurring at the same time. If events A and B are **mutually exclusive** or **disjoint,** which means they cannot occur at the same time:

$$P(A \text{ or } B) = P(A) + P(B)$$

If events A and B are not mutually exclusive, $P(A \text{ or } B) = P(A) + P(B) - P(A \text{ and } B)$ where $P(A \text{ and } B)$ represents the probability of event A and B both occurring at the same time. An example of two events that are mutually exclusive are rolling a 6 on a die and rolling an odd number on a die. The probability of rolling a 6 or rolling an odd number is:

$$\frac{1}{6} + \frac{3}{6} = \frac{4}{6} = \frac{2}{3}$$

Rolling a 6 and rolling an even number are not mutually exclusive because there is some overlap. The probability of rolling a 6 or rolling an even number is:

$$\frac{1}{6} + \frac{3}{6} - \frac{1}{6} = \frac{3}{6} = \frac{1}{2}$$

Conditional Probability

The **multiplication rule** is necessary when finding the probability that an event A occurs in a first trial and event B occurs in a second trial, which is written as $P(A \text{ and } B)$. This rule differs if the events are independent or dependent. Two events A and B are **independent** if the occurrence of one event does not affect the probability that the other will occur. If A and B are not independent, they are **dependent,** and the outcome of the first event somehow affects the outcome of the second. If events A and B are independent, $P(A \text{ and } B) = P(A)P(B)$, and if events A and B are dependent, $P(A \text{ and } B) = P(A)P(B|A)$, where $P(B|A)$ represents the probability event B occurs given that event A has already occurred.

$P(B|A)$ represents **conditional probability,** or the probability of event B occurring given that event A has already occurred. $P(B|A)$ can be found by dividing the probability of events A and B both occurring by the probability of event A occurring using the formula $P(B|A) = \frac{P(A \text{ and } B)}{P(A)}$ and represents the total number of outcomes remaining for B to occur after A occurs. This formula is derived from the multiplication rule with dependent events by dividing both sides by $P(A)$. Note that $P(B|A)$ and $P(A|B)$ are not the same. The first quantity shows that event B has occurred after event A, and the second quantity shows that event A has occurred after event B. To incorrectly interchange these ideas is known as **confusion of the inverse.**

Consider the case of drawing two cards from a deck of fifty-two cards. The probability of pulling two queens would vary based on whether the initial card was placed back in the deck for the second pull. If the card is placed back in, the probability of pulling two queens is:

$$\frac{4}{52} \times \frac{4}{52} = 0.00592$$

If the card is not placed back in, the probability of pulling two queens is:

$$\frac{4}{52} \times \frac{3}{51} = 0.00452$$

When the card is not placed back in, both the numerator and denominator of the second probability decrease by 1. This is due to the fact that, theoretically, there is one less queen in the deck, and there is one less total card in the deck as well.

Conditional probability is used frequently when probabilities are calculated from tables. Two-way frequency tables display data with two variables and highlight the relationships between those two variables. They are often used to summarize survey results and are also known as **contingency tables**. Each cell shows a count pertaining to that individual variable pairing, known as a **joint frequency**, and the totals of each row and column also are in the tables. Consider the following two-way frequency table:

	70 or older	**69 or younger**	**Totals**
Women	20	40	60
Men	5	35	40
Total	25	75	100

This table shows the breakdown of ages and sexes of 100 people in a particular village. Consider a randomly selected villager. The probability of selecting a male 69 years old or younger is $\frac{35}{100}$ because there are 35 males under the age of 70 and 100 total villagers.

Probability Distributions

A **discrete random variable** is a set of values that is either finite or countably infinite. If there are infinitely many values, being **countable** means that each individual value can be paired with a natural number. For example, the number of coin tosses before getting heads could potentially be infinite, but the total number of tosses is countable. Each toss refers to a number, like the first toss, second toss, etc. A **continuous random variable** has infinitely many values that are not countable.

The individual items cannot be enumerated; an example of such a set is any type of measurement. There are infinitely many heights of human beings due to decimals that exist within each inch, centimeter, millimeter, etc. Each type of variable has its own probability distribution, which calculates the probability for each potential value of the random variable. Probability distributions exist in tables, formulas, or graphs. The expected value of a random variable represents what the mean value should be in either a large sample size or after many trials. According to the Law of Large Numbers, after many trials, the actual mean and that of the probability distribution should be very close to the expected value. The **expected value** is a weighted average that is calculated as $E(X) = \sum x_i p_i$, where x_i represents the value of each outcome, and p_i represents the probability of each outcome. The expected value if all of the probabilities are equal is:

$$E(X) = \frac{x_1 + x_2 + \cdots + x_n}{n}$$

Expected value is often called the **mean of the random variable** and is known as a measure of central tendency like mean and mode.

A **binomial probability distribution** is a probability distribution that adheres to some important criteria. The distribution must consist of a fixed number of trials where all trials are independent, each trial has an outcome classified as either success or failure, and the probability of a success is the same in each trial. Within any binomial experiment, x is the number of resulting successes, n is the number of trials, P is the probability of success within each trial, and $Q = 1 - P$ is the probability of failure within each trial. The probability of obtaining x successes within n trials is $\binom{n}{x} P^x (1 - P)^{n-x}$, where $\binom{n}{x} = \frac{n!}{x!(n-x)!}$ is called the **binomial coefficient**. A binomial probability distribution could be used to find the probability of obtaining exactly two heads on five tosses of a coin. In the formula, $x = 2$, $n = 5$, $P = 0.5$, and $Q = 0.5$.

A **uniform probability distribution** exists when there is constant probability. Each random variable has equal probability, and its graph is a rectangle because the height, representing the probability, is constant.

Finally, a **normal probability distribution** has a graph that is symmetric and bell-shaped; an example using body weight is shown here:

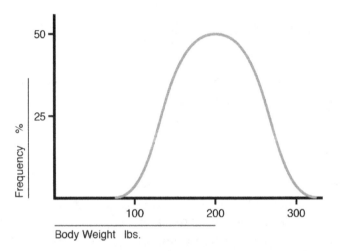

Population percentages can be estimated using normal distributions. For example, the probability that a data point will be less than the mean is 50 percent. The Empirical Rule states that 68 percent of the data falls within 1 standard deviation of the mean, 95 percent falls within 2 standard deviations of the mean, and 99.7 percent falls within 3 standard deviations of the mean. A standard normal distribution is a normal distribution with a mean equal to 0 and standard deviation equal to 1. The area under the entire curve of a standard normal distribution is equal to 1.

Basic Algebra

Adding, Subtracting, Multiplying, and Factoring Linear Expressions

An **algebraic expression** is a mathematical phrase that may contain numbers, variables, and mathematical operations. An expression represents a single quantity. For example, $3x + 2$ is an algebraic expression.

An **algebraic equation** is a mathematical sentence with two expressions that are equal to each other. That is, an equation must contain an equals sign, as in $3x + 2 = 17$. This statement says that the value of the expression on the left side of the equals sign is equivalent to the value of the expression on the right side. In an expression, there are not two sides because there is no equals sign. The equals sign ($=$) is the difference between an expression and an equation.

To distinguish an expression from an equation, just look for the equals sign.

Example: Determine whether each of these is an expression or an equation.

- $16 + 4x = 9x - 7$ Solution: Equation

- $-27x - 42 + 19y$ Solution: Expression

- $4 = x + 3$ Solution: Equation

Adding and Subtracting Linear Algebraic Expressions

To add and subtract linear algebra expressions, you must combine like terms. **Like terms** are described as those terms that have the same variable with the same exponent. In the following example, the x-terms can be added because the variable is the same and the exponent on the variable of one is also the same. These terms add to be $9x$. The other like terms are called **constants** because they have no variable component. These terms will add to be nine.

Example: Add $(3x - 5) + (6x + 14)$

$3x - 5 + 6x + 14$	Rewrite without parentheses
$3x + 6x - 5 + 14$	Commutative property of addition
$9x + 9$	Combine like terms

When subtracting linear expressions, be careful to add the opposite when combining like terms. Do this by distributing -1, which is multiplying each term inside the second parenthesis by negative one. Remember that distributing -1 changes the sign of each term.

Example: Subtract $(17x + 3) - (27x - 8)$

$17x + 3 - 27x + 8$	Distributive Property
$17x - 27x + 3 + 8$	Commutative property of addition
$-10x + 11$	Combine like terms

Example: Simplify by adding or subtracting:

$$(6m + 28z - 9) + (14m + 13) - (-4z + 8m + 12)$$

$6m + 28z - 9 + 14m + 13 + 4z - 8m - 12$	Distributive Property
$6m + 14m - 8m + 28z + 4z - 9 + 13 - 12$	Commutative Property of Addition
$12m + 32z - 8$	Combine like terms

The Distributive Property: $a(b + c) = ab + ac$

The distributive property is a way of taking a factor and multiplying it through a given expression in parentheses. Each term inside the parentheses is multiplied by the outside factor, eliminating the parentheses. The following example shows how to distribute the number 3 to all the terms inside the parentheses.

Example: Use the distributive property to write an equivalent algebraic expression:

$3(2x + 7y + 6)$

$3(2x) + 3(7y) + 3(6)$ Distributive property

$6x + 21y + 18$ Simplify

Because $a - b$ can be written $a + (-b)$, the distributive property can be applied in the example below.

Example: Use the distributive property to write an equivalent algebraic expression.

$7(5m - 8)$

$7[5m + (-8)]$ Rewrite subtraction as addition of -8

$7(5m) + 7(-8)$ Distributive property

$35m - 56$ Simplify

In the following example, note that the factor of 2 is written to the right of the parentheses but is still distributed as before.

Example: Use the distributive property to write an equivalent algebraic expression:

$(3m + 4x - 10)2$

$(3m)2 + (4x)2 + (-10)2$ Distributive property

$6m + 8x - 20$ Simplify

Example: $-(-2m + 6x)$

In this example, the negative sign in front of the parentheses can be interpreted as $-1(-2m + 6x)$

$-1(-2m + 6x)$

$-1(-2m) + (-1)(6x)$ Distributive property

$2m - 6x$ Simplify

Creating an Equivalent Form of an Algebraic Expression

Two algebraic expressions are equivalent if, even though they look different, they represent the same expression. Therefore, plugging in the same values into the variables in each expression will result in the same result in both expressions. To obtain an equivalent form of an algebraic expression, laws of algebra must be followed. For instance, addition and multiplication are both commutative and associative.

Therefore, terms in an algebraic expression can be added in any order and multiplied in any order. For instance, $4x + 2y$ is equivalent to:

$$2y + 4x \text{ and } y \times 2 + x \times 4$$

Also, the distributive law allows a number to be distributed throughout parentheses, as in the following:

$$a(b + c) = ab + ac$$

The two expressions on both sides of the equals sign are equivalent. Also, collecting like terms is important when working with equivalent forms. The simplest version of an expression is always the one easiest to work with, so all like terms (those with the same variables raised to the same powers) must be combined.

Note that an expression is not an equation, and therefore expressions cannot be multiplied times numbers, divided by numbers, or have numbers added to them or subtracted from them and still have equivalent expressions. These processes can only happen in equations when the same step is performed on both sides of the equals sign.

Evaluating Algebraic Expressions

To evaluate an algebra expression for a given value of a variable, replace the variable with the given value. Then perform the given operations to simplify the expression.

Example: Evaluate $12 + x$ for $x = 9$

$12 + (9)$ Replace x with the value of 9 as given in the problem. It is a good idea to always use parentheses when substituting this value. This will be particularly important in the following examples.

21 Add

Now see that when x is 9, the value of the given expression is 21.

Example: Evaluate $4x + 7$ for $x = 3$

$4(3) + 7$ Replace the x in the expression with 3

$12 + 7$ Multiply (remember order of operations)

19 Add

Therefore, when x is 3, the value of the given expression is 19.

Example: Evaluate $-7m - 3r - 18$ for $m = 2$ and $r = -1$

$-7(2) - 3(-1) - 18$ Replace m with 2 and r with -1

$-14 + 3 - 18$ Multiply

-29 Add

So, when m is 2 and r is -1, the value of the given expression is -29.

Creating Algebraic Expressions

A **variable** is a symbol used to represent a number. Letters, like x, y, and z, are often used as variables in algebra.

A **constant** is a number that cannot change its value. For example, 18 is a constant.   3

A **term** is a constant, variable, or the product of constants and variables. In an expression, terms are separated by $+$ and $-$ signs. Examples of terms are $24x$, -32, and $15xyz$.

Like terms are terms that contain the same variables. For example, $6z$ and $-8z$ are like terms, and $9xy$ and $17xy$ are like terms. Constants, like 23 and 51, are like terms as well.

A **factor** is something that is multiplied by something else. A factor may be a constant, a variable, or a sum of constants or variables.

A **coefficient** is the numerical factor in a term that has a variable. In the term $16x$, the coefficient is 16.

Example: Given the expression, $6x - 12y + 18$, answer the following questions.

- How many terms are in the expression?
  - Solution: 3
- Name the terms.
  - Solution: 6x, –12y, and 18
  - (Notice that the minus sign preceding the 12 is interpreted to represent negative 12)
- Name the factors.
  - Solution: 6, x, –12, y
- What are the coefficients in this expression?
  - Solution: 6 and –12
- What is the constant in this expression?
  - Solution: 18

Adding, Subtracting, Multiplying, Dividing, and Factoring Polynomials

Adding, Subtracting, and Multiplying Polynomial Equations
When working with polynomials, like terms are terms that contain exactly the same variables with the same powers. For example, x^4y^5 and $9x^4y^5$ are like terms. The coefficients are different, but the same variables are raised to the same powers. When adding polynomials, only terms that are like can be added. When adding two like terms, just add the coefficients and leave the variables alone. This process uses the distributive property. For example:

$$x^4y^5 + 9x^4y^5$$

$$(1 + 9)x^4y^5$$

$$10x^4y^5$$

Therefore, when adding two polynomials, simply add the like terms together. Unlike terms cannot be combined.

Subtracting polynomials involves adding the opposite of the polynomial being subtracted. Basically, the sign of each term in the polynomial being subtracted is changed, and then the like terms are combined because it is now an addition problem. For example, consider the following:

$$6x^2 - 4x + 2 - (4x^2 - 8x + 1).$$

Add the opposite of the second polynomial to obtain:

$$6x^2 - 4x + 2 + (-4x^2 + 8x - 1)$$

Then, collect like terms to obtain:

$$2x^2 + 4x + 1$$

Multiplying polynomials involves using the product rule for exponents that:

$$b^m b^n = b^{m+n}$$

Basically, when multiplying expressions with the same base, just add the exponents. Multiplying a monomial times a monomial involves multiplying the coefficients together and then multiplying the variables together using the product rule for exponents. For instance:

$$8x^2 y \times 4x^4 y^2 = 32x^6 y^3$$

When multiplying a monomial times a polynomial that is not a monomial, use the distributive property to multiply each term of the polynomial times the monomial. For example:

$$3x(x^2 + 3x - 4)$$

$$3x^3 + 9x^2 - 12x$$

Finally, multiplying two polynomials when neither one is a monomial involves multiplying each term of the first polynomial times each term of the second polynomial. There are some shortcuts, given certain scenarios.

For instance, a binomial multiplied by a binomial can be found by using the **FOIL** (Firsts, Outers, Inners, Lasts) method shown here:

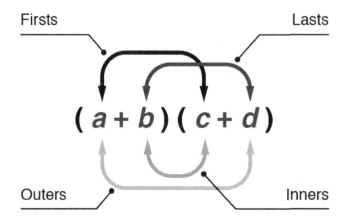

Finding the product of a sum and difference of the same two terms is simple because if it was to be foiled out, the outer and inner terms would cancel out. For instance:

$$(x + y)(x - y) = x^2 + xy - xy - y^2$$

Finally, the square of a binomial can be found using the following formula:

$$(a \pm b)^2 = a^2 \pm 2ab + b^2$$

Expanding Polynomials
A **polynomial** is a mathematical expression containing the sum and difference of one or more terms that are constants multiplied times variables raised to positive powers. A polynomial is considered expanded when there are no variables contained within parentheses, the distributive property has been carried out for any terms that were within parentheses, and like terms have been collected.

Recall that the distributive property is a way of taking a factor and multiplying it through a given expression in parentheses. Each term inside the parentheses is multiplied by the outside factor, eliminating the parentheses. In simple terms, the distributive property is:

$$a(b + c) = ab + ac$$

To exemplify how to use the distributive property and expand a polynomial, consider the equation:

$$3(x + 2) - 5x = 4x + 1$$

Use the distributive property to clear the parentheses. Therefore, multiply each term inside the parentheses by 3. This step results in:

$$3x + 6 - 5x = 4x + 1$$

Next, collect like terms on the left-hand side. Like terms are terms with the same variable or variables raised to the same exponent(s). Only like terms can be combined through addition or subtraction. After collecting like terms, the equation is:

$$-2x + 6 = 4x + 1$$

Finally, apply the addition and multiplication principles. Add $2x$ to both sides to obtain:

$$6 = 6x + 1$$

Then, subtract 1 from both sides to obtain $5 = 6x$. Finally, divide both sides by 6 to obtain the solution:

$$\frac{5}{6} = x$$

Factoring
A factorization of an algebraic expression can be found. Throughout the process, a more complicated expression can be decomposed into products of simpler expressions. To factor a polynomial, first determine if there is a greatest common factor. If there is, factor it out. For example, $2x^2 + 8x$ has a greatest common factor of $2x$ and can be written as $2x(x + 4)$. Once the greatest common monomial factor is factored out, if applicable, count the number of terms in the polynomial. If there are two terms, is it a difference of squares, a sum of cubes, or a difference of cubes?

If so, the following rules can be used:

$$a^2 - b^2 = (a + b)(a - b)$$

$$a^3 + b^3 = (a + b)(a^2 - ab + b^2)$$

$$a^3 - b^3 = (a - b)(a^2 + ab + b^2)$$

If there are three terms, and if the trinomial is a perfect square trinomial, it can be factored into the following:

$$a^2 + 2ab + b^2 = (a + b)^2$$

$$a^2 - 2ab + b^2 = (a - b)^2$$

If not, try factoring into a product of two binomials by trial and error into a form of $(x + p)(x + q)$. For example, to factor $x^2 + 6x + 8$, determine what two numbers have a product of 8 and a sum of 6. Those numbers are 4 and 2, so the trinomial factors into $(x + 2)(x + 4)$.

Finally, if there are four terms, try factoring by grouping. First, group terms together that have a common monomial factor. Then, factor out the common monomial factor from the first two terms. Next, look to see if a common factor can be factored out of the second set of two terms that results in a common binomial factor. Finally, factor out the common binomial factor of each expression, for example:

$$xy - x + 5y - 5$$

$$x(y - 1) + 5(y - 1)$$

$$(y - 1)(x + 5)$$

After the expression is completely factored, check to see if the factorization is correct by multiplying to try to obtain the original expression. Factorizations are helpful in solving equations that consist of a polynomial set equal to zero. If the product of two algebraic expressions equals zero, then at least one of the factors is equal to zero. Therefore, factor the polynomial within the equation, set each factor equal to zero, and solve. For example, $x^2 + 7x - 18 = 0$ can be solved by factoring into:

$$(x + 9)(x - 2) = 0$$

Set each factor equal to zero, and solve to obtain $x = -9$ and $x = 2$.

Creating Polynomials from Written Descriptions

The key to representing problem situations using algebraic expressions, or polynomials, is being able to translate written or verbal phrases and scenarios into correct algebraic expressions. Key words and phrases can be translated into mathematical expressions that involve mathematical operations and variables. For instance, words and phrases such as *plus, sum, more than, added to,* and *increased by* represent addition. Words and phrases such as *less than, decreased by, minus,* and *subtracted from* represent subtraction. Words and phrases such as *times, product, double,* and *triple* represent multiplication, and words and phrases such as *quotient* and *ratio* represent division. Variables are introduced when an unknown quantity needs to be found. Letters such as x, y, and z can be utilized to represent these unknown quantities.

Algebraic expressions, such as polynomials, can be formed by combining variables and numbers with operations such as addition, subtraction, multiplication, and division. For instance, let's say you run a catering business and want to give a formula to your customers that would price their bill for a certain number of steak dinners, pasta dinners, and chicken dinners. If it is known that each steak dinner costs $15, each pasta dinner costs $11, and each chicken dinner costs $12, a formula could be generated. If a customer orders x steak dinners, y pasta dinners, and z chicken dinners, the total bill would be equal to:

$$15x + 11y + 12z$$

The cost of each type of dinner is equal to the number ordered multiplied by the cost per item. The total bill is then the sum of those three amounts. Note that this expression is a polynomial formed from the three variables x, y, and z. Now let's say that this customer wanted a weekly catered dinner for one year. Because there are 52 weeks in a year, the bill for the year would be 52 times the bill, which results in the polynomial:

$$52(15x + 11y + 12z)$$

Note that two individual weeks would have a total bill of:

$$2(15x + 11y + 12z)$$

Which is equivalent to:

$$30x + 22y + 24z$$

Adding, Subtracting, Multiplying, Dividing Rational Expressions

A **rational expression** is a fraction or a ratio in which both the numerator and denominator are polynomials that are not equal to zero. A **polynomial** is a mathematical expression containing the sum and difference of one or more terms that are constants multiplied times variables raised to positive powers. Here are some examples of rational expressions:

$$\frac{2x^2+6x}{x}, \ \frac{x-2}{x^2-6x+8}, \text{ and } \frac{x+2}{x^3-1}$$

Such expressions can be simplified using different forms of division. The first example can be simplified in two ways. First, because the denominator is a monomial, the expression can be split up into two expressions: $\frac{2x^2}{x} + \frac{6x}{x}$, and then simplified using properties of exponents as $2x + 6$. It also can be simplified using factoring and then crossing out common factors in the numerator and denominator. For instance, it can be written as:

$$\frac{2x(x+3)}{x}$$

$$2(x+3) = 2x + 6$$

The second expression above can also be simplified using factoring. It can be written as:

$$\frac{x-2}{(x-2)(x-4)} = \frac{1}{x-4}$$

Finally, the third example can only be simplified using long division, as there are no common factors in the numerator and denominator. First, divide the first term of the denominator by the first term of the numerator, then write that in the quotient. Then, multiply the divisor times that number and write it below the dividend. Subtract, bring down the next term from the dividend, and continue that process with the next first term and first term of the divisor. Continue the process until every term in the divisor is accounted for.

Here is the actual long division:

Simplifying Expressions Using Long Division

$$
\begin{array}{r}
x^2 \quad -2x \quad +4 \\
x+2 \enclose{longdiv}{x^3 \qquad\qquad -1} \\
x^3 +2x^2 \\
\hline
-2x^2 \qquad -1 \\
-2x^2 -4x \\
\hline
4x \quad -1 \\
4x \quad +8 \\
\hline
-9
\end{array}
$$

Writing an Expression from a Written Description

When presented with a real-world problem that must be solved, the first step is always to determine what the unknown quantity is that must be solved for. Use a variable, such as x or t, to represent that unknown quantity. Sometimes, there can be two or more unknown quantities. In this case, either choose an additional variable, or if a relationship exists between the unknown quantities, express the other quantities in terms of the original variable. After choosing the variables, form algebraic expressions and/or equations that represent the verbal statement in the problem. The following table shows examples of vocabulary used to represent the different operations:

Addition	Sum, plus, total, increase, more than, combined, in all
Subtraction	Difference, less than, subtract, reduce, decrease, fewer, remain
Multiplication	Product, multiply, times, part of, twice, triple
Division	Quotient, divide, split, each, equal parts, per, average, shared

The combination of operations and variables form both mathematical expression and equations. As mentioned, the difference between expressions and equations are that there is no equals sign in an expression, and that expressions are **evaluated** to find an unknown quantity, while equations are **solved** to find an unknown quantity. Also, inequalities can exist within verbal mathematical statements. Instead of a statement of equality, expressions state quantities are *less than*, *less than or equal to*, *greater than*, or *greater than or equal to*. Another type of inequality is when a quantity is said to be *not equal to* another quantity. The symbol used to represent "not equal to" is ≠.

The steps for solving inequalities in one variable are the same steps for solving equations in one variable. The addition and multiplication principles are used. However, to maintain a true statement when using the $<$, $\leq$, $>$, and $\geq$ symbols, if a negative number is either multiplied by both sides of an inequality or divided from both sides of an inequality, the sign must be flipped. For instance, consider the following inequality: $3 - 5x \leq 8$. First, 3 is subtracted from each side to obtain $-5x \leq 5$. Then, both sides are divided by -5, while flipping the sign, to obtain $x \geq -1$. Therefore, any real number greater than or equal to -1 satisfies the original inequality.

Using Linear Equations to Solve Real-World Problems

An **equation in one variable** is a mathematical statement where two algebraic expressions in one variable, usually x, are set equal. To solve the equation, the variable must be isolated on one side of the equals sign. The addition and multiplication principles of equality are used to isolate the variable. The **addition principle of equality** states that the same number can be added to or subtracted from both sides of an equation. Because the same value is being used on both sides of the equals sign, equality is maintained. For example, the equation $2x - 3 = 5x$ is equivalent to both:

$$(2x - 3) + 3 = 5x + 3$$

And:

$$(2x - 3) - 5 = 5x - 5$$

This principle can be used to solve the following equation:

$$x + 5 = 4$$

The variable x must be isolated, so to move the 5 from the left side, subtract 5 from both sides of the equals sign. Therefore:

$$x + 5 - 5 = 4 - 5$$

So, the solution is $x = -1$. This process illustrates the idea of an **additive inverse** because subtracting 5 is the same as adding -5. Basically, add the opposite of the number that must be removed to both sides of the equals sign. The **multiplication principle of equality** states that equality is maintained when a number is either multiplied by both expressions on each side of the equals sign, or when both expressions are divided by the same number. For example, $4x = 5$ is equivalent to both $16x = 20$ and $x = \frac{5}{4}$.

Multiplying both sides times 4 and dividing both sides by 4 maintains equality. Solving the equation $6x - 18 = 5$ requires the use of both principles. First, apply the addition principle to add 18 to both sides of the equals sign, which results in $6x = 23$. Then use the multiplication principle to divide both sides by 6, giving the solution $x = \frac{23}{6}$. Using the multiplication principle in the solving process is the same as involving a multiplicative inverse. A **multiplicative inverse** is a value that, when multiplied by a given number, results in 1. Dividing by 6 is the same as multiplying by $\frac{1}{6}$, which is both the reciprocal and multiplicative inverse of 6.

When solving a linear equation in one variable, checking the answer shows if the solution process was performed correctly. Plug the solution into the variable in the original equation. If the result is a false statement, something was done incorrectly during the solution procedure. Checking the example above gives the following:

$$6 \times \frac{23}{6} - 18 = 23 - 18 = 5$$

Therefore, the solution is correct.

Some equations in one variable involve fractions or the use of the **distributive property**. In either case, the goal is to obtain only one variable term and then use the addition and multiplication principles to isolate that variable. Consider the equation $\frac{2}{3}x = 6$. To solve for x, multiply each side of the equation by the reciprocal of $\frac{2}{3}$, which is $\frac{3}{2}$. This step results in:

$$\frac{3}{2} \times \frac{2}{3}x = \frac{3}{2} \times 6$$

which simplifies into the solution $x = 9$. Now consider the equation:

$$3(x + 2) - 5x = 4x + 1$$

Use the distributive property to clear the parentheses. Therefore, multiply each term inside the parentheses by 3. This step results in:

$$3x + 6 - 5x = 4x + 1$$

Next, collect like terms on the left-hand side. **Like terms** are terms with the same variable or variables raised to the same exponent(s). Only like terms can be combined through addition or subtraction. After collecting like terms, the equation is:

$$-2x + 6 = 4x + 1$$

Finally, apply the addition and multiplication principles. Add $2x$ to both sides to obtain:

$$6 = 6x + 1$$

Then, subtract 1 from both sides to obtain $5 = 6x$. Finally, divide both sides by 6 to obtain the solution:

$$\frac{5}{6} = x$$

Two other types of solutions can be obtained when solving an equation in one variable. The final result could be that there is either no solution or that the solution set contains all real numbers. Consider the equation:

$$4x = 6x + 5 - 2x$$

First, the like terms can be combined on the right to obtain:

$$4x = 4x + 5$$

Next, subtract $4x$ from both sides. This step results in the false statement $0 = 5$. There is no value that can be plugged into x that will ever make this equation true. Therefore, there is no solution. The solution procedure contained correct steps, but the result of a false statement means that no value satisfies the equation. The symbolic way to denote that no solution exists is $\emptyset$. Next, consider the equation:

$$5x + 4 + 2x = 9 + 7x - 5$$

Combining the like terms on both sides results in:

$$7x + 4 = 7x + 4$$

The left-hand side is exactly the same as the right-hand side. Using the addition principle to move terms, the result is $0 = 0$, which is always true. Therefore, the original equation is true for any number, and the solution set is all real numbers. The symbolic way to denote such a solution set is $\mathbb{R}$, or in interval notation, $(-\infty, \infty)$.

One-step problems take only one mathematical step to solve. For example, solving the equation $5x = 45$ is a one-step problem because the one step of dividing both sides of the equation by 5 is the only step necessary to obtain the solution $x = 9$. The multiplication principle of equality is the one step used to isolate the variable. The equation is of the form $ax = b$, where a and b are rational numbers. Similarly, the addition principle of equality could be the one step needed to solve a problem. In this case, the equation would be of the form $x + a = b$ or $x - a = b$, for real numbers a and b.

A **multi-step problem** involves more than one step to find the solution, or it could consist of solving more than one equation. An equation that involves both the addition principle and the multiplication principle is a two-step problem, and an example of such an equation is $2x - 4 = 5$. Solving involves adding 4 to both sides and then dividing both sides by 2. An example of a two-step problem involving two separate equations is $y = 3x, 2x + y = 4$. The two equations form a system of two equations that must be solved together in two variables. The system can be solved by the substitution method. Since y is already solved for in terms of x, plug $3x$ in for y into the equation $2x + y = 4$, resulting in $2x + 3x = 4$. Therefore, $5x = 4$ and $x = \frac{4}{5}$. Because there are two variables, the solution consists of a value for both x and for y. Substitute $x = \frac{4}{5}$ into either original equation to find y. The easiest choice is $y = 3x$. Therefore:

$$y = 3 \times \frac{4}{5} = \frac{12}{5}$$

The solution can be written as the ordered pair:

$$\left(\frac{4}{5}, \frac{12}{5}\right)$$

Real-world problems can be translated into both one-step and multi-step problems. In either case, the word problem must be translated from the verbal form into mathematical expressions and equations that can be solved using algebra. An example of a one-step real-world problem is the following: A cat weighs half as much as a dog living in the same house. If the dog weighs 14.5 pounds, how much does the cat weigh? To solve this problem, an equation can be used. In any word problem, the first step is to define variables that represent the unknown quantities. For this problem, let x be equal to the unknown weight of the cat. Because two times the weight of the cat equals 14.5 pounds, the equation to be solved is:

$$2x = 14.5$$

Use the multiplication principle to divide both sides by 2. Therefore, $x = 7.25$. The cat weighs 7.25 pounds.

Most of the time, real-world problems are more difficult than this one and consist of multi-step problems. The following is an example of a multi-step problem: The sum of two consecutive page numbers is equal to 437. What are those page numbers? First, define the unknown quantities. If x is equal to the first page number, then $x + 1$ is equal to the next page number because they are consecutive integers. Their sum is equal to 437, and this statement translates to the equation:

$$x + x + 1 = 437$$

To solve, first collect like terms to obtain:

$$2x + 1 = 437$$

Then, subtract 1 from both sides and then divide by 2. The solution to the equation is $x = 218$. Therefore, the two consecutive page numbers that satisfy the problem are 218 and 219. It is always important to make sure that answers to real-world problems make sense. For instance, it should be a red flag if the solution to this same problem resulted in decimals, which would indicate the need to check the work. Page numbers are whole numbers; therefore, if decimals are found to be answers, the solution process should be double-checked to see where mistakes were made.

Solving a System of Two Linear Equations

An example of a system of two linear equations in two variables is the following:

$$2x + 5y = 8$$

$$5x + 48y = 9$$

A solution to a system of two linear equations is an ordered pair that satisfies both the equations in the system. A system can have one solution, no solution, or infinitely many solutions. The solution can be found through a graphing technique. The solution of a system of equations is actually equal to the point of intersection of both lines. If the lines intersect at one point, there is one solution and the system is said to be **consistent**. However, if the two lines are parallel, they will never intersect and there is no solution. In this case, the system is said to be **inconsistent.** Third, if the two lines are actually the same line, there are infinitely many solutions and the solution set is equal to the entire line. The lines are dependent.

Here is a summary of the three cases:

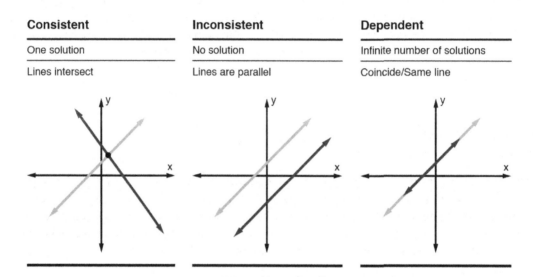

Consistent	Inconsistent	Dependent
One solution	No solution	Infinite number of solutions
Lines intersect	Lines are parallel	Coincide/Same line

Consider the following system of equations:

$$y + x = 3$$

$$y - x = 1$$

To find the solution graphically, graph both lines on the same *xy*-plane. Graph each line using either a table of ordered pairs, the *x*- and *y*-intercepts, or slope and the *y*-intercept. Then, locate the point of intersection.

The graph is shown here:

The System of Equations
$$\begin{cases} y + x = 3 \\ y - x = 1 \end{cases}$$

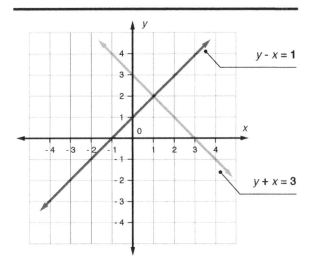

It can be seen that the point of intersection is the ordered pair (1, 2). This solution can be checked by plugging it back into both original equations to make sure it results in true statements. This process results in:

$$2 + 1 = 3$$

$$2 - 1 = 1$$

Both are true equations, so therefore the point of intersection is truly the solution.

The following system has no solution:

$$y = 4x + 1$$

$$y = 4x - 1$$

Both lines have the same slope and different y-Intercepts; therefore, they are parallel. This means that they run alongside each other and never intersect.

Finally, the following solution has infinitely many solutions:

$$2x - 7y = 12$$

$$4x - 14y = 24$$

Note that the second equation is equal to the first equation times 2. Therefore, they are the same line. The solution set can be written in set notation as $\{(x, y) | 2x - 7y = 12\}$, which represents the entire line.

There are two algebraic methods to finding solutions. The first is substitution. This is better suited when one of the equations is already solved for one variable, or it is easy to do so. Then this equation gets substituted into the other equation for that variable, resulting in an equation in one variable. Solve for the given variable, and plug that value into one of the original equations to find the other variable. This last step is known as back-substitution.

Here is an example of solving a system with the substitution method:

$$y = x + 1 \qquad 2y = 3x$$

$$2y = 3x$$

$$2(x + 1) = 3x$$

$$2x + 2 = 3x$$

$$\underline{-2x \qquad\quad -2x}$$

$$2 = x$$

$$y = x + 1$$

$$y = 2 + 1 = 3$$

Solution: (2 , 3)

The other method is known as elimination, or the addition method. This is better suited when the equations are in standard form:

$$Ax + By = C$$

The goal in this method is to multiply one or both equations times numbers that result in opposite coefficients. Then add the equations together to obtain an equation in one variable. Solve for the given variable, and then take that value and back-substitute to obtain the other part of the ordered pair solution.

Here is an example of elimination:

$$\begin{cases} -x + 5y = 8 \\ 3x + 7y = -2 \end{cases} \xRightarrow{\times 3} \begin{cases} -3x + 15y = 24 \\ 3x + 7y = -2 \end{cases}$$

$$-3x + 15y = 24$$
$$\underline{3x + 7y = -2}$$
$$22y = 22$$
$$\frac{22y}{22} = \frac{22}{22}$$
$$y = 1$$

Note that in order to check an answer when solving a system of equations, the solution must be checked in both original equations to show that it solves both equations.

Solving Inequalities and Graphing the Answer on a Number Line

A **linear equation** *in x* can be written in the form $ax + b = 0$. A **linear inequality** is very similar, although the equals sign is replaced by an inequality symbol such as $<, >, \leq$, or $\geq$. In any case, a can never be 0. Some examples of linear inequalities in one variable are $2x + 3 < 0$ and $4x - 2 \leq 0$. Solving an inequality involves finding the set of numbers that when plugged into the variable, make the inequality a true statement. These numbers are known as the **solution set** of the inequality. To solve an inequality, use the same properties that are necessary in solving equations. First, add or subtract variable terms and/or constants to obtain all variable terms on one side of the equals sign and all constant terms on the other side. Then, either multiply both sides times the same number, or divide both sides by the same number, to obtain an inequality that gives the solution set. When multiplying times, or dividing by, a negative number in an inequality, change the direction of the inequality symbol. The solution set can be graphed on a number line. Consider the linear inequality:

$$-2x - 5 > x + 6$$

First, add 5 to both sides and subtract $-x$ off of both sides to obtain $-3x > 11$. Then, divide both sides by -3, making sure to change the direction of the inequality symbol. These steps result in the solution:

$$x < -\frac{11}{3}$$

Therefore, any number less than $-\frac{11}{3}$ satisfies this inequality.

Inequalities can be solved in a similar method as equations. Basically, the goal is to isolate the variable, and this process can be completed by adding numbers onto both sides, subtracting numbers off of both sides, multiplying numbers onto both sides, and dividing numbers off of both sides of the inequality. Basically, if something is done to one side, it has to be done to the other side, just like when solving equations. However, there is one important difference, and that difference occurs when multiplying times negative numbers and dividing by negative numbers. If either one of these steps must be performed in the solution process, the inequality symbol must be reversed. Consider the following inequality: $2 - 3x < 11$. The goal is to isolate the variable x, so first subtract 2 off both sides to obtain $-3x < 9$. Then divide both sides by -3, making sure to "*flip the sign.*" This results in $x > -3$, which is the solution set. This solution set means that all numbers greater than -3 satisfy the original inequality, and therefore any number larger than -3 is a solution. In **set-builder notation,** this set can be written as $\{x | x > -3\}$, which is read "all *x* values such that *x* is greater than -3." In addition to the inequality form of the solution, solutions of inequalities can be expressed by using both a number line and interval notation. Here is a chart that highlights all three types of expressing the solutions:

Interval Notation	Number Line Sketch	Set-builder Notation
(a , b)		$\{\, x \mid a < x < b \}$
(a , b]		$\{\, x \mid a < x \le b \}$
[a , b)		$\{\, x \mid a \le x < b \}$
[a , b]		$\{\, x \mid a \le x \le b \}$
(a , ∞)		$\{\, x \mid x > a \}$
(- ∞ , b)		$\{\, x \mid x < b \}$
[a , ∞)		$\{\, x \mid x \ge a \}$
(- ∞ , b]		$\{\, x \mid x \le b \}$
(- ∞, ∞)		$\mathbb{R}$

Quadratic Equations with One Variable

A quadratic equation in standard form, $ax^2 + bx + c = 0$, can have either two solutions, one solution, or two complex solutions (no real solutions). This is determined using the determinant $b^2 - 4ac$. If the determinant is positive, there are two real solutions. If the determinant is negative, there are no real solutions. If the determinant is equal to 0, there is one real solution. For example, given the quadratic equation $4x^2 - 2x + 1 = 0$, its determinant is:

$$(-2)^2 - 4(4)(1)$$

$$4 - 16 = -12$$

So it has two complex solutions, meaning no real solutions.

There are quite a few ways to solve a quadratic equation. The first is by **factoring.** If the equation is in standard form and the polynomial can be factored, set each factor equal to 0, and solve using the Principle of Zero Products. For example:

$$x^2 - 4x + 3 = (x - 3)(x - 1)$$

Therefore, the solutions of $x^2 - 4x + 3 = 0$ are those that satisfy both $x - 3 = 0$ and $x - 1 = 0$, or $x = 3$ and $x = 1$. This is the simplest method to solve quadratic equations; however, not all polynomials inside the quadratic equations can be factored.

Another method is **completing the square**. The polynomial $x^2 + 10x - 9$ cannot be factored, so the next option is to complete the square in the equation $x^2 + 10x - 9 = 0$ to find its solutions. The first step is to add 9 to both sides, moving the constant over to the right side, resulting in:

$$x^2 + 10x = 9$$

Then the coefficient of x is divided by 2 and squared. This result is then added to both sides of the equation. In this example, $\left(\frac{10}{2}\right)^2 = 25$ is added to both sides of the equation to obtain:

$$x^2 + 10x + 25$$

$$9 + 25 = 34$$

The left-hand side can then be factored into:

$$(x + 5)^2 = 34$$

Solving for x then involves taking the square root of both sides and subtracting 5. This leads to the two solutions:

$$x = \pm\sqrt{34} - 5$$

The third method is the **quadratic formula.** Given a quadratic equation in standard form:

$$ax^2 + bx + c = 0$$

Its solutions always can be found using the formula:

$$x = \frac{-b \pm \sqrt{b^2 - 4ac}}{2a}$$

Graphs and Functions

Locating Points and Graphing Equations

Graphs, equations, and tables are three different ways to represent linear relationships. The following graph shows a linear relationship because the relationship between the two variables is constant. As the distance increases by 25 miles, the time lapses by 1 hour. This pattern continues for the rest of the graph. The line represents a constant rate of 25 miles per hour. This graph can also be used to solve problems involving predictions for a future time. After 8 hours of travel, the rate can be used to predict the distance covered. Eight hours of travel at 25 miles per hour covers a distance of 200 miles. The equation at the top of the graph corresponds to this rate also. The same prediction of distance in a given time can be found using the equation.

For a time of 10 hours, the distance would be 250 miles, as the equation yields:

$$d = 25 \times 10 = 250$$

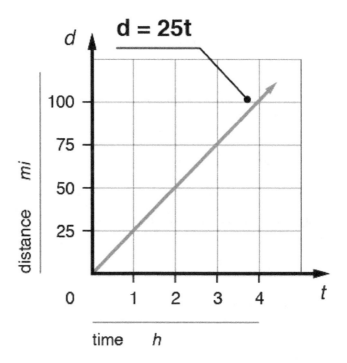

Another representation of a linear relationship can be seen in a table. The first thing to observe from the table is that the y-values increase by the same amount of 3 each time. As the x-values increase by 1, the y-values increase by 3. This pattern shows that the relationship is linear. If this table shows the money earned, y-value, for the hours worked, x-value, then it can be used to predict how much money will be earned for future hours. If 6 hours are worked, then the pay would be $19. For further hours and money to be determined, it would be helpful to have an equation that models this table of values. The equation will show the relationship between x and y. The y-value can each time be determined by multiplying the x-value by 3, then adding 1. The following equation models this relationship:

$$y = 3x + 1$$

Now that there is an equation, any number of hours, *x*, can be substituted into the equation to find the amount of money earned, *y*.

$y = 3x + 1$	
x	y
0	1
1	4
2	7
4	13
5	16

Determining the Slope of a Line from a Graph, Equation, or Table

A linear function that models a linear relationship between two quantities is of the form $y = mx + b$, or in function form $f(x) = mx + b$. In a linear function, the value of *y* depends on the value of *x*, and *y* increases or decreases at a constant rate as *x* increases. Therefore, the independent variable is *x*, and the dependent variable is *y*. The graph of a linear function is a line, and the constant rate can be seen by looking at the steepness, or **slope**, of the line. If the line increases from left to right, the slope is positive. If the line slopes downward from left to right, the slope is negative. Slope is rise over run or how much the *y*-values change for a given change in *x*-values.

In the function, *m* represents slope. Each point on the line is an **ordered pair** (*x, y*), where *x* represents the *x*-coordinate of the point and *y* represents the *y*-coordinate of the point. The point where *x* = 0 is known as the *y*-intercept, and it is the place where the line crosses the *y*-axis. If *x* = 0 is plugged into $f(x) = mx + b$, the result is $f(0) = b$, so therefore, the point (0, *b*) is the *y*-intercept of the line. The derivative of a linear function is its slope. The slope can also be determined by finding the difference in the *y*-values between two ordered pairs and dividing this difference by the difference in the *x*-values of the same two ordered pairs.

Proportional Relationships for Equations and Graphs

Proportional relationships can be seen in equations and graphs. Consider the following situation. A taxicab driver charges a flat fee of $2 per ride and $3 a mile. This statement can be modeled by the function $f(x) = 3x + 2$ where *x* represents the number of miles and $f(x) = y$ represents the total cost of the ride. The total cost increases at a constant rate of $2 per mile, and that is why this situation is a linear relationship. The slope $m = 3$ is equivalent to this rate of change. The flat fee of $2 is the *y*-

intercept. It is the place where the graph crosses the x-axis, and it represents the cost when $x = 0$, or when no miles have been traveled in the cab. The y-intercept in this situation represents the flat fee.

Features of Graphs and Tables for Linear and Nonlinear Relationships

When given data in ordered pairs, choosing an appropriate function or equation to model the data is important. Besides linear relationships, other common relationships that exist are quadratic and exponential. A helpful way to determine what type of function to use is to find the difference between consecutive dependent variables. Basically, find pairs of ordered pairs where the x-values increase by 1, and take the difference of the y-values. If the differences between subsequent y-values are always the same value, then the function is **linear.**

If the differences between subsequent y-values are not the same, the function could be quadratic or exponential. In **quadratic functions**, when the differences between the x-values are the same (for example, increasing by 1), the differences between subsequent y-values will not be the same. Instead, the difference of the differences between subsequent y–values will be the same. For example, in the simplest quadratic function, $y = x^2$, as the x-values increase by 1, the y–values increase by different amounts, but the difference between the difference is constant (2).

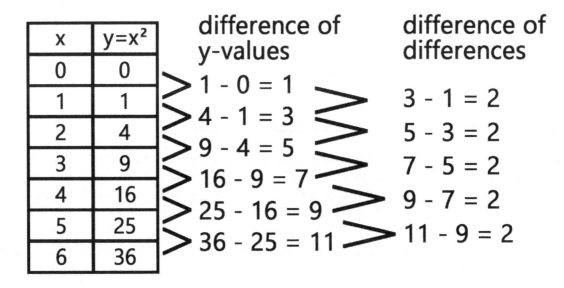

If consecutive differences between the differences are not the same, try taking ratios of consecutive y-values. If the ratios are the same, the data have an **exponential** relationship and an exponential function should be used.

For example, the ordered pairs (1, 4), (2, 6), (3 ,8), and (4,10) have a linear relationship because the difference in y-values is 2 for every increase in x of 1. The ordered pairs (1, 0), (2, 3), (3, 10), and (4, 21) have a nonlinear relationship. The first differences in y-values are 3, 7, and 11, however, consecutive second differences are both 4, so the function is quadratic. Third, the ordered pairs (1, 10), (2, 30), (3, 90), and (4, 270) have an exponential relationship. Taking ratios of consecutive y-values leads to a common ratio of 4.

The general form of a **quadratic equation** is $y = ax^2 + bx + c$, and its **vertex form** is $y = a(x - h)^2 + k$, with vertex (h, k). If the vertex and one other point are known, the vertex form should be used to solve for *a*. If three points, not the vertex, are known, the general form should be used. The three points create a system of three equations in three unknowns that can be solved for.

The general form of an **exponential function** is $y = b \times a^x$, where *a* is the **base** and *b* is the **y-intercept**.

Recall that a linear relationship exists between two variables if they are proportional to each other. Basically, if one quantity increases, the other quantity increases or decreases at a constant rate. The graph of a linear relationship is a straight line. If the line goes through the *origin*, the point (0, 0), then there is direct variation between the two quantities and the equation for direct variation is $y = kx$, where *k* is known as the constant of variation.

Recall also that a nonlinear relationship exists between two variables if an increase in one quantity does not correspond with a constant change in the other quantity. The graphs of nonlinear relationships are not straight lines. Often, the graphs of nonlinear relationships involve curves, where the change in the variable is predictable, but not constant.

A common nonlinear relationship between two variables involves inverse variation where one quantity varies inversely with respect to another. The equation for inverse variation is $y = \frac{k}{x}$, where *k* is still known as the constant of variation. Here is the graph of the curve $y = \frac{3}{x}$:

The Graph of $y = \dfrac{3}{x}$

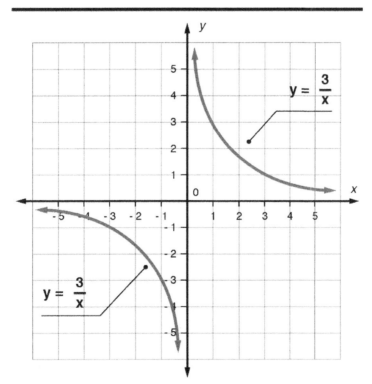

Other common nonlinear functions involve polynomial functions. The squaring function $f(x) = x^2$ can be seen here:

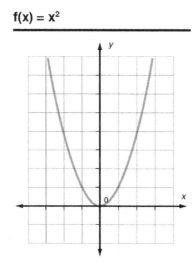

f(x) = x²

Notice that as the independent variable x increases when $x > 0$, the dependent variable y also increases. However, y does not increase at a constant rate.

Exponential growth involves a quantity, the dependent variable, changing by a common ratio every unit increase or equal interval. The equation of exponential growth is $y = a^x$ for $a > 0$, $a \neq 1$. The value a is known as the **base**. Consider the exponential equation $y = 2^x$. When x equals 1, y equals 2, and when x equals 2, y equals 4. For every unit increase in x, the value of the output variable doubles. Here is the graph of $y = 2^x$. Notice that as the dependent variable, y, gets very large, x increases slightly. This characteristic of this graph is why sometimes a quantity is said to be blowing up exponentially.

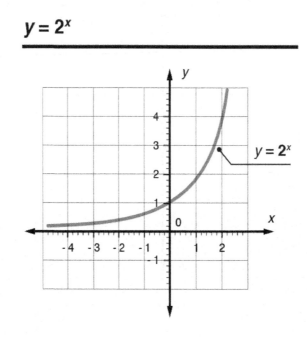

y = 2ˣ

Equation of a Line from the Slope and a Point on a Line

A **linear function** of the form $f(x) = mx + b$ has two important quantities: m and b. The quantity m represents the slope of the line, and the quantity b represents the y-intercept of the line. When the function represents an actual real-life situation, or mathematical model, these two quantities are very meaningful. The slope, m, represents the rate of change, or the amount y increases or decreases given an increase in x. If m is positive, the rate of change is positive, and if m is negative, the rate of change is negative. The y-intercept, b, represents the amount of the quantity y when x is 0. In many applications, if the x-variable is never a negative quantity, the y-intercept represents the initial amount of the quantity y. Often the x-variable represents time, so it makes sense that the x-variable is never negative.

Consider the following example. These two equations represent the cost, C, of t-shirts, x, at two different printing companies:

$$C(x) = 7x$$

$$C(x) = 5x + 25$$

The first equation represents a scenario that shows the cost per t-shirt is $7. In this equation, x varies directly with y. There is no y-intercept, which means that there is no initial cost for using that printing company. The rate of change is 7, which is price per shirt. The second equation represents a scenario that has both an initial cost and a cost per t-shirt. The slope 5 shows that each shirt is $5. The y-intercept 25 shows that there is an initial cost of using that company. Therefore, it makes sense to use the first company at $7 a shirt when only purchasing a small number of t-shirts. However, any large orders would be cheaper by going with the second company because eventually that initial cost will be negligible.

If a line has slope m and y-intercept b, then the equation of the line is written as $y = mx + b$, which is defined as the slope-intercept form of the equation of a line. For instance, if a line has slope -2 and the y-intercept has coordinate (0, -4), its corresponding equation is $y = -2x - 4$ because $m = -2$ and $b = -4$. To determine the equation of the line, the point given does not have to be a y-intercept. If another point on the line is given along with the slope, the point-slope form of the equation of the line can be used. This form states that for any line with slope m and point (x_1, y_1), its corresponding equation is:

$$y - y_1 = m(x - x_1)$$

For example, if a line has slope 2 and passes through the point (2, -1), its corresponding point-slope form is:

$$y - (-1) = 2(x - 2)$$

Distribute the 2 and then solve for y to determine that its slope-intercept form is:

$$y = 2x - 5$$

Another case in which an equation of a line can be found involves horizontal lines. Horizontal lines have slope equal to zero. Therefore, if a horizontal line has a y-intercept corresponding to the point (0, -2), then the equation of the line is $y = 0x - 2$, which simplifies to $y = -2$. Similarly, equations of vertical lines can be found. If a vertical line passes through the point (2, 4), then the equation of the line is $x = 2$. Vertical lines have an undefined slope.

Equation of a Line from Two Points

In order to find the equation of a line, the slope of that line and a point on that line need to be found. If two points are given on the line, the slope can be calculated as the rise over the run. In other words, if points (x_1, y_1) and (x_2, y_2) are given, the slope formula is:

$$m = \frac{y_2 - y_1}{x_2 - y_1}$$

Once the slope is calculated, its value and either one of the given two points can be plugged into the point-slope form of the line:

$$y - y_1 = m(x - x_1)$$

For example, consider the line which contains the two points (-5, -2) and (-4, 11). To find the equation of the line, first the slope is calculated as:

$$m = \frac{11 - (-2)}{-4 - (-5)}$$

$$\frac{11 + 2}{-4 + 5} = \frac{13}{1} = 13$$

Then, choose either point to plug into the point-slope form. Usually, the point that is easier to work with is chosen. That point might mean the one with less fractions, decimals, or negative signs. In this case, choose (-4, 11) to be plugged into the point-slope formula. Therefore, we obtain $y - 11 = 13(x - (-4))$, which can be simplified into slope-intercept form by distributing the 13 and solving for y. The slope-intercept form is obtained here:

$$y - 11 = 13(x + 4)$$

$$y - 11 = 13x + 52$$

$$y = 13x + 63$$

Using Slope of a Line

Two lines are **parallel** if they never intersect. Given the equation of two lines, they are parallel if they have the same slope and different y-intercepts. If they had the same slope and same y-intercept, they would be the same line. Therefore, in order to show two lines are parallel, put them in slope-intercept form, $y = mx + b$, to find m and b. The two lines $y = 2x + 6$ and $4x - 2y = 6$ are parallel. The second line in slope intercept is $y = 2x - 3$. Both lines have the same slope, 2, and different y-intercepts.

Two lines are **perpendicular** if they intersect at a right angle. Given the equation of two lines, they are perpendicular if their slopes are negative reciprocals. Therefore, the product of both slopes is equal to -1. For example, the lines $y = 4x + 1$ and $y = -\frac{1}{4}x + 1$ are perpendicular because their slopes are negative reciprocals. The product of 4 and $-\frac{1}{4}$ is -1.

Functions Shown in Different Ways

A **relation** is any set of ordered pairs (x, y). The first set of points, known as the x-coordinates, make up the domain of the relation. The second set of points, known as the y-coordinates, make up the range of the relation. A relation in which every member of the domain corresponds to only one member of the range is known as a **function.** A function cannot have a member of the domain corresponding to two members of the range. Functions are most often given in terms of equations instead of ordered pairs. For instance, here is an equation of a line:

$$y = 2x + 4$$

In function notation, this can be written as:

$$f(x) = 2x + 4$$

The expression $f(x)$ is read "f of x" and it shows that the inputs, the x-values, get plugged into the function and the output is $y = f(x)$. The set of all inputs are in the domain and the set of all outputs are in the range.

The x-values are known as the **independent variables** of the function and the y-values are known as the **dependent variables** of the function. The y-values depend on the x-values. For instance, if $x = 2$ is plugged into the function shown above, the y-value depends on that input.

$$f(2) = 2 \times 2 + 4 = 8.$$

Therefore, $f(2) = 8$, which is the same as writing the ordered pair (2, 8). To graph a function, graph it in equation form. Therefore, replace $f(x)$ with h and plot ordered pairs.

Due to the definition of a function, the graph of a function cannot have two of the same x-components paired to different y-component. For example, the ordered pairs (3, 4) and (3, -1) cannot be in a valid function. Therefore, all graphs of functions pass the **vertical line test**. If any vertical line intersects a graph in more than one place, the graph is not that of a function. For instance, the graph of a circle is not a function because one can draw a vertical line through a circle and the line would intersect the circle twice. Common functions include lines and polynomials, and they all pass the vertical line test.

Functions in Tables and Graphs

As mentioned, in math, a relation is a relationship between two sets of numbers. By using a rule, it takes a number from the first set and matches it to a number in the second set. A relation consists of a set of inputs, known as the **domain,** and a set of outputs, known as the **range.** A function is a relation in which each member of the domain is paired to only one other member of the range. In other words, each input has only one output.

Here is an example of a relation that is not a function:

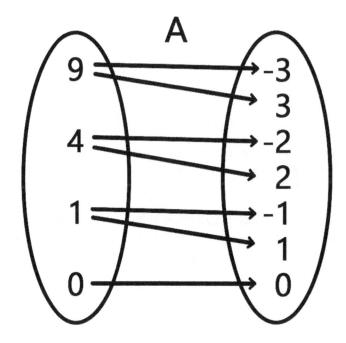

Every member of the first set, the domain, is mapped to two members of the second set, the range. Therefore, this relation is not a function.

In addition to a diagram representing sets, a function can be represented by a table of ordered pairs, a graph of ordered pairs (a scatterplot), or a set of ordered pairs as shown in the following:

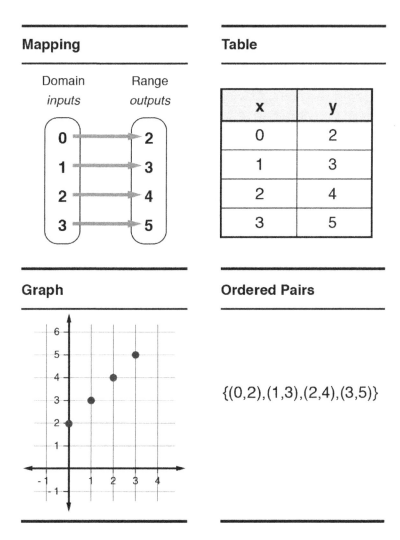

Mapping

Domain — inputs

Range — outputs

Table

x	y
0	2
1	3
2	4
3	5

Graph

Ordered Pairs

$\{(0,2),(1,3),(2,4),(3,5)\}$

Note that this relation is a function because every member of the domain is mapped to exactly one member of the range.

An equation occurs when two algebraic expressions are set equal to one another. Functions can be represented in equation form. Given an equation in two variables, x and y, it can be expressed in function form if solved for y. For example, the linear equation $2x + y = 5$ can be solved for y to obtain $y = -2x + 5$, otherwise known as **slope-intercept** form. To place the equation in function form, replace y with $f(x)$, which is read "f of x." Therefore:

$$f(x) = -2x + 5$$

This notation clarifies the input–output relationship of the function. The function f is a function of x, so an x value can be plugged into the function to obtain an output. For example:

$$f(2) = -2 \times 2 + 5 = 1$$

Therefore, an input of 2 corresponds to an output of 1.

A function can be graphed by plotting ordered pairs in the *xy*-plane in the same way that the equation form is graphed. The graph of a function always passes the Vertical Line Test. Basically, for any graph, if a vertical line can be drawn through any part of the graph and it hits the graph in more than one place, the graph is not a function. For example, the graph of a circle is not a function. The Vertical Line Test shows that with these relationships, the same *x* value has more than one *y* value, which goes against the definition of a function.

Inequalities look like equations, but instead of an equals sign, $<, >, \leq, \geq$, or $\neq$ are used. Here are some examples of inequalities: $2x + 7 < y$, $3x^2 \geq 5$, and $x \neq 4$. Inequalities show relationships between algebraic expressions when the quantities are different. Inequalities can also be expressed in function form if they are solved for *y*. For instance, the first inequality listed above can be written as:

$$2x + y < f(x)$$

Evaluating Functions

Functions are usually denoted by a letter, and the most common letter used is f. If x is defined to be the input variable of the function, then $f(x)$ is the output of the function and is read as "f of x." There usually is a rule associated with the function, typically in terms of a formula. For instance, consider the function:

$$f(x) = x^2 + 1$$

The function can be thought of as directions describing what is done to the input variable when it is inputted into the function. In this case, the input variable is squared and then added to 1. For example:

$$f(2) = 2^2 + 1 = 4 + 1 = 5$$

Therefore, the input is 2, its output is 5, and the corresponding ordered pair is (2, 5). In other words, we have evaluated the function at 2 to obtain 5. Input values can also be negative numbers. Using the same function, if the input variable is -2, its corresponding output is:

$$f(-2) = (-2)^2 + 1 = 4 + 1 = 5$$

In this case, both -2 and 2 correspond to the same output of 5. Within functions, two input variables can have the same output variables but the same input variable cannot have two separate output variables.

Reasoning Through Language Arts

Reading for Meaning

Events, Plots, Characters, Settings, and Ideas

Putting Events in Order

Sequence structure is the order of events in which a story or information is presented to the audience. Sometimes the text will be presented in chronological order, or sometimes it will be presented by displaying the most recent information first, then moving backwards in time. The sequence structure depends on the author, the context, and the audience. The structure of a text also depends on the genre in which the text is written. Is it literary fiction? Is it a magazine article? Is it instructions for how to complete a certain task? Different genres will have different purposes for switching up the sequence.

The structure presented in literary fiction, called **narrative structure**, is the foundation on which the text moves. The narrative structure comes from the plot and setting. The plot is the sequence of events in the narrative that move the text forward through cause and effect. The setting is the place or time period in which the story takes place. Narrative structure has two main categories: linear and nonlinear.

Linear Narrative

A narrative is linear when it is told in chronological order. Traditional linear narratives will follow the plot diagram below depicting the narrative arc. The narrative arc consists of the exposition, conflict, rising action, climax, falling action, and resolution.

- **Exposition**: The exposition is in the beginning of a narrative and introduces the characters, setting, and background information of the story. The exposition provides the context for the upcoming narrative. Exposition literally means "a showing forth" in Latin.

- **Conflict**: In a traditional narrative, the conflict appears toward the beginning of the story after the audience becomes familiar with the characters and setting. The conflict is a single instance between characters, nature, or the self, in which the central character is forced to make a decision or move forward with some kind of action. The conflict presents something for the main character, or protagonist, to overcome.

- **Rising Action**: The rising action is the part of the story that leads into the climax. The rising action will develop the characters and plot while creating tension and suspense that eventually lead to the climax.

- **Climax**: The climax is the part of the story where the tension produced in the rising action comes to a culmination. The climax is the peak of the story. In a traditional structure, everything before the climax builds up to it, and everything after the climax falls from it. It is the height of the narrative, and it is usually either the most exciting part of the story or a turning point in the character's journey.

- **Falling Action**: The falling action happens as a result of the climax. Characters continue to develop, although there is a wrapping up of loose ends here. The falling action leads to the resolution.

- **Resolution**: The resolution is where the story comes to an end and usually leaves the reader with the satisfaction of knowing what happened within the story and why. However, stories do not always end in this fashion. Sometimes readers can be confused or frustrated at the end from lack of information or the absence of a happy ending.

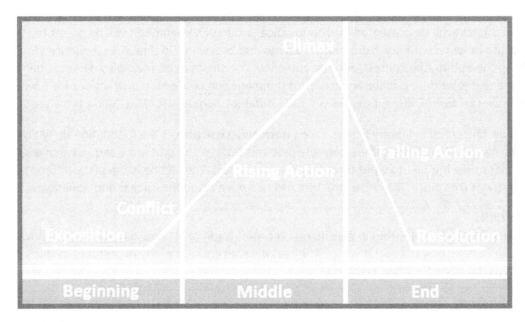

Nonlinear Narrative

A nonlinear narrative deviates from the traditional narrative because it does not always follow the traditional plot structure of the narrative arc. Nonlinear narratives may include structures that are disjointed, circular, or disruptive, in the sense that they do not follow chronological order. **In medias res** is an example of a nonlinear structure. *In medias res* is Latin for "in the middle of things," which is how many ancient texts, especially epic poems, began their story, such as Homer's *Iliad*. Instead of having a clear exposition with a full development of characters, they would begin right in the middle of the action.

Many modernist texts in the late nineteenth and early twentieth centuries experimented with disjointed narratives, moving away from traditional linear narrative. Disjointed narratives are depicted in novels like *Catch 22*, where the author, Joseph Heller, structures the narrative based on free association of ideas rather than chronology. Another nonlinear narrative can be seen in the novel *Wuthering Heights*, written by Emily Brontë; after the first chapter, the narrative progresses retrospectively instead of chronologically. There seem to be two narratives in *Wuthering Heights* working at the same time: a present narrative as well as a past narrative. Authors employ disrupting narratives for various reasons; some use it for the purpose of creating situational irony for the readers, while some use it to create a certain effect, such as excitement, discomfort, or fear.

Sequence Structure in Technical Documents

The purpose of technical documents, such as instructions manuals, cookbooks, or "user-friendly" documents, is to provide information to users as clearly and efficiently as possible. In order to do this, the sequence structure in technical documents should be as straightforward as possible. This usually involves some kind of chronological order or a direct sequence of events. For example, someone who is reading an instruction manual on how to set up their Smart TV wants directions in a clear, simple, straightforward manner that does not confuse them or leave them guessing about the proper sequence.

Sequence Structure in Informational Texts

The structure of informational texts depends on the specific genre. For example, a newspaper article may start by stating an exciting event that happened, then talk about that event in chronological order. Many informational texts also use cause and effect structure, which describes an event and then identifies reasons for why that event occurred. Some essays may write about their subjects by way of comparison and contrast, which is a structure that compares two things or contrasts them to highlight their differences. Other documents, such as proposals, will have a problem to solution structure, where the document highlights some kind of problem and then offers a solution. Finally, some informational texts are written with lush details and description in order to captivate the audience, allowing them to visualize the information presented to them. This type of structure is known as descriptive.

Making Inferences or Drawing Conclusions about Plots, Sequence of Events, Characters, Settings, and Ideas in Passages

One technique authors often use to make their fictional stories more interesting is not giving away too much information by providing hints and description. It is then up to the reader to draw a conclusion about the author's meaning by connecting textual clues with the reader's own pre-existing experiences and knowledge. Drawing conclusions is important as a reading strategy for understanding what is occurring in a text. Rather than directly stating who, what, where, when, or why, authors often describe story elements. Then, readers must draw conclusions to understand significant story components. As they go through a text, readers can think about the setting, characters, plot, problem, and solution; whether the author provided any clues for consideration; and combine any story clues with their existing knowledge and experiences to draw conclusions about what occurs in the text.

Making Predictions

Before and during reading, readers can apply the reading strategy of making predictions about what they think may happen next. For example, what plot and character developments will occur in fiction? What points will the author discuss in nonfiction? Making predictions about portions of text they have not yet read prepares readers mentally for reading, and also gives them a purpose for reading. To inform and make predictions about text, the reader can do the following:

- Consider the title of the text and what it implies
- Look at the cover of the book
- Look at any illustrations or diagrams for additional visual information
- Analyze the structure of the text
- Apply outside experience and knowledge to the text

Readers may adjust their predictions as they read. Reader predictions may or may not come true in text.

Making Inferences

An inference is an educated guess or conclusion based on sound evidence and reasoning within the text. When multiple-choice test questions ask about the logical conclusion that can be drawn from reading text, the test-taker must identify which choice will unavoidably lead to that conclusion. In order to eliminate the incorrect choices, the test-taker should come up with a hypothetical situation wherein an answer choice is true, but the conclusion is not true.

Here is an example:

> Fred purchased the newest PC available on the market. Therefore, he purchased the most expensive PC in the computer store.

> What can one assume for this conclusion to follow logically?

> a. Fred enjoys purchasing expensive items.
> b. PCs are some of the most expensive personal technology products available.
> c. The newest PC is the most expensive one.

The premise of the text is the first sentence: Fred purchased the newest PC. The conclusion is the second sentence: Fred purchased the most expensive PC. Recent release and price are two different factors; the difference between them is the logical gap. To eliminate the gap, one must connect the new information from the conclusion with the pertinent information from the premise. In this example, there must be a connection between product recency and product price. Therefore, a possible bridge to the logical gap could be a sentence stating that the newest PCs always cost the most.

Analyzing Relationships within Passages, Including How People, Events, and Ideas are Connected

In a passage on the test, the relationships between people, events, and ideas may be clearly stated, or the reader might have to infer the relationships based on clues in the passage. To infer means to arrive at a conclusion based on evidence, clues, or facts.

People might be related through connections like family or friendship, or through events that link them directly or indirectly. In the passage, relationships may be described in background information or dialogue, or they may be implied through interactions between the characters.

Events and ideas in a passage can be related through sequence, comparison, or cause and effect. Relating events and ideas through sequence means ordering them chronologically, alphabetically, or geographically. An author can also relate ideas and events through comparison, showing the similarities and differences.

Sequence

When ideas are related through sequence, as in a series of events, the author typically uses signal words like *after, and then, while,* and *before.*

Cause and Effect

In cause and effect relationships, the cause is an event or circumstance that occurs before and is directly responsible for the effect. Signal words such as *because, due to,* and *as a result of* indicate a cause and effect relationship. When the relationship is implied, the reader must use clues in the passage to infer that the effect resulted from the cause.

Compare and Contrast

Ideas are often connected through comparisons of their similarities using the signal words such as *like* or *and.*

Understanding Main Ideas and Details

Determining the Relationship Between Ideas

The main idea of a passage is related to the details that support it. These details demonstrate specific examples that support the main idea. For example, suppose a passage's main idea is "*Schools should be privatized.*" The passage supports this main idea with this detail: "*Students who attend private schools perform better.*" The main idea is further supported by this statistical detail: "*65% of students with the highest national GPAs are from private schools.*"

Like the connections between events, these relationships may be clearly stated, or they may be implied. The reader can infer these relationships by analyzing signal words and the stated or implied connections. Below is a list of signal words.

Time Order
Time order can be inferred through words like *before* and *after,* or *next, then, first, lastly,* which signal that events occur one after the other.

Examples
Specific examples support a main idea or definitions.

Comparisons
The words like *and, also, like* signal a comparison of the similarities between ideas

Contrasts
Words like *but, however* signal a contrast of the differences between ideas.

Cause and Effect
Words like *because, so, therefore* signal that one event or circumstance is the cause of another.

Spatial
Ideas or objects are described by their location with respect to each other by using words such as *behind* and *to the right.*

Analyzing How Details Develop the Main Idea
A paragraph is made up of a series of related sentences that support a single main idea or message. The paragraph might clearly state the main idea or imply it through supporting details, guiding the reader to the desired conclusion.

The main idea is what the passage is about. When this main idea is clearly stated, the reader can easily determine how the evidence or details that follow support this main idea. Suppose the author states, "*Because of their inexperience with predators, dodos went extinct in the 17ᵗʰ century.*" The subsequent sentences will support this claim with details. The author will describe the sequence of events, each one leading the reader to agree with the statement based on the evidence and reach the conclusion the author wants them to reach.

If the main idea is not directly stated, the author must imply the main idea through the use of strong supporting details. These details could be comparisons of like ideas or contrasts of different ideas, factual evidence in graphs and statistics, quotes from experts, or vivid descriptions that evoke the desired emotional response.

Point of View and Purpose

Identifying the Author's Point of View and Purpose
Authors may have many purposes for writing a specific text. Their purposes may be to try and convince readers to agree with their position on a subject, to impart information, or to entertain. Other writers are motivated to write from a desire to express their own feelings. Authors' purposes are their reasons for

writing something. A single author may have one overriding purpose for writing or multiple reasons. An author may explicitly state their intention in the text, or the reader may need to infer that intention. Those who read reflectively benefit from identifying the purpose because it enables them to analyze information in the text. By knowing why the author wrote the text, readers can glean ideas for how to approach it.

The following is a list of questions readers can ask in order to discern an author's purpose for writing a text:

- From the title of the text, why do you think the author wrote it?
- Was the purpose of the text to give information to readers?
- Did the author want to describe an event, issue, or individual?
- Was it written to express emotions and thoughts?
- Did the author want to convince readers to consider a particular issue?
- Was the author primarily motivated to write the text to entertain?
- Why do you think the author wrote this text from a certain point of view?
- What is your response to the text as a reader?
- Did the author state their purpose for writing it?

Students should read to interpret information rather than simply content themselves with roles as text consumers. Being able to identify an author's purpose efficiently improves reading comprehension, develops critical thinking, and makes students more likely to consider issues in depth before accepting writer viewpoints. Authors of fiction frequently write to entertain readers. Another purpose for writing fiction is making a political statement; for example, Jonathan Swift wrote "A Modest Proposal" (1729) as a political satire. Another purpose for writing fiction as well as nonfiction is to persuade readers to take some action or further a particular cause. Fiction authors and poets both frequently write to evoke certain moods; for example, Edgar Allan Poe wrote novels, short stories, and poems that evoke moods of gloom, guilt, terror, and dread. Another purpose of poets is evoking certain emotions: love is popular, as in Shakespeare's sonnets and numerous others. In "The Waste Land" (1922), T.S. Eliot evokes society's alienation, disaffection, sterility, and fragmentation.

Authors seldom directly state their purposes in texts. Some students may be confronted with nonfiction texts such as biographies, histories, magazine and newspaper articles, and instruction manuals, among others. To identify the purpose in nonfiction texts, students can ask the following questions:

- Is the author trying to teach something?
- Is the author trying to persuade the reader?
- Is the author imparting factual information only?
- Is this a reliable source?
- Does the author have some kind of hidden agenda?

To apply author purpose in nonfictional passages, students can also analyze sentence structure, word choice, and transitions to answer the aforementioned questions and to make inferences. For example, authors wanting to convince readers to view a topic negatively often choose words with negative connotations.

Narrative Writing

Narrative writing tells a story. The most prominent examples of narrative writing are fictional novels. Here are some examples:

- Mark Twain's The Adventures of Tom Sawyer and The Adventures of Huckleberry Finn
- Victor Hugo's *Les Misérables*
- Charles Dickens' Great Expectations, David Copperfield, and A Tale of Two Cities
- Jane Austen's Northanger Abbey, Mansfield Park, Pride and Prejudice, Sense and Sensibility, and Emma
- Toni Morrison's Beloved, The Bluest Eye, and Song of Solomon
- Gabriel García Márquez's One Hundred Years of Solitude and Love in the Time of Cholera

Some nonfiction works are also written in narrative form. For example, some authors choose a narrative style to convey factual information about a topic, such as a specific animal, country, geographic region, and scientific or natural phenomenon.

Since narrative is the type of writing that tells a story, it must be told by someone, who is the narrator. The narrator may be a fictional character telling the story from their own viewpoint. This narrator uses the first person (*I, me, my, mine* and *we, us, our,* and *ours*). The narrator may simply be the author; for example, when Louisa May Alcott writes "Dear reader" in *Little Women*, she (the author) addresses us as readers. In this case, the novel is typically told in third person, referring to the characters as he, she, they, or them. Another more common technique is the omniscient narrator; i.e. the story is told by an unidentified individual who sees and knows everything about the events and characters—not only their externalized actions, but also their internalized feelings and thoughts. Second person, i.e. writing the story by addressing readers as "you" throughout, is less frequently used.

Expository Writing

Expository writing is also known as informational writing. Its purpose is not to tell a story as in narrative writing, to paint a picture as in descriptive writing, or to persuade readers to agree with something as in argumentative writing. Rather, its point is to communicate information to the reader. As such, the point of view of the author will necessarily be more objective. Whereas other types of writing appeal to the reader's emotions, appeal to the reader's reason by using logic, or use subjective descriptions to sway the reader's opinion or thinking, expository writing seeks to do none of these but simply to provide facts, evidence, observations, and objective descriptions of the subject matter. Some examples of expository writing include research reports, journal articles, articles and books about historical events or periods, academic subject textbooks, news articles and other factual journalistic reports, essays, how-to articles, and user instruction manuals.

Technical Writing

Technical writing is similar to expository writing in that it is factual, objective, and intended to provide information to the reader. Indeed, it may even be considered a subcategory of expository writing. However, technical writing differs from expository writing in that (1) it is specific to a particular field, discipline, or subject; and (2) it uses the specific technical terminology that belongs only to that area. Writing that uses technical terms is intended only for an audience familiar with those terms. A primary example of technical writing today is writing related to computer programming and use.

Persuasive Writing

Persuasive writing is intended to persuade the reader to agree with the author's position. It is also known as argumentative writing. Some writers may be responding to other writers' arguments, in which case they make reference to those authors or text and then disagree with them. However, another common technique is for the author to anticipate opposing viewpoints in general, both from other authors and from the author's own readers. The author brings up these opposing viewpoints, and then refutes them before they can even be raised, strengthening the author's argument. Writers persuade readers by appealing to the readers' reason and emotion, as well as to their own character and credibility. Aristotle called these appeals *logos*, *pathos*, and *ethos*, respectively.

Determining How the Author Explains a Position and Responds to Different Viewpoints
An author should clearly state his or her opinion and use evidence, such as research studies, statistics, and examples, to support his or her stance. Although somewhat counterintuitive, raising opposing viewpoints or presenting the counterargument actually strengthens in argument (if done correctly). This is because it gives the author the platform to provide evidence against that viewpoint—to disprove it—instead of leaving readers wondering if another viewpoint is more rational.

Inferring the Author's Purpose in the Passage When it is Not Stated
The author's attitude toward a certain person or idea, or his or her purpose, may not always be stated. While it may seem impossible to know exactly what the author felt toward their subject, there are clues to indicate the emotion, or lack thereof, of the author. Clues like word choice or style will alert readers to the author's attitude. Some possible words that name the author's attitude are listed below:

- Admiring
- Angry
- Critical
- Defensive
- Enthusiastic
- Humorous
- Moralizing
- Neutral
- Objective
- Patriotic
- Persuasive
- Playful
- Sentimental
- Serious
- Supportive
- Sympathetic
- Unsupportive

An author's tone is the author's attitude toward their subject and is usually indicated by word choice. If an author's attitude toward their subject is one of disdain, the author will show the subject in a negative light, using deflating words or words that are negatively charged. If an author's attitude toward their subject is one of praise, the author will use agreeable words and show the subject in a positive light. If an author takes a neutral tone towards their subject, their words will be neutral as well, and they probably will show all sides of their subject, not just the negative or positive side.

Style is another indication of the author's attitude and includes aspects such as sentence structure, type of language, and formatting. Sentence structure is how a sentence is put together. Sometimes, short, choppy sentences will indicate a certain tone given the surrounding context, while longer sentences may serve to create a buffer to avoid being too harsh, or may be used to explain additional information. Style may also include formal or informal language. Using formal language to talk about a subject may indicate a level of respect. Using informal language may be used to create an atmosphere of friendliness or familiarity with a subject. Again, it depends on the surrounding context whether or not language is used in a negative or positive way. Style may also include formatting, such as determining the length of paragraphs or figuring out how to address the reader at the very beginning of the text.

The following is a passage from *The Florentine Painters of the Renaissance* by Bernhard Berenson. Following the passage is a question stem regarding the author's attitude toward their subject:

Let us look now at an even greater triumph of movement than the Nudes, Pollaiuolo's "Hercules Strangling Antæus." As you realise the suction of Hercules' grip on the earth, the swelling of his calves with the pressure that falls on them, the violent throwing back of his chest, the stifling force of his embrace; as you realise the supreme effort of Antæus, with one hand crushing down upon the head and the other tearing at the arm of Hercules, you feel as if a fountain of energy had sprung up under your feet and were playing through your veins. I cannot refrain from mentioning still another masterpiece, this time not only of movement, but of tactile values and personal beauty as well—Pollaiuolo's "David" at Berlin. The young warrior has sped his stone, cut off the giant's head, and now he strides over it, his graceful, slender figure still vibrating with the rapidity of his triumph, expectant, as if fearing the ease of it. What lightness, what buoyancy we feel as we realise the movement of this wonderful youth!

Which one of the following best captures the author's attitude toward the paintings depicted in the passage?
  a. Neutrality towards the subject in this passage.
  b. Disdain for the violence found in the paintings.
  c. Excitement for the physical beauty found within the paintings.
  d. Passion for the movement and energy of the paintings.

Choice *D* is the best answer. We know that the author feels positively about the subject because of the word choice. Berenson uses words and phrases like "supreme," "fountain of energy," "graceful," "figure still vibrating," "lightness," "buoyancy," and "wonderful youth." Notice also the exclamation mark at the end of the paragraph. These words and style depict an author full of passion, especially for the movement and energy found within the paintings.

Choice *A* is incorrect because the author is biased towards the subject due to the energy he writes with—he calls the movement in the paintings "wonderful" and by the other word choices and phrases, readers can tell that this is not an objective analysis of these paintings. Choice *B* is incorrect because, although the author does mention the "violence" in the stance of Hercules, he does not exude disdain towards this. Choice *C* is incorrect. There is excitement in the author's tone, and some of this excitement is directed towards the paintings' physical beauty. However, this is not the *best* answer choice. Choice *D* is more accurate when stating the passion is for the movement and energy of the paintings, of which physical beauty is included.

Tone and Figurative Language

Understanding How Words Affect Tone
Words can be very powerful. When written words are used with the intent to make an argument or support a position, the words used—and the way in which they are arranged—can have a dramatic effect on the readers. Clichés, colloquialisms, run-on sentences, and misused words are all examples of ways that word choice can negatively affect writing quality. Unless the writer carefully considers word choice, a written work stands to lose credibility.

If a writer's overall intent is to provide a clear meaning on a subject, he or she must consider not only the exact words to use, but also their placement, repetition, and suitability. Academic writing should be intentional and clear, and it should be devoid of awkward or vague descriptions that can easily lead to misunderstandings. When readers find themselves reading and rereading just to gain a clear understanding of the writer's intent, there may be an issue with word choice. Although the words used in academic writing are different from those used in a casual conversation, they shouldn't necessarily be overly academic either. It may be relevant to employ key words that are associated with the subject, but struggling to inject these words into a paper just to sound academic may defeat the purpose. If the message cannot be clearly understood the first time, word choice may be the culprit.

Word choice also conveys the author's attitude and sets a tone. Although each word in a sentence carries a specific **denotation**, it might also carry positive or negative **connotations**—and it is the connotations that set the tone and convey the author's attitude. Consider the following similar sentences:

> It was the same old routine that happens every Saturday morning—eat, exercise, chores.
> The Saturday morning routine went off without a hitch—eat, exercise, chores.

The first sentence carries a negative connotation with the author's "same old routine" word choice. The feelings and attitudes associated with this phrase suggest that the author is bored or annoyed at the Saturday morning routine. Although the second sentence carries the same topic—explaining the Saturday morning routine—the choice to use the expression "without a hitch" conveys a positive or cheery attitude.

An author's writing style can likewise be greatly affected by word choice. When writing for an academic audience, for example, it is necessary for the author to consider how to convey the message by carefully considering word choice. If the author interchanges between third-person formal writing and second-person informal writing, the author's writing quality and credibility are at risk. Formal writing involves complex sentences, an objective viewpoint, and the use of full words as opposed to the use of a subjective viewpoint, contractions, and first- or second-person usage commonly found in informal writing.

Content validity, the author's ability to support the argument, and the audience's ability to comprehend the written work are all affected by the author's word choice.

Understanding How Figurative Language Affects the Meaning of Words or Phrases
Authors of a text use language with multiple levels of meaning for many different reasons. When the meaning of a text requires directness, the author will use literal language to provide clarity. On the other hand, an author will use figurative language to produce an emotional effect or facilitate a deeper understanding of a word or passage. For example, a set of instructions on how to use a computer would require literal language. However, a commentary on the social implications of immigration bans might contain a wide range of figurative language to elicit an empathetic response. A single text can have a mixture of both literal and figurative language.

Literal Language
Literal language uses words in accordance with their actual definition. Many informational texts employ literal language because it is straightforward and precise. Documents such as instructions, proposals, technical documents, and workplace documents use literal language for the majority of their writing, so there is no confusion or complexity of meaning for readers to decipher. The information is best communicated through clear and precise language.

The following are brief examples of literal language:

- I cook with olive oil.
- There are 365 days in a year.
- My grandma's name is Barbara.
- Yesterday we had some scattered thunderstorms.
- World War II began in 1939.
- Blue whales are the largest species of whale.

Figurative Language

Not meant to be taken literal, figurative language is useful when the author of a text wants to produce an emotional effect in the reader or add a heightened complexity to the meaning of the text. Figurative language is used more heavily in texts such as literary fiction, poetry, critical theory, and speeches. Figurative language goes beyond literal language, allowing readers to form associations they wouldn't normally form with literal language. Using language in a figurative sense appeals to the imagination of the reader. It is important to remember that words signify objects and ideas and are not the objects and ideas themselves. Figurative language can highlight this detachment by creating multiple associations, but it also points to the fact that language is fluid and capable of creating a world full of linguistic possibilities. It can be argued that figurative language is the heart of communication even outside of fiction and poetry. People connect through humor, metaphors, cultural allusions, puns, and symbolism in their everyday rhetoric. The following are terms associated with figurative language:

A **simile** is a comparison of two things using *like, than,* or *as.* A simile usually takes objects that have no apparent connection, such as a mind and an orchid, and compares them:

> His mind was as complex and rare as a field of ghost orchids.

Similes encourage new, fresh perspectives on objects or ideas that would not otherwise occur. Unlike similes, metaphors are comparisons that do not use *like, than,* or *as.* So, a metaphor from the above example would be:

> His mind was a field of ghost orchids.

Thus, similes highlight the comparison by focusing on the figurative side of the language, elucidating the author's intent. Metaphors, however, provide a beautiful yet somewhat equivocal comparison.

A popular use of figurative language, **metaphors** compare objects or ideas directly, asserting that something *is* a certain thing, even if it isn't. The following is an example of a metaphor used by writer Virginia Woolf:

> Books are the mirrors of the soul.

Metaphors consist of two parts: a tenor and a vehicle. Tenor refers to the object being described, and vehicle refers to the figurative language making the comparison. In this example, the tenor is "books" and the vehicle is "mirrors of the soul." Perhaps the author meant to say that written language (books) reflect a person's most inner thoughts and desires.

Dead metaphors are phrases that have been overused to the point where the figurative language has taken on a literal meaning, like "crystal clear." This phrase is in such popular use that the meaning seems literal ("perfectly clear") even when it is not. al clear." However, since the latter part of the phrase is in such popular use, the meaning seems literal ("I understand what you're saying") even when it's not.

89

Finally, an Finally, an extended metaphor is one that goes on for several paragraphs, is one that goes on for several paragraphs, or even an entire text. "On First Looking into Chapman's Homer," a poem by John Keats, begins, "Much have I travell'd in the realms of gold," and goes on to explain the first time he hears Chapman's translation of Homer's writing. The extended metaphor begins in the first line as Keats compares travelling into "realms of gold" to the act of hearing a certain kind of literature for the first time. The metaphor continues through the end of the poem where Keats stands "Silent, upon a peak in Darien," having heard the end of Chapman's translation. Keats gained insight into new lands (new text) and was richer because of it.

The following are brief definitions and examples of popular figurative language:

- **Onomatopoeia**: A word that, when spoken, imitates the sound to which it refers. Ex: "We heard a loud *boom* while driving to the beach yesterday."

- **Personification**: When human characteristics are given to animals, inanimate objects, or abstractions. An example would be in William Wordsworth's poem "Daffodils" where he sees a "crowd . . . / of golden daffodils . . . / Fluttering and dancing in the breeze." Dancing is usually a characteristic attributed solely to humans, but Wordsworth personifies the daffodils here as a crowd of people dancing.

- **Juxtaposition**: Juxtaposition places two objects side by side for comparison or contrast. For example, Milton juxtaposes God and Satan in "Paradise Lost."

- **Paradox**: A paradox is a statement that appears self-contradictory but is actually true. One example of a paradox is when Socrates said, "I know one thing; that I know nothing." Seemingly, if Socrates knew nothing, he wouldn't know that he knew nothing. However, he is using figurative language not to say that he literally knows nothing, but that true wisdom begins with casting all presuppositions about the world aside.

- **Hyperbole**: A hyperbole is an exaggeration. Ex: "I'm so tired I could sleep for centuries."

- **Allusion**: An allusion is a reference to a character or event that happened in the past. T.S. Eliot's "The Waste Land" is a poem littered with allusions, including, "I will show you fear in a handful of dust," alluding to Genesis 3:19: "For you are dust, and to dust you shall return."

- **Pun**: Puns are used in popular culture to invoke humor by exploiting the meanings of words. They can also be used in literature to give hints of meaning in unexpected places. In "Romeo and Juliet," Mercutio makes a pun after he is stabbed by Tybalt: "look for me tomorrow and you will find me a grave man."

- **Imagery**: This is a collection of images given to the reader by the author. If a text is rich in imagery, it is easier for the reader to imagine themselves in the author's world.

  One example of a poem that relies on *imagery* is William Carlos Williams' "The Red Wheelbarrow":

  > so much depends
  > upon
  >
  > a red wheel
  > barrow

glazed with rain
water

beside the white
chickens

The starkness of the imagery and the placement of the words in this poem bring to life the images of a purely simple world. Through its imagery, this poem tells a story in just sixteen words.

- **Symbolism**: A symbol is used to represent an idea or belief system. For example, poets in Western civilization have been using the symbol of a rose for hundreds of years to represent love. In Japan, poets have used the firefly to symbolize passionate love, and sometimes even spirits of those who have died. Symbols can also express powerful political commentary and can be used in propaganda.

- **Irony**: There are three types of irony: verbal, dramatic, and situational. Verbal irony is when a person states one thing and means the opposite. For example, a person is probably using irony when they say, "I can't wait to study for this exam next week." Dramatic irony occurs in a narrative and happens when the audience knows something that the characters do not. In the modern TV series *Hannibal*, the audience knows that Hannibal Lecter is a serial killer, but most of the main characters do not. This is dramatic irony. Finally, situational irony is when one expects something to happen, and the opposite occurs. For example, we can say that a fire station burning down would be an instance of situational irony.

Understanding How the Use of Words, Phrases, or Figurative Language Influences the Author's Purpose
As mentioned, denotation refers to a word's explicit definition, like that found in the dictionary. Denotation is often set in comparison to connotation. Connotation is the emotional, cultural, social, or personal implication associated with a word. Denotation is more of an objective definition, whereas connotation can be more subjective, although many connotative meanings of words are similar for certain cultures. The denotative meanings of words are usually based on facts, and the connotative meanings of words are usually based on emotion.

Here are some examples of words and their denotative and connotative meanings in Western culture:

Word	Denotative Meaning	Connotative Meaning
Home	A permanent place where one lives, usually as a member of a family.	A place of warmth; a place of familiarity; comforting; a place of safety and security. "Home" usually has a positive connotation.
Snake	A long reptile with no limbs and strong jaws that moves along the ground; some snakes have a poisonous bite.	An evil omen; a slithery creature (human or nonhuman) that is deceitful or unwelcome. "Snake" usually has a negative connotation.
Winter	A season of the year that is the coldest, usually from December to February in the northern hemisphere and from June to August in the southern hemisphere.	Circle of life, especially that of death and dying; cold or icy; dark and gloomy; hibernation, sleep, or rest. Winter can have a negative connotation, although many who have access to heat may enjoy the snowy season from their homes.

Additionally, depending on the author's purpose with a text, he or she will either use formal language or more informal language. Formal language is less personal than informal language. It is more "buttoned-up" and business-like, adhering to proper grammatical rules. It is used in professional or academic contexts, to convey respect or authority. For example, one would use formal language to write an informative or argumentative essay for school or to address a superior. Formal language avoids contractions, slang, colloquialisms, and first-person pronouns. Formal language uses sentences that are usually more complex and often in passive voice. Punctuation can differ as well. For example, exclamation points (!) are used to show strong emotion or can be used as an interjection but should be used sparingly in formal writing situations.

Informal language is often used when communicating with family members, friends, peers, and those known more personally. It is more casual, spontaneous, and forgiving in its conformity to grammatical rules and conventions. Informal language is used for personal emails and correspondence between coworkers or other familial relationships. The tone is more relaxed. In informal writing, slang, contractions, clichés, and the first- and second-person are often used.

Organizing Ideas

Determining How a Section Fits into a Passage and Helps Develop the Ideas
To determine whether details support the main idea of a passage, determine the relationships between ideas. For example, suppose the main idea of the passage is "*Due to inflation, the value of a dollar is different today than it was 100 years ago and different than it will be 100 years from now.*" This idea is then directly supported by the statements like this: "*In 1918, an item that cost a dollar would cost almost $17 today.*" Another detail supporting it is this: "*Experts estimate that the expected rate of inflation over the next 10 years is about 2%.*"

To determine whether a detail supports the main idea, a reader must determine whether there is a connection. Then they must decide whether the detail directly supports the main idea, one of the supporting ideas, or none of the ideas. Suppose a supporting idea in a passage about inflation is "*In 1918, an item that cost a dollar would cost almost $17 today.*" The reader then reads this detail: "*That means a coffee that costs $5 today would have cost about 30 cents back then.*" The detail about the coffee price is an example of the supporting idea about prices in 1918. However, the statement *The value of an item is determined by how popular the item is* does not support the main idea ("*Due to inflation, the value of a dollar is different today than it was 100 years ago and different than it will be 100 years from now.*") or any of the supporting ideas.

Analyzing How a Text is Organized
Text structure is the way in which the author organizes and presents textual information so readers can follow and comprehend it. One kind of text structure is sequence. This means the author arranges the text in a logical order from beginning to middle to end. There are three types of sequences:

- **Chronological**: ordering events in time from earliest to latest
- **Spatial**: describing objects, people, or spaces according to their relationships to one another in space
- **Order of Importance**: addressing topics, characters, or ideas according to how important they are, from either least important to most important

Chronological sequence is the most common sequential text structure. Readers can identify sequential structure by looking for words that signal it, like *first, earlier, meanwhile, next, then, later, finally;* and specific times and dates the author includes as chronological references.

Problem-Solution Text Structure

The problem-solution text structure organizes textual information by presenting readers with a problem and then developing its solution throughout the course of the text. The author may present a variety of alternatives as possible solutions, eliminating each as they are found unsuccessful, or gradually leading up to the ultimate solution. For example, in fiction, an author might write a murder mystery novel and have the character(s) solve it through investigating various clues or character alibis until the killer is identified. In nonfiction, an author writing an essay or book on a real-world problem might discuss various alternatives and explain their disadvantages or why they would not work before identifying the best solution. For scientific research, an author reporting and discussing scientific experiment results would explain why various alternatives failed or succeeded.

Comparison-Contrast Text Structure

Comparison identifies similarities between two or more things. **Contrast** identifies differences between two or more things. Authors typically employ both to illustrate relationships between things by highlighting their commonalities and deviations. For example, a writer might compare Windows and Linux as operating systems, and contrast Linux as free and open-source vs. Windows as proprietary. When writing an essay, sometimes it is useful to create an image of the two objects or events you are comparing or contrasting. Venn diagrams are useful because they show the differences as well as the similarities between two things. Once you've seen the similarities and differences on paper, it might be helpful to create an outline of the essay with both comparison and contrast. Every outline will look different, because every two or more things will have a different number of comparisons and contrasts. Say you are trying to compare and contrast carrots with sweet potatoes.

Here is an example of a compare/contrast outline using those topics:

- *Introduction:* Talk about why you are comparing and contrasting carrots and sweet potatoes. Give the thesis statement.
- *Body paragraph 1:* Sweet potatoes and carrots are both root vegetables (similarity)
- *Body paragraph 2:* Sweet potatoes and carrots are both orange (similarity)
- *Body paragraph 3:* Sweet potatoes and carrots have different nutritional components (difference)
- *Conclusion:* Restate the purpose of your comparison/contrast essay.

Of course, if there is only one similarity between your topics and two differences, you will want to rearrange your outline. Always tailor your essay to what works best with your topic.

Descriptive Text Structure

Description can be both a type of text structure and a type of text. Some texts are descriptive throughout entire books. For example, a book may describe the geography of a certain country, state, or region, or tell readers all about dolphins by describing many of their characteristics. Many other texts are not descriptive throughout, but use descriptive passages within the overall text. The following are a few examples of descriptive text:

- When the author describes a character in a novel
- When the author sets the scene for an event by describing the setting
- When a biographer describes the personality and behaviors of a real-life individual

- When a historian describes the details of a particular battle within a book about a specific war
- When a travel writer describes the climate, people, foods, and/or customs of a certain place

A hallmark of description is using sensory details, painting a vivid picture so readers can imagine it almost as if they were experiencing it personally.

Cause and Effect Text Structure
When using cause and effect to extrapolate meaning from text, readers must determine the cause when the author only communicates effects. For example, if a description of a child eating an ice cream cone includes details like beads of sweat forming on the child's face and the ice cream dripping down her hand faster than she can lick it off, the reader can infer or conclude it must be hot outside. A useful technique for making such decisions is wording them in "If...then" form, e.g. "If the child is perspiring and the ice cream melting, it may be a hot day." Cause and effect text structures explain why certain events or actions resulted in particular outcomes. For example, an author might describe America's historical large flocks of dodo birds, the fact that gunshots did not startle/frighten dodos, and that because dodos did not flee,

Understanding the Meaning and Purpose of Transition Words
In writing, some sentences naturally lead to others, whereas in other cases, a new sentence expresses a new idea. Transitional phrases connect sentences and the ideas they convey, which makes the writing coherent. Transitional language also guides the reader from one thought to the next. For example, when pointing out an objection to the previous idea, starting a sentence with "However," "But," or "On the other hand" is transitional. When adding another idea or detail, writers use "Also," "In addition," "Furthermore," "Further," "Moreover," "Not only," etc. Readers have difficulty perceiving connections between ideas without such transitional wording.

Analyzing How the Organization of a Paragraph or Passage Supports the Author's Ideas
The author's purpose determines the organization of a passage. The organization guides the connections between the ideas that support the main idea and establishes relationships between supporting ideas.

For example, suppose an author wants to explain why Henry VIII created the Church of England. This author might choose a sequential structure with time order relationships to describe the events. She would also include the cause and effect structure between the ideas to establish the relationship between the result and the circumstances that caused the result: "*Henry VIII created the Church of England because he was excommunicated from the Catholic church.*"

If the author's purpose is to convince someone to leave the Church of England to return to Catholicism, they might use comparison and contrast to describe the ideals of the two churches. The same series of events and cause and effect relationships contribute to the main idea, but the conclusion the author wants the reader to make is different: "*The foundation of the Church of England is sinful.*" The author could also use order of importance to provide a list of reasons why this event occurred, or the reasons the reader should leave the Church of England.

Comparing Different Ways of Presenting Ideas

Evaluating Two Different Texts and How They Address Scope, Purpose, Emphasis, Audience, and Impact
The author's purpose for writing the passage guides how they address the audience, the scope or level of detail of the passage, what to emphasize, and the impact they want the passage to have on the reader.

If the author wants to persuade an audience of voters to vote for an independent candidate, they might show parallels between the sentiments that divided the North and South leading up to the Civil War and the division between the two major parties in the US government today. The author would emphasize the opposing views of the two sides in the Civil War and their inability to compromise, and the author would demonstrate a similar lack of compromise in politics today. The author might compare the issues that divided the sides and the possible reasons behind each side's position, such as the economic power of the voters or human rights, and they might emphasize the outcome of the conflict then. The success of or failure of the author to convince their audience to vote for the independent candidate would determine the impact of the passage.

Using the same information, a passage about voting and the Civil War could be used to persuade voters to vote for one of the two major parties. A passage like that would have a different scope, emphasis, and impact than one about independent candidates. The author in favor of a major party would focus the scope of the argument on different details, drawing positive comparisons to their party and negative comparisons to the opposing party. The author might emphasize the result of the conflict, rather than the events that led to war; or the author might minimize the impact of the division between the two sides. This author might emphasize the great deeds the party has accomplished, despite the opposition. The success of this author in convincing voters to back a particular side would determine the impact of the passage.

Evaluating Two Different Passages, Focusing on Point of View, Tone, Style, Organization, Purpose, or Impact
The author's purpose affects the way they address the point of view from which they are writing, the tone and style of the passage, and the organization they use.

Suppose a passage has this main idea: "Nuclear power is safe." If the author is an investor in nuclear power plant construction, their purpose might be change the public's perception of nuclear power and smooth the way for any permits. The author might use a persuasive style and a friendly tone to convince people to feel good about nuclear power.

Suppose a passage has this main idea: "All high school graduates should go to college because it's the way to get ahead in life." If the author is the dean of students at a college, he might believe in college for and also believe his college is the best. His secondary purpose might be to recruit more students to attend his college. From his point of view, students should attend his school. He might use a persuasive style and a friendly tone to convince graduates using examples and testimonials from students that his college offers the best academics and the most exciting campus life.

Lastly, suppose a passage were written from the point of view of a low-income high school graduate who can't afford college. This passage's main idea might be: "*All students should be able to attend college.*" Instead of potential college students, the audience for this passage might be a state Congressperson, and the author's purpose might be to advocate better funding for state universities. The style of this passage might also include persuasive language. The tone would be serious, and the author might use statistics to contrast the lives of those who attended college and those who did not.

Identifying and Creating Arguments

The Relationship Between Evidence and Main Ideas and Details

Summarizing Information from a Passage
An important skill is the ability to read a complex text and then reduce its length and complexity by focusing on the key events and details. A summary is a shortened version of the original text, written by

the reader in their own words. The summary should be shorter than the original text, and it must include the most critical points.

In order to effectively summarize a complex text, it's necessary to understand the original source and identify the major points covered. It may be helpful to outline the original text to get the big picture and avoid getting bogged down in the minor details. For example, a summary wouldn't include a statistic from the original source unless it was the major focus of the text. It is also important for readers to use their own words but still retain the original meaning of the passage. The key to a good summary is emphasizing the main idea without changing the focus of the original information.

Complex texts will likely be more difficult to summarize. Readers must evaluate all points from the original source, filter out the unnecessary details, and maintain only the essential ideas. The summary often mirrors the original text's organizational structure. For example, in a problem-solution text structure, the author typically presents readers with a problem and then develops solutions through the course of the text. An effective summary would likely retain this general structure, rephrasing the problem and then reporting the most useful or plausible solutions.

Paraphrasing is somewhat similar to summarizing. It calls for the reader to take a small part of the passage and list or describe its main points. Paraphrasing is more than rewording the original passage, though. As with summary, a paraphrase should be written in the reader's own words, while still retaining the meaning of the original source. The main difference between summarizing and paraphrasing is that a summary would be appropriate for a much larger text, while paraphrase might focus on just a few lines of text. Effective paraphrasing will indicate an understanding of the original source, yet still help the reader expand on their interpretation. A paraphrase should neither add new information nor remove essential facts that change the meaning of the source.

Identifying the Relationship Between the Main Idea and Details of a Passage
The topic of a text is the general subject matter. Text topics can usually be expressed in one word, or a few words at most. Additionally, readers should ask themselves what point the author is trying to make. This point is the main idea of the text, the one thing the author wants readers to know concerning the topic. Once the author has established the main idea, they will support the main idea by supporting details. Supporting details are evidence that support the main idea and include personal testimonies, examples, or statistics.

One analogy for these components and their relationships is that a text is like a well-designed house. The topic is the roof, covering all rooms. The main idea is the frame. The supporting details are the various rooms. To identify the topic of a text, readers can ask themselves what or who the author is writing about in the paragraph. To locate the main idea, readers can ask themselves what one idea the author wants readers to know about the topic. To identify supporting details, readers can put the main idea into question form and ask "what does the author use to prove or explain their main idea?"

Let's look at an example. An author is writing an essay about the Amazon rainforest and trying to convince the audience that more funding should go into protecting the area from deforestation. The author makes the argument stronger by including evidence of the benefits of the rainforest: it provides habitats to a variety of species, it provides much of the earth's oxygen which in turn cleans the atmosphere, and it is the home to medicinal plants that may be the answer to some of the world's deadliest diseases.

Here is an outline of the essay looking at topic, main idea, and supporting details:

Topic: Amazon rainforest
Main Idea: The Amazon rainforest should receive more funding in order to protect it from deforestation.
Supporting Details:
    1. It provides habitats to a variety of species
    2. It provides much of the earth's oxygen which in turn cleans the atmosphere
    3. It is home to medicinal plants that may be the answer to some of the world's deadliest diseases.

Notice that the topic of the essay is listed in a few key words: "Amazon rainforest". The main idea tells us what about the topic is important: that the topic should be funded in order to prevent deforestation. Finally, the supporting details are what author relies on to convince the audience to act or to believe in the truth of the main idea.

Determining the Main idea of a Passage
The main idea a paragraph is sometimes clearly stated at the beginning or the end of the paragraph, or it might be implied so the reader must infer it from supporting details.

To determine the main idea of a paragraph, the reader must determine the topic. The topic is different than the main idea. The main idea is the author's message; the topic is who or what the passage is about. To determine the topic when it is not explicitly stated, the reader examines the supporting details within the paragraph.

These details might include answers to questions raised throughout the passage, events or ideas that move a story along, or evidence that leads the reader to a specific conclusion. These details might include vivid descriptions of events, quotes that offer an opinion, comparisons of ideas, and statistics and graphs.

Based on the topic and the supporting details or evidence, the reader can form a conclusion about the author's purpose to infer the main idea of the paragraph.

Determining Which Details Support a Main Idea
Authors use both major and minor details to support the main idea of a passage.

Major details are general statements that contribute directly to supporting the main idea. For example, suppose the main idea of a passage is *"Many variables determine what type of car you should buy."* The following major detail supports this main idea: *"One of the variables is where you live."*

Minor details are specific examples of the major details. For example, consider these details:

> *If it snows a lot, you might need AWD. If you live in the city and have to commute, you might want good gas mileage.*

These are minor details that support one of the passage's major details: *One of the variables is where you live.* Minor details contribute to the overall message of the main idea, but they do not directly support the main idea. Instead, they develop and support the major details that support the main idea.

Identifying the Theme and Supportive Elements in Fiction and Nonfiction
The theme of a piece of text is the central idea the author communicates. Whereas the topic of a passage of text may be concrete in nature, by contrast the theme is always conceptual. For example, while the topic of Mark Twain's novel *The Adventures of Huckleberry Finn* might be described as something like the

coming-of-age experiences of a poor, illiterate, functionally orphaned boy around and on the Mississippi River in 19th-century Missouri, one theme of the book might be that human beings are corrupted by society. Another might be that slavery and "civilized" society itself are hypocritical. Whereas the main idea in a text is the most important single point that the author wants to make, the theme is the concept or view around which the author centers the text.

Throughout time, humans have told stories with similar themes. Some themes are universal across time, space, and culture. These include themes of the individual as a hero, conflicts of the individual against nature, the individual against society, change vs. tradition, the circle of life, coming-of-age, and the complexities of love. Themes involving war and peace have featured prominently in diverse works, like Homer's *Iliad*, Tolstoy's *War and Peace* (1869), Stephen Crane's *The Red Badge of Courage* (1895), Hemingway's *A Farewell to Arms* (1929), and Margaret Mitchell's *Gone with the Wind* (1936). Another universal literary theme is that of the quest. These appear in folklore from countries and cultures worldwide, including the Gilgamesh Epic, Arthurian legend's Holy Grail quest, Virgil's *Aeneid*, Homer's *Odyssey*, and the *Argonautica*. Cervantes' *Don Quixote* is a parody of chivalric quests. J.R.R. Tolkien's *The Lord of the Rings* trilogy (1954) also features a quest.

One instance of similar themes across cultures is when those cultures are in countries that are geographically close to each other. For example, a folklore story of a rabbit in the moon using a mortar and pestle is shared among China, Japan, Korea, and Thailand—making medicine in China, making rice cakes in Japan and Korea, and hulling rice in Thailand. Another instance is when cultures are more distant geographically, but their languages are related. For example, East Turkestan's Uighurs and people in Turkey share tales of folk hero Effendi Nasreddin Hodja. Another instance, which may either be called cultural diffusion or simply reflect commonalities in the human imagination, involves shared themes among geographically and linguistically different cultures: both Cameroon's and Greece's folklore tell of centaurs; Cameroon, India, Malaysia, Thailand, and Japan, of mermaids; Brazil, Peru, China, Japan, Malaysia, Indonesia, and Cameroon, of underwater civilizations; and China, Japan, Thailand, Vietnam, Malaysia, Brazil, and Peru, of shape-shifters.

Two prevalent literary themes are love and friendship, which can end happily, sadly, or both. William Shakespeare's *Romeo and Juliet*, Emily Brontë's *Wuthering Heights*, Leo Tolstoy's *Anna Karenina*, and both *Pride and Prejudice* and *Sense and Sensibility* by Jane Austen are famous examples. Another theme recurring in popular literature is of revenge, an old theme in dramatic literature, e.g. Elizabethans Thomas Kyd's *The Spanish Tragedy* and Thomas Middleton's *The Revenger's Tragedy*. Some more well-known instances include Shakespeare's tragedies *Hamlet* and *Macbeth*, Alexandre Dumas' *The Count of Monte Cristo*, John Grisham's *A Time to Kill*, and Stieg Larsson's *The Girl Who Kicked the Hornet's Nest*.

Themes are underlying meanings in literature. For example, if a story's main idea is a character succeeding against all odds, the theme is overcoming obstacles. If a story's main idea is one character wanting what another character has, the theme is jealousy. If a story's main idea is a character doing something they were afraid to do, the theme is courage. Themes differ from topics in that a topic is a subject matter; a theme is the author's opinion about it. For example, a work could have a topic of war and a theme that war is a curse. Authors present themes through characters' feelings, thoughts, experiences, dialogue, plot actions, and events. Themes function as "glue" holding other essential story elements together. They offer readers insights into characters' experiences, the author's philosophy, and how the world works.

Drawing Conclusions, Making Inferences, and Evaluating Evidence

Making Generalizations Based on Evidence
Readers form conclusions about the main idea of a passage based on the evidence provided. As the reader gathers more details from the text, their conclusion could change or be proven by the evidence. They start to form generalizations about the author's message.

Generalizations are broad conclusions based on the way a reader interprets details. These generalizations might be correct, or the reader might have made inferences from the evidence that the author did not intend.

The reader should closely examine the evidence in the text to determine if the generalizations were actually implied by the author, or if they are incorrect.

For example:

1. Humans cannot live without water.
2. People use thousands of gallons of water a year to keep their lawns green.

Based on these details, the reader concludes that *people who water their lawns are wasting water.*

The author has not offered evidence that watering lawns is wasteful, so the generalization is not supported. If the passage had included the detail that scientists say water will soon be dangerously scarce, then the generalization would be supported. On the other hand, the following statement would contradict the reader's generalization: *Lawns also need water to grow and sustain life.*

Using Main Ideas to Draw Conclusions
Authors describe settings, characters, character emotions, and events. Readers must infer to understand text fully. Inferring enables readers to figure out meanings of unfamiliar words, make predictions about upcoming text, draw conclusions, and reflect on reading. Readers can infer about text before, during, and after reading. In everyday life, we use sensory information to infer. Readers can do the same with text. When authors do not answer all reader questions, readers must infer by saying "I think....This could be....This is because....Maybe....This means....I guess..." etc. Looking at illustrations, considering characters' behaviors, and asking questions during reading facilitate inference. Taking clues from text and connecting text to prior knowledge help to draw conclusions. Readers can infer word meanings, settings, reasons for occurrences, character emotions, pronoun referents, author messages, and answers to questions unstated in text. To practice inference, students can read sentences written/selected by the instructor, discuss the setting and character, draw conclusions, and make predictions.

Making inferences and drawing conclusions involve skills that are quite similar: both require readers to fill in information the author has omitted. Authors may omit information as a technique for inducing readers to discover the outcomes themselves; or they may consider certain information unimportant; or they may assume their reading audience already knows certain information. To make an inference or draw a conclusion about text, readers should observe all facts and arguments the author has presented and consider what they already know from their own personal experiences. Reading students taking multiple-choice tests that refer to text passages can determine correct and incorrect choices based on the information in the passage. For example, from a text passage describing an individual's signs of anxiety while unloading groceries and nervously clutching their wallet at a grocery store checkout, readers can infer or conclude that the individual may not have enough money to pay for everything.

Describing the Steps of an Argument

When authors write text for the purpose of persuading others to agree with them, they assume a position with the subject matter about which they are writing. Rather than presenting information objectively, the author treats the subject matter subjectively so that the information presented supports his or her position. In their argumentation, the author presents information that refutes or weakens opposing positions. Another technique authors use in persuasive writing is to anticipate arguments against the position. When students learn to read subjectively, they gain experience with the concept of persuasion in writing, and learn to identify positions taken by authors. This enhances their reading comprehension and develops their skills for identifying pro and con arguments and biases.

There are five main parts of the classical argument that writers employ in a well-designed stance:

- Introduction: In the introduction to a classical argument, the author establishes goodwill and rapport with the reading audience, warms up the readers, and states the thesis or general theme of the argument.

- Narration: In the narration portion, the author gives a summary of pertinent background information, informs the readers of anything they need to know regarding the circumstances and environment surrounding and/or stimulating the argument, and establishes what is at risk or the stakes in the issue or topic. Literature reviews are common examples of narrations in academic writing.

- Confirmation: The confirmation states all claims supporting the thesis and furnishes evidence for each claim, arranging this material in logical order—e.g. from most obvious to most subtle or strongest to weakest.

- Refutation and Concession: The refutation and concession discuss opposing views and anticipate reader objections without weakening the thesis, yet permitting as many oppositions as possible.

- Summation: The summation strengthens the argument while summarizing it, supplying a strong conclusion and showing readers the superiority of the author's solution.

Introduction

A classical argument's introduction must pique reader interest, get readers to perceive the author as a writer, and establish the author's position. Shocking statistics, new ways of restating issues, or quotations or anecdotes focusing the text can pique reader interest. Personal statements, parallel instances, or analogies can also begin introductions—so can bold thesis statements if the author believes readers will agree. Word choice is also important for establishing author image with readers.

The introduction should typically narrow down to a clear, sound thesis statement. If readers cannot locate one sentence in the introduction explicitly stating the writer's position or the point they support, the writer probably has not refined the introduction sufficiently.

Narration and Confirmation

The narration part of a classical argument should create a context for the argument by explaining the issue to which the argument is responding, and by supplying any background information that influences the issue. Readers should understand the issues, alternatives, and stakes in the argument by the end of the narration to enable them to evaluate the author's claims equitably. The confirmation part of the classical argument enables the author to explain why they believe in the argument's thesis. The author builds a chain of reasoning by developing several individual supporting claims and explaining why that evidence supports each claim and also supports the overall thesis of the argument.

Refutation and Concession and Summation

The classical argument is the model for argumentative/persuasive writing, so authors often use it to establish, promote, and defend their positions. In the refutation aspect of the refutation and concession part of the argument, authors disarm reader opposition by anticipating and answering their possible objections, persuading them to accept the author's viewpoint. In the concession aspect, authors can concede those opposing viewpoints with which they agree.

This can avoid weakening the author's thesis while establishing reader respect and goodwill for the author: all refutation and no concession can antagonize readers who disagree with the author's position. In the conclusion part of the classical argument, a less skilled writer might simply summarize or restate the thesis and related claims; however, this does not provide the argument with either momentum or closure. More skilled authors revisit the issues and the narration part of the argument, reminding readers of what is at stake.

Identifying Evidence Used to Support a Claim or Conclusion

In the GED Reasoning Through Language Arts Section, there will likely be questions that ask the test taker about evidence or principles expressed in the selection that pertain to the argument. A principle functions as a fundamental truth used as a basis for a scenario or system of reasoning. Principles or evidence might serve as the cause of something or as a final cause; alternatively, principles may serve as moral, juridical, or scientific law.

Principles as Cause

Principles can function in different ways according to the way we express the principle. For the circumstance of cause and effect, principle refers to the cause that was efficient for the effect to come into existence. The principle as cause is traced back to Aristotelian reasoning, which surmises that every event is moved by something prior to it, or has a cause.

Principles as Law

We see principles as law at work in moral law, juridical law, and scientific law. In moral law, principles are what our predecessors teach us as children. "Do unto others," or the "golden rule," is a principle that society has embedded in us so that we are able to function as a civilized people. Principles as moral law are restrictive to the individual as a way of protecting the other person, the whole, or society.

Principles in juridical law are created by the State and also function to limit the liberty of individuals in order to protect the masses. The principles formed in juridical law are written rules that seek to establish a foundation which people can adhere to. The "homestead principle" is an example of a principle in juridical law. The "homestead principle" would function as someone gaining ownership of land because they have made it into a farm, or utilized some resource that has been unused on the land prior to their cultivation of it.

Principles as scientific law function as natural laws, including the Laws of Thermodynamics or natural selection. Principles in scientific law function as laws used to predict certain phenomena that happen in nature. In this context, principles are able to predict results of future experiments. They are developed from facts and also have the ability to be strongly supported by observational evidence.

Determining Whether Evidence is Relevant and Sufficient

Selecting the most relevant material to support a written text is a necessity in producing quality writing and for the credibility of an author. Arguments lacking in reasons or examples won't work in persuading the audience later on, because their hearts have not been pulled. Using examples to support ideas also gives the writing rhetorical effects such as pathos (emotion), logos (logic), or ethos (credibility), all three of which are necessary for a successful text.

An author needs to think about the audience. Are they indifferent? An author might use a personal story or example in order stir empathy. Are they resistant? If so, an author might use logical reasoning based in factual evidence so they will be convinced. Personal stories or testimonials, statistics, or documentary evidence are various types of examples that one can use for their writing.

Determining Whether a Statement Is or Is Not Supported

Evidence used in arguments must be credible and valid, such as that from peer-reviewed scholarly journals. Peer-reviewed sources are sources that have been reviewed by other experts in the field. must also be relevant by being up-to-date, especially those within the science or technology fields.

For example, let's a passage is discussion pesticides and the collapse of bees. An argument without relevant examples would look like this:

> With the use of the world's most popular pesticides, bees are becoming extinct. This is also causing ecological devastation. We must do something soon about the bee population, or else we will chase bees to extinction and lose valuable resources as a result.

Here is the same argument with examples. The added examples are in italics:

> With the use of the world's most popular pesticides, bees are becoming extinct. *Beekeepers have reported losing 55 to 95 percent of their colony in just two short years due to toxic poisoning.* This is also causing ecological devastation. *Bees are known for pollinating more than two-thirds of the world's most essential crops.* We must do something soon about the bee population, or else we will chase bees to extinction and lose valuable resources as a result.

Adding examples to the above argument brings life to the bees—they are living, dying, pollinating—and readers feel more compelled to act as a result of adding relevant examples to the argument.

Assessing Whether an Argument is Valid

Not all arguments are valid. Authors sometimes have one or more flaws in their argument's reasoning. Some GED questions may ask you to provide a description of that error. In order for you to be able to describe what flaw is occurring in the argument, it will help to know of various argumentative flaws, such as red herring, false choice, and correlation vs. causation.

Here are some examples of what this type of question looks like:

- The reasoning in the argument is flawed because the argument . . .
- The argument is most vulnerable to criticism on the grounds that it . . .
- Which one of the following is an error in the argument's reasoning?
- A flaw in the reasoning of the argument is that . . .
- Which one of the following most accurately describes X's criticism of the argument made by Y?

Bait/Switch

One common flaw that is good to know is called the "bait and switch." It occurs when the test makers will provide an argument that offers evidence about X, and ends the argument with a conclusion about Y. A "bait and switch" answer choice will look like this:

> The argument assumes that X does in fact address Y without providing justification.

Let's look at an example:

> Hannah will most likely always work out and maintain a healthy physique. After all, Hannah's IQ is extremely high.

The correct answer will look like this:

> The argument assumes that Hannah's high IQ addresses her likelihood of always working out without providing justification.

Ascriptive Error

The ascriptive argument will begin the argument with something a third party has claimed. Usually, it will be something very general, like "Some people say that . . ." or "Generally, it has been said . . ." Then, the arguer will follow up that claim with a refutation or opposing view. The problem here is that when the arguer phrases something in this general sense without a credible source, their refutation of that evidence doesn't really matter. Here's an example:

It has been said that peppermint oil has been proven to relieve stomach issues and, in some cases, prevent cancer. I can attest to the relief in stomach issues; however, there is just not enough evidence to prove whether or not peppermint oil has the ability to prevent any kind of cancerous cells from forming in the body.

The correct answer will look like this:

> The argument assumes that the refuting evidence matters to the position that is being challenged.

> We have no credible source in this argument, so the refutation is senseless.

Prescriptive Error

First, let's take a look at what "prescriptive" means. Prescriptive means to give directions, or to say something *ought to* or *should* do something else. Sometimes an argument will be a descriptive premise (simply describing) that leads to a prescriptive conclusion, which makes for a very weak argument. This is like saying "There is a hurricane coming; therefore, we should leave the state." Even though this seems like common sense, the logical soundness of this argument is missing. A valid argument is when the truth of the premise leads absolutely to the truth of the conclusion. It's when the conclusion *is* something, not when the conclusion *should* be or do something. The flaw here is the assumption that the conclusion is going to work out; something prescriptive is not ever guaranteed to work out in a logical argument.

False Choice

A false choice, or false dilemma, flaw is a statement that assumes only the object it lists in the statement is the solution, or the only options that exist, for that problem. Here is an example:

> I didn't get the grade I want in Chemistry class. I must either be really stupid, I didn't get enough sleep, or I didn't eat enough that day.

This is a false choice error. We are offered only three options for why the speaker did not get the grade he or she wanted in Chemistry class. However, there is potentially more options why the grade was not achieved other than the three listed. The speaker could have been fighting a cold, or the professor may not have taught the material in a comprehensive way. It is our job as test takers to recognize that there are more options other than the choices we are given, although it appears that the only three choices are listed in the example.

Red Herring

A red herring is a point offered in an argument that is only meant to distract or mislead. A red herring will throw something out after the argument that is unrelated to the argument, although it still commands attention, thus taking attention away from the relevant issue. The following is an example of a red herring fallacy:

> Kirby: It seems like therapy is moving toward a more holistic model rather than something prescriptive, where the space between a therapist and client is seen more organic rather than a controlled space. This helps empower the client to reach their own conclusions about what should be done rather than having someone tell them what to do.

> Barlock: What's the point of therapy anyway? It seems like "talking out" problems with a stranger is a waste of time and always has been. Is it even successful as a profession?

We see Kirby present an argument about the route therapy is taking toward the future. Instead of responding to the argument by presenting their own side regarding where therapy is headed, Barlock questions the overall point of therapy. Barlock throws out a red herring here: Kirby cannot proceed with the argument because now Kirby must defend the existence of therapy instead of its future.

Correlation Versus Causality

Test takers should be careful when reviewing causal conclusions because the reasoning is often flawed, incorrectly classifying correlation as causality. Two events that may or may not be associated with one another are said to be linked such that one was the cause or reason for the other, which is considered the effect. To be a true "cause-and-effect" relationship, one factor or event must occur first (the cause) and be the sole reason (unless others are also listed) that the other occurred (the effect). The cause serves as the initiator of the relationship between the two events.

For example, consider the following argument:

> Last weekend, the local bakery ran out of blueberry muffins and some customers had to select something else instead. This week, the bakery's sales have fallen. Therefore, the blueberry muffin shortage last weekend resulted in fewer sales this week.

In this argument, the author states that the decline in sales this week (the effect) was caused by the shortage of blueberry muffins last weekend. However, there are other viable alternate causes for the decline in sales this week besides the blueberry muffin shortage. Perhaps it is summer and many normal patrons are away this week on vacation, or maybe another local bakery just opened or is running a special sale this week. There might be a large construction project or road work in town near the bakery, deterring customers from navigating the detours or busy roads. It is entirely possible that the decline this week is just a random coincidence and not attributable to any factor other than chance, and that next week, sales will return to normal or even exceed typical sales. Insufficient evidence exists to confidently assert that the blueberry muffin shortage was the sole reason for the decline in sales, thus mistaking correlation for causation.

Identifying Assumptions in an Argument and Determining if they are Supported by the Evidence

In the structure of an argument, an **assumption** is an unstated premise. To identify the assumption within arguments, you must find something that the argument is relying on that the author is not stating explicitly. Many strengthening and weakening questions deal with unstated assumptions, as well as necessary and sufficient assumption questions. Let's take a look at what an unsated assumption looks like:

> All restaurants in the Seattle area serve vegan food. *Haile's Seafood* must serve vegan food.

Let's identify all parts of the argument, including the unstated assumption. The conclusion of this argument is the last sentence: *Haile's Seafood* must serve vegan food. The premise we are given is the first sentence: All restaurants in the Seattle area serve vegan food. Now let's ask ourselves if there's a missing link. How did the author reach this conclusion? The author reached this conclusion with an unstated assumption, which might look like this: *Haile's Seafood* is in the Seattle area. Now we have the argument:

> Premise: All restaurants in the Seattle area serve vegan food.
> Unstated Assumption: *Haile's Seafood* is in the Seattle area.
> Conclusion: *Haile's Seafood* must serve vegan food

Another way to look at the missing link is like this: there is a connection between "Seattle area" and "vegan food," and one between *"Haile's Seafood"* and "vegan food", but there is no connection between "Seattle area" and *"Haile's Seafood."* The unstated assumption identifies this connection.

Analyzing Two Arguments and Evaluating the Types of Evidence Used to Support Each Claim

Authors use arguments to persuade the reader to agree with their claim. When analyzing the strength of an argument, the reader should summarize the author's message and determine whether or not they agree or disagree with the claim and whether, overall, they believe the evidence the author presented to support it. Next the reader should identify each individual detail and examine it to determine if it supports the claim, fails to support the claim, or even distract from the author's goal.

As the reader examines each piece of evidence, they should consider its type and its effectiveness in proving the author's claim both individually and in context. Readers should pay attention to details that contradict the evidence, and readers should question the way the author uses details.

Facts can often be interpreted in ways that mislead the reader, so they will believe a claim that is unproven or untrue. If the claim is supported by facts such as statistics or empirical evidence, the reader should examine them. The reader should question whether the facts are relevant or, if they were interpreted differently, they would lead to a different conclusion that would fail to support the argument or even contradict it. They should consider whether the scope of the data is appropriate, or if it is too broad or narrow, and they should question whether the source is credible, and whether a source has been cited at all.

If the author uses expert opinions, the reader should consider whether the expert is credible and appropriate. If a doctor offers an opinion about pediatric medicine, the reader should ask if the doctor is an expert in that field or if they practice a different form of medicine.

Finally, the reader should consider whether the author has anticipated and adequately responded to any potential counterarguments to the claim and decide if the argument overall is strong, or if the evidence fails to persuade them to accept the author's claim.

Extending Your Understanding to New Situations

Combining Information from Different Sources
When a student has an assignment to research and write a paper, one of the first steps after determining the topic is to select research sources. The student may begin by conducting an Internet or library search of the topic, may refer to a reading list provided by the instructor, or may use an annotated bibliography of works related to the topic. Once appropriate, valid, and relevant sources are found, the information gleaned from them must be synthesized and incorporated together to form a coherent, logical argument or paper. The following include descriptions of some common sources:

Books as Resources
To evaluate the worth of the book for the research paper, the student first considers the book title to get an idea of its content. Then the student can scan the book's table of contents for chapter titles and topics to get further ideas of their applicability to the topic. The student may also turn to the end of the book to look for an alphabetized index. Most academic textbooks and scholarly works have these; students can look up key topic terms to see how many are included and how many pages are devoted to them.

Textbooks that are designed well employ varied text features for organizing their main ideas, illustrating central concepts, spotlighting significant details, and signaling evidence that supports the ideas and points conveyed. When a textbook uses these features in recurrent patterns that are predictable, it makes it easier for readers to locate information and come up with connections. When readers comprehend how to make use of text features, they will take less time and effort deciphering how the text is organized, leaving them more time and energy for focusing on the actual content in the text. Instructional activities can include not only previewing text through observing main text features, but moreover through examining and deconstructing the text and ascertaining how the text features can aid them in locating and applying text information for learning.

Included among various text features are a table of contents, headings, subheadings, an index, a glossary, a foreword, a preface, paragraphing spaces, bullet lists, footnotes, sidebars, diagrams, graphs, charts, pictures, illustrations, captions, italics, boldface, colors, and symbols. A glossary is a list of key vocabulary words and/or technical terminology and definitions. This helps readers recognize or learn specialized terms used in the text before reading it. A foreword is typically written by someone other than the text author and appears at the beginning to introduce, inform, recommend, and/or praise the work. A preface is often written by the author and also appears at the beginning, to introduce or explain something about the text, like new additions. A sidebar is a box with text and sometimes graphics at the left or right side of a page, typically focusing on a more specific issue, example, or aspect of the subject. Footnotes are additional comments/notes at the bottom of the page, signaled by superscript numbers in the text.

When examining a book, a journal article, a monograph, or other publication, the table of contents is in the front. In books, it is typically found following the title page, publication information (often on the facing side of the title page), and dedication page, when one is included. In shorter publications, the table of contents may follow the title page, or the title on the same page. The table of contents in a book lists the number and title of each chapter and its beginning page number. An index, which is most common in books but may also be included in shorter works, is at the back of the publication. Books, especially academic texts, frequently have two: a subject index and an author index. Readers can look alphabetically for specific subjects in the subject index. Likewise, they can look up specific authors cited, quoted, discussed, or mentioned in the author index.

The index in a book offers particular advantages to students. For example, college course instructors typically assign certain textbooks, but do not expect students to read the entire book from cover to cover immediately. They usually assign specific chapters to read in preparation for specific lectures and/or discussions in certain upcoming classes. Reading portions at a time, some students may find references they either do not fully understand or want to know more about. They can look these topics up in the book's subject index to find them in later chapters. When a text author refers to another author, students can also look up the name in the book's author index to find all page numbers of all other references to that author. College students also typically are assigned research papers to write. A book's subject and author indexes can guide students to pages that may help inform them of other books to use for researching paper topics.

Journal Articles

Like books, journal articles are primary or secondary sources the student may need to use for researching any topic. To assess whether a journal article will be a useful source for a particular paper topic, a student can first get some idea about the content of the article by reading its title and subtitle, if any exists. Many journal articles, particularly scientific ones, include abstracts. These are brief summaries of the content. The student should read the abstract to get a more specific idea of whether the experiment, literature review, or other work documented is applicable to the paper topic. Students should also check the references at the end of the article, which today often contain links to related works for exploring the topic further.

Headings and subheadings concisely inform readers what each section of a paper contains, as well as showing how its information is organized both visually and verbally. Headings are typically up to about five words long. They are not meant to give in-depth analytical information about the topic of their section, but rather an idea of its subject matter. Text authors should maintain consistent style across all headings. Readers should not expect headings if there is not material for more than one heading at each level, just as a list is unnecessary for a single item. Subheadings may be a bit longer than headings because they expand upon them. Readers should skim the subheadings in a paper to use them as a map of how content is arranged. Subheadings are in smaller fonts than headings to mirror relative importance. Subheadings are not necessary for every paragraph. They should enhance content, not substitute for topic sentences.

When a heading is brief, simple, and written in the form of a question, it can have the effect of further drawing readers into the text. An effective author will also answer the question in the heading soon in the following text. Question headings and their text answers are particularly helpful for engaging readers with average reading skills. Both headings and subheadings are most effective with more readers when they are obvious, simple, and get to their points immediately. Simple headings attract readers; simple subheadings allow readers a break, during which they also inform reader decisions whether to continue reading or not. Headings stand out from other text through boldface, but also italicizing and underlining them would be excessive. Uppercase-lowercase headings are easier for readers to comprehend than all capitals. More legible fonts are better. Some experts prefer serif fonts in text, but sans-serif fonts in headings. Brief subheadings that preview upcoming chunks of information reach more readers.

Encyclopedias and Dictionaries

Dictionaries and encyclopedias are both reference books for looking up information alphabetically. Dictionaries are more exclusively focused on vocabulary words. They include each word's correct spelling, pronunciation, variants, part(s) of speech, definitions of one or more meanings, and examples used in a sentence. Some dictionaries provide illustrations of certain words when these inform the meaning. Some dictionaries also offer synonyms, antonyms, and related words under a word's entry. Encyclopedias, like dictionaries, often provide word pronunciations and definitions. However, they have broader scopes: one can look up entire subjects in encyclopedias, not just words, and find comprehensive, detailed information about historical events, famous people, countries, disciplines of study, and many other things. Dictionaries are for finding word meanings, pronunciations, and spellings; encyclopedias are for finding breadth and depth of information on a variety of topics.

Card Catalogs

A card catalog is a means of organizing, classifying, and locating the large numbers of books found in libraries. Without being able to look up books in library card catalogs, it would be virtually impossible to find them on library shelves. Card catalogs may be on traditional paper cards filed in drawers, or electronic catalogs accessible online; some libraries combine both. Books are shelved by subject area; subjects are coded using formal classification systems—standardized sets of rules for identifying and labeling books by subject and author. These assign each book a call number: a code indicating the classification system, subject, author, and title. Call numbers also function as bookshelf "addresses" where books can be located. Most public libraries use the Dewey Decimal Classification System. Most university, college, and research libraries use the Library of Congress Classification. Nursing students will also encounter the National Institute of Health's National Library of Medicine Classification System, which major collections of health sciences publications utilize.

Databases

A database is a collection of digital information organized for easy access, updating, and management. Users can sort and search databases for information. One way of classifying databases is by content, i.e. full-text, numerical, bibliographical, or images. Another classification method used in computing is by organizational approach. The most common approach is a relational database, which is tabular and defines data so they can be accessed and reorganized in various ways. A distributed database can be reproduced or interspersed among different locations within a network. An object-oriented database is organized to be aligned with object classes and subclasses defining the data. Databases usually collect files like product inventories, catalogs, customer profiles, sales transactions, student bodies, and resources. An associated set of application programs is a database management system or database manager. It enables users to specify which reports to generate, control access to reading and writing data, and analyze database usage. Structured Query Language (SQL) is a standard computer language for updating, querying, and otherwise interfacing with databases.

Websites

On the Internet or in computer software programs, text features include URLs, home pages, pop-up menus, drop-down menus, bookmarks, buttons, links, navigation bars, text boxes, arrows, symbols, colors, graphics, logos, and abbreviations. URLs (Universal Resource Locators) indicate the internet "address" or location of a website or web page. They often start with www. (world wide web) or http:// (hypertext transfer protocol) or https:// (the "s" indicates a secure site) and appear in the Internet browser's top address bar. Clickable buttons are often links to specific pages on a website or other external sites. Users can click on some buttons to open pop-up or drop-down menus, which offer a list of actions or departments from which to select. Bookmarks are the electronic versions of physical bookmarks. When users bookmark a website/page, a link is established to the site URL and saved, enabling returning to the site in the future without having to remember its name or URL by clicking the bookmark.

Readers can more easily navigate websites and read their information by observing and utilizing their various text features. For example, most fully developed websites include search bars, where users can type in topics, questions, titles, or names to locate specific information within the large amounts stored on many sites. Navigation bars (software developers frequently use the abbreviation term "navbar") are graphical user interfaces (GUIs) that facilitate visiting different sections, departments, or pages within a website, which can be difficult or impossible to find without these. Typically, they appear as a series of links running horizontally across the top of each page. Navigation bars displayed vertically along the left side of the page are also called sidebars. Links, i.e. hyperlinks, enable hyperspeed browsing by allowing readers to jump to new pages/sites. They may be URLs, words, phrases, images, buttons, etc. They are often but not always underlined and/or blue, or other colors.

Drawing Conclusions
Making inferences and drawing conclusions both require readers to fill in information the writer has omitted. To make an inference or draw a conclusion about the text, test takers should observe all facts and arguments the test writer has presented. The best way to understand ways to drawing well-supported conclusions and generalizations is by practice. Consider the following example:

Nutritionist: "Many bodybuilders turn to whey protein as a way to repair their muscle tissue after working out. Recent studies are showing that using whey as a source of protein is linked to prostate cancer in men. Bodybuilders who use whey protein may consider switching to a plant-based protein source in order to avoid the negative effects that could come with whey protein consumption."

Which of the following most accurately expresses the conclusion of the nutritionist's argument?

> a. Whey protein is an excellent way to repair muscles after a workout.
> b. Bodybuilders should switch from whey to plant-based protein.
> c. Whey protein causes every single instance of prostate cancer in men.
> d. We still don't know the causes of prostate cancer in men.

The correct answer choice is *B*: bodybuilders should switch from whey to a plant-based protein. We can gather this from the entirety of the passage, as it begins with what kind of protein bodybuilders consume, the dangers of that protein, and what kind of protein to switch to. Choice *A* is incorrect; this is the opposite of what the passage states. When reading through answer choices, it's important to look for choices that include the words "every," "always," or "all." In many instances, absolute answer choices will not be the correct answer. Take Choice *C* for example: the passage does not state that whey protein causes "every single instance" of prostate cancer in men, only that it is *linked* to prostate cancer in men. Choice *D* is incorrect; although the nutritionist does not list all causes of prostate cancer in men, he/she does not conclude that we do not know any causes of prostate cancer in men.

The key to drawing well-supported conclusions is to read the question a few times and then paraphrase the passage. This will help you get an idea of the passage's conclusion before reading the different answer choices. Remember that drawing a conclusion is different than making an assumption. Conclusions must rely solely on the facts of the passage. Making an assumption goes beyond the facts of the passage, so be careful of answer choices depicting assumptions instead of passage-based conclusions.

There will be questions on the GED that give a scenario with a general conclusion and ask you to apply that general conclusion to a new context. Skills for making inferences and drawing conclusions will be helpful in the first portion of this question. Reading the initial scenario carefully and finding the general concept, or the bigger picture, is necessary for when the test taker attempts to apply this general concept to the new context the question provides. Here is an example of a test question that asks the test taker to apply information in a selection to a new context:

The placebo effect is a phenomenon used in clinical trial studies to test the effectiveness of new medications. A group of people are given either the new medication or the placebo, but are not told which they receive. Interestingly, about one-third of people who are given the placebo in clinical trials will report a cessation of their symptoms. In one trial in 1925, a group of people were given sugar pills and told their migraines should dissipate as a result of the pills. Forty-two percent noticed that in the following six months, their weekly migraines evaporated. Researchers believe that human belief and expectation might be a reason that the placebo will work in some patients.

Considering the phenomenon of the placebo effect, what would probably happen to someone who is given a shot with no medication and told their arm should go numb from it?

    a. The patient might experience some burning in their arm, but then they would feel nothing.
    b. The patient would feel their arm going numb, as the placebo effect is certain to work.
    c. Nothing would happen, because the shot does not actually have any medication in it.
    d. The individual might actually experience a numbing sensation in their arm, as the placebo works on some people by simply being told the placebo will have certain effects.

The answer is Choice *D*. The individual might actually experience a numbing sensation in their arm, as the placebo works on some people by simply being told the placebo will have certain effects. Choices *B* and *C* are too absolute to be considered correct—watch out for words like "never" or "always" in the answer choices so you can rule them out if possible. Choice *A* is incorrect because we don't know what the initial sensation of the shot would feel like for this individual. The placebo effect would have a chance of working with the shot, just like it would have a chance of working in the above example with the pill. The patient's belief in an effect is what can possibly manifest the desired result of the placebo.

Grammar and Language

Word Usage

Correcting Errors with Frequently Confused Words
Affect Versus Effect
The word *affect* is a verb, and it means to have an effect on something or to influence something. *Affect* is sometimes used as a noun, and in this way, it means an emotional response or disposition. The word *effect* is mostly used as a noun, and it refers to the impact or result of something. However, *effect* can also be used as a verb, meaning to cause something to come into being. Replacing the verb *effect* with *bring about* is useful in determining which word to use. Here are four examples of these words used correctly:

- Affect as verb: Her bravery the night before *affected* the way her peers treated her.
- Affect as noun: He had a moody *affect* during his everyday routines.
- Effect as noun: The breakfast I ate this morning had a negative *effect* on me.
- Effect as verb: He was able to *effect* change in his life after he became well again.

Among Versus Between

When choosing between the words *between* and *among*, usually we would use *between* if it involved two choices, and *among* if it involved multiple choices. Here is an example:

> I had to choose *between* purple and blue.
> She distributed the ice cream *among* her four children.

This rule, however, isn't absolute. The word *between* can usually be used when talking about two or more things, if those things are distinct. For example:

> He chose between vanilla, chocolate, and strawberry.

But if we are choosing among a group of things or people, or about items collectively, it is best to use *among*.

> He chose among many ice cream flavors.

Amount Versus Number

Amount is used with nouns that cannot be counted; *number* is used with nouns that can be counted. Here is an example:

> No one knew the amount of bravery she was capable of.
> The number of items on the list came to be twenty-seven.

Good Versus Well

The word *good* is an adjective. We use the word *good* before nouns and after linking verbs. The word *well* is usually an adverb, although sometimes *well* can be an adjective when pertaining to someone's health.

> Before a noun: Tina did a good job giving her speech in class today.
> After linking verb: Tina's lunch smells good.

In the first example, we see that *good* is an adjective describing the word *job*. In the second example, the word *good* is an adjective to modify the subject *lunch*.

> Tina did well giving her speech in class today.

In the above example, *well* does not describe a noun; it answers the question *how did Tina do?* making *well* an adverb.

> Tina began to get well again after her surgery.

In the above example, we see *well* as an adjective pertaining to Tina and her health.

Bad Versus Badly

Similar rules are applied to *bad* and *badly* just as in *good* and *well*. *Bad* is used to describe a noun or used after a linking verb. The word *badly* is an adverb that modifies an action verb. Let's look at the same example above using the word *bad* and *badly:*

> Before a noun: Tina did a bad job giving her speech in class today.
> After a linking verb: Tina's lunch smells bad.
> Badly as an adverb: Tina did badly giving her speech in class today.

Bring Versus Take

The word *bring* implies carrying something toward the speaker, like *Please bring your pencils to class tomorrow*. The word *take* implies carrying something away from the speaker, like *Please take your pencils home with you tonight*.

Can versus May (Could versus Might)

In formal English, the word *may* is used to ask or grant permission, and the word *can* refers to ability or capability, like the following examples:

> May I go to the restroom before class starts?
> Can he ride his bicycle yet?

Keep these differences in mind while writing. With informal language, we often use *can* to ask permission, but in formal writing, it is more appropriate to use the word *may* when granting or asking for permission.

Farther Versus Further

The word *farther* refers to a measurable or physical distance, like the following:

> How much farther is the walk to your house?

In the above context, the word *farther* is used because it is referring to a physical distance from one point to another.

The word *further* implies a metaphorical or figurative distance. *Further* may also mean "in addition to," like *I have nothing further to write.*

> We drifted further apart as the years went by.

Fewer Versus Less

The word *fewer* is used with plural nouns and when discussing countable things. The word *less* is used with singular mass nouns. For example, we might have *fewer* coins, groceries, restaurants, or tablets, but we have *less* money, food, security, or light.

Hear Versus Here

The word *hear* and *here* have distinctly different meanings, but they sound the same, so they are often misspelled. The word *hear* means to perceive with the ear, like *I hear the train coming this way*. The word *here* refers to a place or position, like *I put the money here a few seconds ago*.

I.E. Versus E.G.

The abbreviation *i.e.* stands for "in other words." *I.e.* does not have to do with listing examples, but gives an alternate point of view of a statement, denoting the phrase "in other words."

The abbreviation *e.g.* stands for "for example." Use *e.g.* when listing one or more examples in your writing.

Learn Versus Teach

The verb *learn* means to receive knowledge or a skill in a subject. The verb *teach* means to give knowledge or a skill to someone. Below are two examples of how to correctly use *learn* and *teach*.

> Today, I learned how to count blocks in school.
> Today, the teacher taught us how to count blocks in school.

Lie Versus Lay

Excluding the definition of "to tell an untruth" for *lie*, the words *lay* and *lie* mean setting/reclining. The word *lay* requires a direct object. Here's an example:

> Lay the book down on my desk please.

The word *book* in the sentence above is the direct object. If one were to lie down on the couch, it would be:

> I want to lie down on the couch.

There is no direct object, so the verb *lie* is used.

What gets confusing between *lay* and *lie* is that *lay* is also a past tense of lie. In this case, we have:

> Yesterday, Nina lay on the countertop and fell asleep.

Even more bizarre, the past tense of *lay* is *laid*. Therefore, we have:

> Last week I laid the book down on your desk.

For the purposes of the present tense, remember that *lay* requires a direct object (you lay something down) and *lie* does not (you lie down by yourself).

Which Versus That

Sometimes, it can be confusing when to use *which* or when to use *that* in a sentence. Basically, the word *that* is used for restrictive clauses (aka essential clauses). Restrictive clauses are clauses that restrict the information in another part of the sentence, so they cannot be taken out. Here's an example:

> The door *that you broke* is being *fixed*.

The restrictive clause is in italics. If we took this information out, it would change the meaning of the sentence. It's a *very specific door* that is being fixed—the one *that you broke*. This information is pertinent to the sentence.

The word *which* is used for nonrestrictive clauses (aka nonessential clauses). This means that the clause beginning with *which* can be taken out of the sentence, and the information in the main clause will not be changed. Here are some examples:

> I only got five hours of sleep last night, which isn't good.
> Mangos, which are cheap, are my favorite food.

Notice in the second sentence, if we took out the middle clause *which are cheap*, mangos would still be the speaker's favorite food. Clauses that begin with *which* are almost always set apart by commas or preceded by commas.

Who Versus Whom

The words *who* and *whom* serve two different purposes within a sentence. *Who* refers to the subject of a clause, which is the noun that is acting. *Whom* refers to the direct object of a clause, which is the noun that is having something done to it. Here are two examples using both *who* and *whom*:

> *Who* left the door open last night?
> *Whom* did you leave that letter with?

Let's answer these questions with the proper noun, *Marie. Marie left the door open last night.* In this sentence, we can see that *Marie* is the subject of the sentence. Since the *who* is referring to the subject of the sentence, the first example is correct. Let's answer the second question. *I left that letter with Marie.* Now, Marie is the *object* of the sentence. This sentence is correct because it uses *whom* to refer to the object of the sentence.

One trick is to replace the word *who* or *whom* with *she* or *her*. When using the word *who*, we can answer the question with *she, he,* or *they*. When using the word whom, we can answer the question with *her, him,* or *them*. Plugging the pronouns in is an easy way to tell if you've used the correct word:

> *She* left the door open last night. (Who?)
> I left the letter with *her*. (Whom?)

Notice again that the word *she* is the subject of the first sentence (so we use *who*), and the word *her* is the object of the second sentence (so we use *whom*).

Homonyms, Homophones, and Homographs

Homophones are words that sound the same in speech, but have different spellings and meanings. For example, *to, too,* and *two* all sound alike, but have three different spellings and meanings. Homophones with different spellings are also called heterographs. Homographs are words that are spelled identically, but have different meanings. If they also have different pronunciations, they are heteronyms. For instance, *tear* pronounced one way means a drop of liquid formed by the eye; pronounced another way, it means to rip. Homophones that are also homographs are homonyms. For example, *bark* can mean the outside of a tree or a dog's vocalization; both meanings have the same spelling. *Stalk* can mean a plant stem or to pursue and/or harass somebody; these are spelled and pronounced the same. *Rose* can mean a flower or the past tense of *rise*. Many non-linguists confuse things by using "homonym" to mean sets of words that are homophones but not homographs, and also those that are homographs but not homophones.

The word *row* can mean to use oars to propel a boat; a linear arrangement of objects or print; or an argument. It is pronounced the same with the first two meanings, but differently with the third. Because it is spelled identically regardless, all three meanings are homographs. However, the two meanings pronounced the same are homophones, whereas the one with the different pronunciation is a heteronym. By contrast, the word *read* means to peruse language, whereas the word *reed* refers to a marsh plant. Because these are pronounced the same way, they are homophones; because they are spelled differently, they are heterographs. Homonyms are both homophones and homographs—pronounced and spelled identically, but with different meanings. One distinction between homonyms is of those with separate, unrelated etymologies, called "true" homonyms, e.g. *skate* meaning a fish or *skate* meaning to glide over ice/water. Those with common origins are called polysemes or polysemous homonyms, e.g. the *mouth* of an animal/human or of a river.

The subject of the sentence is the word or phrase that is being discussed or described relating to the verb. The **complete subject** is a subject with all its parts, like the following: *The stormy weather* ruined their vacation. A **simple subject** is the subject with all the modifiers removed. In the previous example the simple subject would be *weather*. There are various forms of subjects listed below:

- Noun (phrase) or pronoun: *The tiny bird* sang all morning long.
- A to-infinitive clause: *To hike the Appalachian Trail* was her lifelong goal.
- A gerund: *Running* was his new favorite sport.
- A that-clause: *That she was old* did not stop her from living her life.
- A direct quotation: *"Here comes the sun"* is a quote from a Beatle's song.
- A free relative clause: *Whatever she said* is none of my business.
- Implied subject: *(You)* Shut the door!

A **verb** is a word or phrase that expresses action, feeling, or state of being. Verbs explain what their subject is *doing*. Three different types of verbs used in a sentence are action verbs, linking verbs, and helping verbs.

Action verbs show a physical or mental action. Some examples of action verbs are *play, type, jump, write, examine, study, invent, develop,* and *taste*. The following example uses an action verb:

Kat *imagines* that she is a mermaid in the ocean.

The verb *imagines* explains what Kat is doing: she is imagining being a mermaid.

Linking verbs connect the subject to the predicate without expressing an action. The following sentence shows an example of a linking verb:

The mango *tastes* sweet.

The verb *tastes* is a linking verb. The mango doesn't *do* the tasting, but the word *taste* links the mango to its predicate, sweet. Most linking verbs can also be used as action verbs, such as *smell, taste, look, seem, grow,* and *sound*. Saying something *is* something else is also an example of a linking verb. For example, if we were to say, "Peaches is a dog," the verb *is* would be a linking verb in this sentence, since it links the subject to its predicate.

Helping verbs are verbs that help the main verb in a sentence. Examples of helping verbs are *be, am, is, was, have, has, do, did, can, could, may, might, should,* and *must,* among others. The following are examples of helping verbs:

Jessica *is* planning a trip to Hawaii.
Brenda *does* not like camping.
Xavier *should* go to the dance tonight.

Notice that after each of these helping verbs is the main verb of the sentence: *planning, like,* and *go*. Helping verbs usually show an aspect of time.

Errors in subject-verb agreement are very common grammatical. One of the most common instances is when people use a series of nouns as a compound subject with a singular instead of a plural verb. Here is an example:

> Waiting in traffic, walking from the parking lot across campus, and navigating to the classroom in the building *is* time-consuming

instead of saying "*are* time-consuming." Additionally, when a sentence subject is compound, the verb is plural:

> She and her friends *were* laughing.

However, if the conjunction connecting two or more singular nouns or pronouns is "or" or "nor," the verb must be singular to agree:

> That book or another one on that shelf should be helpful.

If a compound subject includes both a singular noun and a plural one, and they are connected by "or" or "nor," the verb must agree with the subject closest to the verb: "Juan or his sisters go out dancing daily"; but "His sisters or Juan goes out dancing daily."

Simply put, singular subjects require singular verbs and plural subjects require plural verbs.

Correcting Pronoun Errors

Pronouns take the place of nouns in sentences. For example, in "John saw Mary, and he waved to her," "he" replaces "John," and "her" replaces "Mary." This reads much better than "John saw Mary, and John waved to Mary," which sounds repetitious. There are several different types of pronouns:

- **Personal pronouns** show the differences in person, gender, number, and case: he, she, they, it, we, him, her, us, them, I, you

- **Demonstrative pronouns** replace a noun phrase: this, these, that, those

- **Interrogative pronouns** are used to ask questions: which, who, whom, what, whose

- **Indefinite pronouns** do not refer to any one thing in particular: none, several, anything, something, anyone, everyone

- **Possessive pronouns** indicate possession: his, hers, theirs, mine, yours

- **Reciprocal pronouns** indicate that two or more people are acting in the same way towards the other: each other, one another

- **Relative pronouns** are used to connect phrases to a noun or pronoun: that, who whom, whose, which, whichever, whatsoever

- **Reflexive pronouns** are preceded by the adverb, adjective, or noun to which it refers: myself, yourself, himself, herself, oneself, itself, ourselves, themselves

Pronoun Case

There are three pronoun cases: subjective case, objective case, and possessive case. Pronouns as subjects are pronouns that replace the subject of the sentence, such as *I, you, he, she, it, we, they* and *who*.

Pronouns as objects replace the object of the sentence, such as *me, you, him, her, it, us, them,* and *whom.* Pronouns that show possession are *mine, yours, hers, its, ours, theirs,* and *whose.*

The following are examples of different pronoun cases:

- **Subject pronoun**: *She* ate the cake for her birthday. *I* saw the movie.
- **Object pronoun**: You gave *me* the card last weekend. She gave the picture to *him.*
- **Possessive pronoun**: That bracelet you found yesterday is *mine. His* name was Casey.

Eliminating Non-Standard English Words or Phrases

Non-standard English words are ones that are used in common speech or informal writing, but which are considered incorrect or improper in formal writing. Non-standard words are used in a way that differs from the classic definition of the word, words derived from other words to form a new word, or slang words. Non-standard words also include commonly misused subject-verb agreement for the verb *to be.* For example, *we was* should be *we were.*

For example, though the word *irregardless* is commonly used in speech and even writing, this is an incorrect form of the word *regardless.* Likewise, the word *ain't* is used to replace *am not,* but *ain't* is not appropriate in formal writing.

Other non-standard English words include abbreviated phrases like *dunno* for *I don't know* or *innit* for *isn't it.* Incorrect contractions such as *gonna, wanna,* and *gotta* should be eliminated from formal writing, as well as purposely misspelled words like *lite* or *sez.*

Sentence Structure

Eliminating Dangling or Misplaced Modifiers

A misplaced modifier is a word, clause, or phrase that refers to an unintended word. Most modifiers can be edited to where they clearly modify the intended word instead of remaining ambiguous. Examples of misplaced modifiers are misplaced adjectives, adverb placement, misplaced phrases, and misplaced clauses. In the examples below, the modifiers are in bold, and the words they modify are in italics.

Misplaced adjectives are adjectives that are intended to modify nouns, but are actually separated from them. Misplaced adjectives can be easily corrected by positioning the adjective beside the noun it modifies. The following is an example followed by a correction:

Incorrect: Grandma ate a **warm** bowl of *soup* for dinner.
Correct: Grandma ate a bowl of **warm** *soup* for dinner.

If spoken in colloquial language, the first sentence would sound pretty accurate. We use misplaced adjectives all the time and they don't *seem* incorrect. However, in the above example, the soup is the thing that is warm, first and foremost, so the adjective should reflect that.

Adverb placement within a sentence can also turn into a misplaced modifier. Depending on where certain adverbs are placed, the meaning of the whole sentence can change. Here are adverb placements that can result in two different meanings:

Meaning 1: She **quickly** *vacuumed* the rug she had bought.
Meaning 2: She vacuumed the rug she had *bought* **quickly.**

117

The placement of the adverb *quickly* determines the meaning of the sentence. In the first example, we are told that she vacuumed the rug quickly. In the second sentence, we are told that she bought the rug quickly. Test takers should be aware of where they place their adverbs so to avoid confusion.

Misplaced phrases are phrases within a sentence that should be rearranged to create clarity. The following is an example of a misplaced phrase with the corrected phrase underneath:

Incorrect: The man gave the *vacation house* to his daughter **with the orange trees**.
Correct: The man gave the *vacation house* **with the orange trees** to his daughter.

The correct sentence places the modifying phrase next to the noun it modifies. In the first sentence, the logic of the sentence tells us that the daughter had the orange trees; the second sentence is corrected to tell us that it's the house that has the orange trees, not the daughter.

Misplaced clauses occur when a clause is modifying an incorrect noun. Below is an example of a misplaced clause:

Incorrect: The restaurant prepared a *dish* for the man **that was cooked in steamed milk**.
Correct: The restaurant prepared a *dish* **that was cooked in steamed milk** for the man.

In the first sentence, the logic of the sentence tells us that there was a man who was cooked in steamed milk, and the restaurant prepared a dish for him! To correct this error, place the clause next to the noun it is supposed to modify. In the corrected version, we see that it's the *dish* that was cooked in steamed milk, not the man.

Editing Sentences for Parallel Structure and Correct Use of Conjunctions
Parallel Sentence Structures
Parallel structure in a sentence matches the forms of sentence components. Any sentence containing more than one description or phrase should keep them consistent in wording and form. Readers can easily follow writers' ideas when they are written in parallel structure, making it an important element of correct sentence construction. For example, this sentence lacks parallelism: "Our coach is a skilled manager, a clever strategist, and works hard." The first two phrases are parallel, but the third is not. Correction: "Our coach is a skilled manager, a clever strategist, and a hard worker." Now all three phrases match in form. Here is another example:

Fred intercepted the ball, escaped tacklers, and a touchdown was scored.

This is also non-parallel. Here is the sentence corrected:

Fred intercepted the ball, escaped tacklers, and scored a touchdown.

Conjunctions
A **conjunction** is a word used to connect clauses or sentences, or the words used to connect words to other words. The words *and, but, or, nor, yet,* and *so* are conjunctions. Two types of conjunctions are called coordinating conjunctions and subordinating conjunctions.

Coordinating conjunctions join two or more items of equal linguistic importance. These conjunctions are *for, and, nor, but, or, yet,* and *so,* also called *FANBOYS* as an acronym. Here are each of the words used as coordinating conjunctions:

- They did not attend the reception that evening, for they were all sick from lunch.
- She bought a chocolate, vanilla, and raspberry cupcake.
- We do not eat pork, nor do we eat fish.
- I would bring her to the circus, but she is afraid of the clowns.
- She went to the grocery store, or she went to the park.
- Grandpa wanted to buy a house, yet he did not want to pay for it.
- We had a baby, so we bought a house on a lake.

Subordinating conjunctions are conjunctions that join an independent clause with a dependent clause. Sometimes they introduce adverbial clauses as well. The most common subordinating conjunctions in the English language are the following: *after, although, as, as far as, as if, as long as, as soon as, as though, because, before, even if, even though, every time, if, in order that, since, so, so that, than, though, unless, until, when, whenever, where, whereas, wherever,* and *while.* Here's an example of the subordinating conjunction, *unless.*

> Theresa was going to go kayaking on the river unless it started to rain.

Notice that "unless it started to rain" is the dependent clause, because it cannot stand by itself as a sentence. "Theresa was going to go kayaking on the river" is the independent clause. Here, *unless* acts as the subordinating conjunction, because the two clauses are not syntactically equal.

Editing for Subject-Verb and Pronoun-Antecedent Agreement
Subject-Verb Agreement
Lack of subject-verb agreement is a very common grammatical error. One of the most common instances is when people use a series of nouns as a compound subject with a singular instead of a plural verb. Here is an example:

> Identifying the best books, locating the sellers with the lowest prices, and paying for them *is* difficult

instead of saying "*are* difficult." Additionally, when a sentence subject is compound, the verb is plural:

> He and his cousins *were* at the reunion.

However, if the conjunction connecting two or more singular nouns or pronouns is "or" or "nor," the verb must be singular to agree:

> That pen or another one like it is in the desk drawer.

If a compound subject includes both a singular noun and a plural one, and they are connected by "or" or "nor," the verb must agree with the subject closest to the verb: "Sally or her sisters go jogging daily"; but "Her sisters or Sally goes jogging daily."

Simply put, singular subjects require singular verbs and plural subjects require plural verbs. A common source of agreement errors is not identifying the sentence subject correctly. For example, people often write sentences incorrectly like, "The group of students *were* complaining about the test." The subject is not the plural "students" but the singular "group." Therefore, the correct sentence should read, "The group of students *was* complaining about the test." The converse also applies, for example, in this incorrect sentence: "The facts in that complicated court case *is* open to question." The subject of the sentence is not the singular "case" but the plural "facts." Hence the sentence would correctly be written: "The facts in that complicated court case *are* open to question." New writers should not be misled by the distance between the subject and verb, especially when another noun with a different number intervenes as in these examples. The verb must agree with the subject, not the noun closest to it.

Pronoun-Antecedent Agreement
Pronouns within a sentence must refer specifically to one noun, known as the **antecedent.** Sometimes, if there are multiple nouns within a sentence, it may be difficult to ascertain which noun belongs to the pronoun. It's important that the pronouns always clearly reference the nouns in the sentence so as not to confuse the reader. Here's an example of an unclear pronoun reference:

> After Catherine cut Libby's hair, David bought her some lunch.

The pronoun in the examples above is *her.* The pronoun could either be referring to *Catherine* or *Libby.* Here are some ways to write the above sentence with a clear pronoun reference:

> After Catherine cut Libby's hair, David bought Libby some lunch.
> David bought Libby some lunch after Catherine cut Libby's hair.

But many times the pronoun will clearly refer to its antecedent, like the following:

> After David cut Catherine's hair, he bought her some lunch.

Eliminating Wordiness or Awkward Sentence Structure
Wordiness means using too many words. It creates awkward sentences that include unnecessarily complex or obscure language. Wordiness makes sentences harder to read and understand.

To eliminate wordiness, eliminate verbs used as nouns; unneeded filler words; qualifiers like *really, very,* and *totally*; words that mean the same thing and are therefore redundant; and abstract language.

Using Verbs as Nouns
Example: A thorough review of the situation is being conducted *for determination of* which party is responsible for *the theft* of the last Easter egg.
Solution: We will review the situation to determine who stole the last Easter egg.

Filler Words
Example: It is important to understand that wordiness does not make you sound smarter.
Solution: Wordiness does not make you sound smarter.

Qualifiers and Redundancy
Example: You sound *really intelligent and smart* when you eliminate wordiness from sentences.
Solution: You sound smart when you eliminate wordiness.

Double Negatives
Example: None of my classes don't assign homework every night.
Solution: Double negatives cause confusion. The above sentence has the following meaning: All of my classes assign homework every night. Therefore, this version that avoids the double negatives is clearer and preferable.

Eliminating Run-On Sentences and Sentence Fragments
Run-On Sentences

A **run-on sentence** combines two or more complete sentences without punctuating them correctly or separating them. For example, a run-on sentence caused by a lack of punctuation is the following:

> There is a malfunction in the computer system however there is nobody available right now who knows how to troubleshoot it.

One correction is, "There is a malfunction in the computer system; however, there is nobody available right now who knows how to troubleshoot it." Another is, "There is a malfunction in the computer system. However, there is nobody available right now who knows how to troubleshoot it."

An example of a comma splice of two sentences is the following:

> Jim decided not to take the bus, he walked home.

Replacing the comma with a period or a semicolon corrects this. Commas that try and separate two independent clauses without a contraction are considered comma splices.

Sentence Fragments

Sentence fragments are caused by absent subjects, absent verbs, or dangling/uncompleted dependent clauses. Every sentence must have a subject and a verb to be complete. An example of a fragment is the following:

> Raining all night long.

In the above example, there is no subject present. "It was raining all night long" is one correction. Another example of a sentence fragment is the second part in the following statement:

> Many scientists think in unusual ways. Einstein, for instance.

The second phrase is a fragment because it has no verb. One correction is "Many scientists, like Einstein, think in unusual ways." Finally, look for "cliffhanger" words like *if, when, because,* or *although* that introduce dependent clauses, which cannot stand alone without an independent clause. For example, to correct the sentence fragment "If you get home early," add an independent clause: "If you get home early, we can go dancing."

Transition Words

In connected writing, some sentences naturally lead to others, whereas in other cases, a new sentence expresses a new idea. Transitional phrases connect sentences and the ideas they convey, which makes the writing coherent. Transitional language also guides the reader from one thought to the next. For example, when pointing out an objection to the previous idea, starting a sentence with "However," "But," or "On the other hand" is transitional. When adding another idea or detail, writers use "Also," "In addition," "Furthermore," "Further," "Moreover," "Not only," etc. Readers have difficulty perceiving connections between ideas without such transitional wording.

Capitalization, Punctuation, and Apostrophes

Using Correct Capitalization

The first word of any document, and of each new sentence, is capitalized. Proper nouns, like names and adjectives derived from proper nouns, should also be capitalized. Here are some examples:

- Grand Canyon
- Pacific Palisades
- Golden Gate Bridge
- Freudian slip
- Shakespearian, Spenserian, or Petrarchan sonnet
- Irish song

Some exceptions are adjectives, originally derived from proper nouns, which through time and usage are no longer capitalized, like *quixotic, herculean*, or *draconian*. Capitals draw attention to specific instances of people, places, and things. Some categories that should be capitalized include the following:

- Brand names
- Companies
- Weekdays
- Months
- Governmental divisions or agencies
- Historical eras
- Major historical events
- Holidays
- Institutions
- Famous buildings
- Ships and other manmade constructions
- Natural and manmade landmarks
- Territories
- Nicknames
- Epithets
- Organizations
- Planets
- Nationalities
- Tribes
- Religions
- Names of religious deities
- Roads
- Special occasions, like the Cannes Film Festival or the Olympic Games

Exceptions

Related to American government, capitalize the noun Congress but not the related adjective congressional. Capitalize the noun U.S. Constitution, but not the related adjective constitutional. Many experts advise leaving the adjectives federal and state in lowercase, as in federal regulations or state water board, and only capitalizing these when they are parts of official titles or names, like Federal Communications Commission or State Water Resources Control Board. While the names of the other planets in the solar system are capitalized as names, Earth is more often capitalized only when being described specifically as a planet, like Earth's orbit, but lowercase otherwise since it is used not only as a proper noun but also to mean *land, ground, soil*, etc. While the name of the Bible should be capitalized, the adjective biblical should not. Regarding biblical subjects, the words heaven, hell, devil, and satanic should not be capitalized. Although race names like Caucasian, African-American, Navajo, Eskimo, East Indian, etc. are capitalized, white and black as races are not.

Names of animal species or breeds are not capitalized unless they include a proper noun. Then, only the proper noun is capitalized. Antelope, black bear, and yellow-bellied sapsucker are not capitalized. However, Bengal tiger, German shepherd, Australian shepherd, French poodle, and Russian blue cat are capitalized.

Other than planets, celestial bodies like the sun, moon, and stars are not capitalized. Medical conditions like tuberculosis or diabetes are lowercase; again, exceptions are proper nouns, like Epstein-Barr syndrome, Alzheimer's disease, and Down syndrome. Seasons and related terms like winter solstice or autumnal equinox are lowercase. Plants, including fruits and vegetables, like poinsettia, celery, or avocados, are not capitalized unless they include proper names, like Douglas fir, Jerusalem artichoke, Damson plums, or Golden Delicious apples.

Titles and Names
When official titles precede names, they should be capitalized, except when there is a comma between the title and name. But if a title follows or replaces a name, it should not be capitalized. For example, "the president" without a name is not capitalized, as in "The president addressed Congress." But with a name it is capitalized, like "President Obama addressed Congress." Or, "Chair of the Board Janet Yellen was appointed by President Obama." One exception is that some publishers and writers nevertheless capitalize President, Queen, Pope, etc., when these are not accompanied by names to show respect for these high offices. However, many writers in America object to this practice for violating democratic principles of equality. Occupations before full names are not capitalized, like owner Mark Cuban, director Martin Scorsese, or coach Roger McDowell.

Some universal rules for capitalization in composition titles include capitalizing the following:

- The first and last words of the title
- Forms of the verb *to be* and all other verbs
- Pronouns
- The word *not*

Universal rules for NOT capitalizing in titles include the articles *the, a,* or *an;* the conjunctions *and, or,* or *nor,* and the preposition *to,* or *to* as part of the infinitive form of a verb. The exception to all of these is UNLESS any of them is the first or last word in the title, in which case they are capitalized. Other words are subject to differences of opinion and differences among various stylebooks or methods. These include *as, but, if,* and *or,* which some capitalize and others do not. Some authorities say no preposition should ever be capitalized; some say prepositions five or more letters long should be capitalized. The *Associated Press Stylebook* advises capitalizing prepositions longer than three letters (like *about, across,* or *with*).

Using Apostrophes with Possessive Nouns Correctly
Possessive forms indicate possession, i.e. that something belongs to or is owned by someone or something. As such, the most common parts of speech to be used in possessive form are adjectives, nouns, and pronouns. The rule for correctly spelling/punctuating possessive nouns and proper nouns is with - *'s,* like "the woman's briefcase" or "Frank's hat." With possessive adjectives, however, apostrophes are not used: these include *my, your, his, her, its, our,* and *their,* like "my book," "your friend," "his car," "her house," "its contents," "our family," or "their property." Possessive pronouns include *mine, yours, his,*

hers, its, ours, and *theirs.* These also have no apostrophes. The difference is that possessive adjectives take direct objects, whereas possessive pronouns replace them. For example, instead of using two possessive adjectives in a row, as in "I forgot my book, so Blanca let me use her book," which reads monotonously, replacing the second one with a possessive pronoun reads better: "I forgot my book, so Blanca let me use hers."

Using Correct Punctuation

Ellipses

Ellipses (. . .) signal omitted text when quoting. Some writers also use them to show a thought trailing off, but this should not be overused outside of dialogue. An example of an ellipsis would be if someone is quoting a phrase out of a professional source but wants to omit part of the phrase that isn't needed: "Dr. Skim's analysis of pollen inside the body is clearly a myth . . . that speaks to the environmental guilt of our society."

Commas

Commas separate words or phrases in a series of three or more. The Oxford comma is the last comma in a series. Many people omit this last comma, but many times it causes confusion. Here is an example:

I love my sisters, the Queen of England and Madonna.

This example without the comma implies that the "Queen of England and Madonna" are the speaker's sisters. However, if the speaker was trying to say that they love their sisters, the Queen of England, as well as Madonna, there should be a comma after "Queen of England" to signify this.

Commas also separate two coordinate adjectives ("big, heavy dog") but not cumulative ones, which should be arranged in a particular order for them to make sense ("beautiful ancient ruins").

A comma ends the first of two independent clauses connected by conjunctions. Here is an example:

I ate a bowl of tomato soup, and I was hungry very shortly after.

Here are some brief rules for commas:

- Commas follow introductory words like *however, furthermore, well, why,* and *actually,* among others.

- Commas go between city and state: Houston, Texas.

- If using a comma between a surname and Jr. or Sr. or a degree like M.D., also follow the whole name with a comma: "Martin Luther King, Jr., wrote that."

- A comma follows a dependent clause beginning a sentence: "Although she was very small, . . ."

- Nonessential modifying words/phrases/clauses are enclosed by commas: "Wendy, who is Peter's sister, closed the window."

- Commas introduce or interrupt direct quotations: "She said, 'I hate him.' 'Why,' I asked, 'do you hate him?'"

Semicolons

Semicolons are used to connect two independent clauses but should never be used in the place of a comma. They can replace periods between two closely connected sentences: "Call back tomorrow; it can wait until then." When writing items in a series and one or more of them contains internal commas, separate them with semicolons, like the following:

> People came from Springfield, Illinois; Alamo, Tennessee; Moscow, Idaho; and other locations.

Hyphens

Here are some rules concerning hyphens:

- Compound adjectives like state-of-the-art or off-campus are hyphenated.

- Original compound verbs and nouns are often hyphenated, like "throne-sat," "video-gamed," "no-meater."

- Adjectives ending in *–ly* are often hyphenated, like "family-owned" or "friendly-looking."

- "Five years old" is not hyphenated, but singular ages like "five-year-old" are.

- Hyphens can clarify. For example, in "stolen vehicle report," "stolen-vehicle report" clarifies that "stolen" modifies "vehicle," not "report."

- Compound numbers twenty-one through ninety-nine are spelled with hyphens.

- Prefixes before proper nouns/adjectives are hyphenated, like "mid-September" and "trans-Pacific."

Parentheses

Parentheses enclose information such as an aside or more clarifying information: "She ultimately replied (after deliberating for an hour) that she was undecided." They are also used to insert short, in-text definitions or acronyms: "His FBS (fasting blood sugar) was higher than normal." When parenthetical information ends the sentence, the period follows the parentheses: "We received new funds ($25,000)." Only put periods within parentheses if the whole sentence is inside them: "Look at this. (You'll be astonished.)" However, this can also be acceptable as a clause: "Look at this (you'll be astonished)." Although parentheses appear to be part of the sentence subject, they are not, and do not change subject-verb agreement: "Will (and his dog) was there."

Quotation Marks

Quotation marks are typically used when someone is quoting a direct word or phrase someone else writes or says. Additionally, quotation marks should be used for the titles of poems, short stories, songs, articles, chapters, and other shorter works. When quotations include punctuation, periods and commas should *always* be placed inside of the quotation marks.

When a quotation contains another quotation inside of it, the outer quotation should be enclosed in double quotation marks and the inner quotation should be enclosed in single quotation marks. For example: "Timmy was begging, 'Don't go! Don't leave!'" When using both double and single quotation marks, writers will find that many word-processing programs may automatically insert enough space between the single and double quotation marks to be visible for clearer reading. But if this is not the case, the writer should write/type them with enough space between to keep them from looking like three single quotation marks. Additionally, non-standard usages, terms used in an unusual fashion, and technical terms are often clarified by quotation marks. Here are some examples:

> My "friend," Dr. Sims, has been micromanaging me again.
> This way of extracting oil has been dubbed "fracking."

Apostrophes

One use of the apostrophe is followed by an *s* to indicate possession, like *Mrs. White's home* or *our neighbor's dog*. When using the *'s* after names or nouns that also end in the letter *s*, no single rule applies: some experts advise adding both the apostrophe and the *s*, like "the Jones's house," while others prefer using only the apostrophe and omitting the additional *s*, like "the Jones' house." The wisest expert advice is to pick one formula or the other and then apply it consistently. Newspapers and magazines often use *'s* after common nouns ending with *s*, but add only the apostrophe after proper nouns or names ending with *s*. One common error is to place the apostrophe before a name's final *s* instead of after it: "Ms. Hasting's book" is incorrect if the name is Ms. Hastings.

Plural nouns should not include apostrophes (e.g. "apostrophe's"). Exceptions are to clarify atypical plurals, like verbs used as nouns: "These are the do's and don'ts." Irregular plurals that do not end in *s* always take apostrophe-*s*, not *s*-apostrophe—a common error, as in "childrens' toys," which should be "children's toys." Compound nouns like mother-in-law, when they are singular and possessive, are followed by apostrophe-*s*, like "your mother-in-law's coat." When a compound noun is plural and possessive, the plural is formed before the apostrophe-*s*, like "your sisters-in-laws' coats." When two people named possess the same thing, use apostrophe-*s* after the second name only, like "Dennis and Pam's house."

Science

Reading for Meaning in Science

Claims and Evidence in Science

Finding Evidence that Supports a Finding

Science passages can contain a lot of information about a specific subject. Generally, there is one main idea and then supporting details. The main idea is the message that the whole passage is conveying, often stated in a complete sentence. The topic of the passage is described in just one word or a short phrase but also conveys a message about the passage. The supporting details provide facts and evidence to support the claim of the main idea. Consider the following passage about earthquakes:

Earthquakes can be described as either a surface earthquake or a deep-focus earthquake. Surface earthquakes occur when the Earth's crust cracks to relieve stress, a fact agreed upon by most scientists. They occur at depths less than 70 km. The origin of deep-focus earthquakes is debated among scientists. However, most scientists believe they are caused by Earth's tectonic plates sliding toward each other to release pressure from fluids inside the tectonic plates. They occur much deeper than surface earthquakes, at depths of 300 to 700 km below the Earth's surface.

The main idea of this passage is about earthquakes and how they are classified into two categories. The supporting details tell us how the earthquakes are classified into each category. The surface earthquakes occur within 70 km of the Earth's surface and happen as a result of a fracture in the Earth's crust. Deep-focus earthquakes occur between 300 and 700 km below the Earth's surface and happen as a result of Earth's tectonic plates sliding toward each other.

Making Sense of Information that differs Between Various Science Sources

Various scientific sources can present information in different ways. It is important to be able to draw conclusions using information from different sources, such as journal articles written by different scientists on the same topic, to better understand the topic being discussed. For example, when discussing the breeding of flowering plants, one scientist may examine color and the other may investigate genotype. Both sets of research will reveal something about future generations of the plants but they use different methods to look at different aspects of genetics.

Looking at this example in more detail, Scientist A decides to crossbreed one white flowering plant with one red flowering plant. The result is one red flowering plant. She repeats this crossbreeding experiment three more times and gets three more red flowering plants. Scientist B analyzes the genotypes of the parent generation red and white flowering plants and finds that both have homozygous genotypes for the color of their flowers. Putting this information together, it is clear that the red flowers have a dominant phenotype based on the results of the crossbreeding experiment. However, the plants that resulted from the crossbreeding will all have a heterozygous genotype.

Science Vocabulary, Terms, and Phrases

Understanding and Explaining Information from Passages

Scientific information can be presented in many different ways. It can be described using words in a passage, by summarizing or paraphrasing the area of research. Summaries involve putting the main idea of a passage into your own words. Paraphrasing involves condensing the original passage without changing the wording too much. The paraphrased version must be attributed to the original source because the words are not newly generated. It can also be described using graphs, charts, and tables. Chemical reactions can be described using equations and formulas. Putting together information from all of these different formats can create a better understanding of the scientific information being portrayed.

For example, the methods of a scientific experiment can be described in a written passage: Scientist A wants to find out if using a fertilizer containing nitrogen improves plant growth, as nitrogen can help plants produce chlorophyll and other proteins. She plants seeds of the same plant into six pots with soil. She adds fertilizer containing nitrogen to three of the plants and fertilizer without nitrogen to the other three plants. She waters each plant daily and records how tall the plants grow over the course of one month. The results of this experiment can be described in a table with exact numbers:

	Inches	
Day	Average growth for plants getting fertilizer with nitrogen	Average growth for plants getting fertilizer without nitrogen
0	0	0
7	1	0
14	2	0.5
21	3	0.75
28	5	1

The data can also be portrayed using a graph for a more visual representation of the difference in plant growth. The conclusion that the plant receiving fertilizer with nitrogen grew much faster can be drawn immediately from looking at this graph:

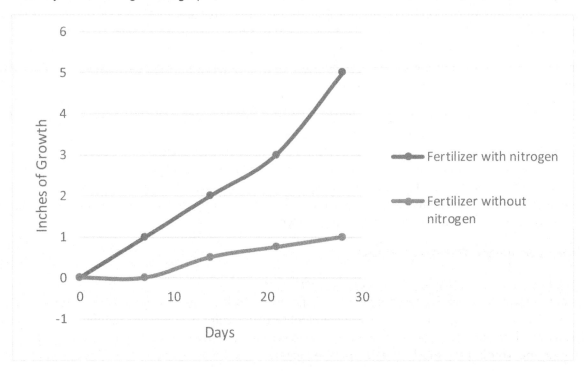

Understanding Symbols, Terms, and Phrases in Science

There are many different symbols, terms, and phrases used when presenting scientific information. When doing a scientific experiment, it is important to first develop a hypothesis about the phenomenon you are observing and the question you are asking. A **hypothesis** is an educated guess about the answer to the question. It is always a declarative statement about the phenomenon, such as "An area with full sunlight will cause the plant to grow faster than an area without sunlight." The experiment is the methods that are used to test whether or not the hypothesis is true. The phenomenon being observed is the dependent variable in the experiment. The outside factor that may be affecting the phenomenon is the independent variable. Data include all of the measurements that are taken during the experimental process. The data collected should be relevant to the phenomenon being observed and the question being asked. Once the data are collected and analyzed, a conclusion can be drawn. The conclusion will either support or disprove the hypothesis.

Chemical reactions are used to describe many different phenomena. They have special symbols in them and are written as follows:

$$C + O_2 \rightarrow CO_2$$

This reaction represents what happens when carbon is burned, which is the addition of oxygen to carbon. The "+" symbol separates two different molecules, either as reactants or products. The "$\rightarrow$" is a yield symbol that shows the direction of the reaction and separates the reactants from the products. The reactants are written on the left side of the yield symbol and the products are written on the right side of the yield symbol.

Different areas of science also have specific, specialized terms that are used for that subspecialty. It is useful to use context clues within a passage or through the graphic elements to decipher the meaning of these terms. The three main branches of science are physical science, Earth science, and life science. The tables below contain some of the major terms used in each branch.

Physical Science Terms	
Term	**Definition**
Aqueous	A solution made with water
Condensation	The changing of a substance from a gas to a liquid
Element	A substance in which all of the atoms in the sample are alike
Evaporation	The changing of a substance from a liquid to a gas
Half Life	The amount of time it takes for one half of a radioactive isotope to decay
Polymer	A large molecule made up of many smaller molecules linked together
Sublimation	The changing of a substance from a solid to a gas without forming a liquid

Earth Science Terms	
Term	**Definition**
Ablation	The loss of ice or snow from a glacier due to melting, evaporation, or erosion
Delta	A flat, low landform near the mouth of a river
Earthquake	The shaking of the ground caused by a sudden movement of the Earth's crust
Erosion	The process of land being worn down by water or wind
Fossil	The preserved remains or traces of a living organism from the past
Mantle	The layer of earth between the outer crust and the core of the Earth
Topography	A description of the physical features of land

Life Science Terms	
Term	**Definition**
Abiotic	The nonliving features of the environment, including air, water, and soil
Antibody	A protein made to attack an antigen
Antigen	A foreign substance in the body
Cell	The smallest unit of a living thing that can perform the functions of life
DNA	The genetic material of all organisms
Homeostasis	The regulation of an organism's internal environment
Nucleus	The organelle of a cell that controls the cell's activities and contains its genetic information
Tissue	A group of similar cells that work together to perform one job

Using Scientific Words to Express Science Information
It is important to use scientific words to express scientific information so that the information is portrayed as technical, accurate, and useful, as opposed to speculation. Scientific law is different than scientific theory. Scientific law is a phenomenon that has been observed repeatedly. Scientific theory is the explanation or reason behind an observed phenomenon. Hypotheses are used to prove whether or not a theory is true. When collecting and discussing data, measurements should be exact and include units so that comparisons are easy to make. A portion of a whole population or other thing being studied is a sample. Samples are representative of the whole population. Trends are patterns or tendencies that are

observed within the data. Predictions are claims or conclusions that can be made based on the trends that are seen. They are a projection of whether or not the hypothesis will be proven to be true. Statistics is a subspecialty of mathematics that helps organize and analyze data to see if differences truly exist between the items that are being compared in the experiment.

Designing and Interpreting Science Experiments

Science Investigations

Designing a Science Investigation
How to design a scientific investigation:

1. Make an observation about a phenomenon.

2. Perform a literature review. This involves researching studies that have already been done and finding out what their results were. This can help form the hypothesis for your scientific investigation.

3. State a clear hypothesis, which is a declarative sentence about what you think the results will show at the end of the experiment.

4. Set up the experimental design. This should include a detailed description of how the data will be collected. There should be a control group, which does not include the variable being tested, in addition to the experimental groups, which have only one variable being changed between them. The design should also include a description about the tools that are being used to make the measurements required. The tools should be capable of making quantitative measurements so that objective comparisons can be made between the groups. Qualitative data does not involve numerical data and can be subjective according to the person making the observations. The objects involved in the experiment should be randomized between all of the groups to remove bias and ensure equal probability of each object receiving a particular treatment. The design should also include a repetition of collecting the measurements to improve accuracy of the treatment results for the groups.

5. After the experiment is completed, the data should be interpreted, which may include the use of statistics to determine the relevance of differences seen between the experimental groups.

6. The investigation should end with a finite conclusion that either supports or disproves the hypothesis.

Identifying and Explaining Independent and Dependent Variables
Independent and dependent are two types of variables that describe how they relate to each other. The **independent variable** is the variable controlled by the experimenter. It stands alone and isn't changed by other parts of the experiment. This variable is normally represented by x and is found on the horizontal, or x-axis, of a graph. The **dependent variable** changes in response to the independent variable. It reacts to, or depends on, the independent variable. This variable is normally represented by y and is found on the vertical, or y-axis of the graph.

The relationship between two variables, x and y, can be seen on a scatterplot.

The following scatterplot shows the relationship between weight and height. The graph shows the weight as x and the height as y. The first dot on the left represents a person who is 45 kg and approximately 150 cm tall. The other dots correspond in the same way. As the dots move to the right and weight increases, height also increases. A line could be drawn through the middle of the dots to move from bottom left to top right. This line would indicate a **positive correlation** between the variables. If the variables had a **negative correlation**, then the dots would move from the top left to the bottom right.

Height and Weight

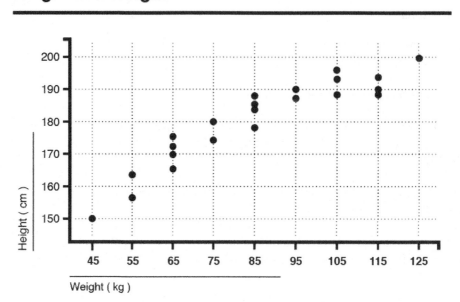

A scatterplot is useful in determining the relationship between two variables, but it's not required. Consider an example where a student scores a different grade on his math test for each week of the month. The independent variable would be the weeks of the month. The dependent variable would be the grades because they change depending on the week. If the grades trended up as the weeks passed, then the relationship between grades and time would be positive. If the grades decreased as the time passed, then the relationship would be negative. (As the number of weeks went up, the grades went down.)

The relationship between two variables can further be described as strong or weak. The relationship between age and height shows a strong positive correlation because children grow taller as they grow up. In adulthood, the relationship between age and height becomes weak, and the dots will spread out. People stop growing in adulthood, and their final heights vary depending on factors like genetics and health. The closer the dots on the graph, the stronger the relationship. As they spread apart, the relationship becomes weaker. If they are too spread out to determine a correlation up or down, then the variables are said to have no correlation.

Variables are values that change, so determining the relationship between them requires an evaluation of who changes them. If the variable changes because of a result in the experiment, then it's dependent. If the variable changes before the experiment, or is changed by the person controlling the experiment, then it's the independent variable. As they interact, one is manipulated by the other. The manipulator is the independent, and the manipulated is the dependent. Once the independent and dependent variable are determined, they can be evaluated to have a positive, negative, or no correlation.

Identifying and Improving Hypotheses for Science Investigations

Sometimes, the original hypothesis of a scientific investigation is weak and needs to be improved. There are several factors to consider when making a hypothesis. For a hypothesis to be strong, the phenomenon being observed and question being asked need to be clear. The language of the hypothesis should be simple and concisely convey what you thought was going to happen when the experiment began. The variables of the experiment should be included, as well as what change to the variables is expected. Finally, the hypothesis needs to be testable. It should be approachable through an experiment where a conclusion can be drawn in defense of the hypothesis or in negation of it.

A weak hypothesis may read as, "A train carrying more passengers might travel slower." This hypothesis is not declarative and does not give enough detail about what is being tested. It does not use scientific terms, either. This hypothesis can be improved by including details and stating it as: "When using the same amount of electricity, a train carrying the maximum weight of passengers will travel at a slower velocity than a train carrying less than the maximum weight of passengers." More detail is included and several variables are mentioned in the hypothesis for a clear comparison. It is also a declarative statement.

Examining Hypotheses

A hypothesis is a well-defined research statement. An experiment then follows, usually using quantitative research. Quantitative research is research based on empirical data.

The results are then analyzed to determine whether the hypothesis was proven or disproven. Examining a hypothesis is also called hypothesis testing. Examining a hypothesis happens most often in science, and it isn't really appropriate for social sciences such as social studies and history. However, qualitative hypotheses can be made in these disciplines to further examine a social or historical event. A hypothesis, in this light, should clearly state the argument that the writer wishes to examine, and the reason or reasons why the author feels it is relevant. This type of hypothesis statement generally requires the "what" and the "why." Consider the following qualitative hypothesis:

> "The Métis in Canada were less discriminated against than were Canada's First Nations since they were partly descendants of European fur traders."

The first half of the hypothesis—"The Métis in Canada were less discriminated against than were Canada's First Nations"—reveals the "what," and the second part—"since they were partly descendants of European fur traders"—is the "why."

In science, hypotheses are generally written as "if, then" statements that require the collection of unbiased, empirical, and quantitative data to either prove or disprove the hypothesis. Consider the following:

> "**If** a hibiscus flower is placed in direct sunlight and watered twice a day, **then** it will thrive."

The basic steps that lead to the formation of a hypothesis and ultimately, a conclusion, include:

Step 1	Making an observation
Step 2	Forming a question based on the observation
Step 3	Forming a hypothesis (a possible answer to the question)
Step 4	Conducting a study (social studies and history) or an experiment (science)
Step 5	Analyzing the data
Step 6	Drawing a conclusion

In order for conclusions to be accepted as valid and credible, it is extremely important that the data collected isn't biased. The researchers must consider all possible angles of the study or experiment, and they must refrain from collecting the data in such a way as to purposely prove the hypothesis. Conducting studies and experiments of this nature helps to advance the different disciplines, challenge widely accepted beliefs, and broaden a global understanding of the fields of social studies, history, and the sciences.

Identifying Possible Errors in a Science Investigation and Changing the Design to Correct Them
Scientific investigations are subject to either random or systematic errors. **Random errors** are the differences in data measurements created by the precision limitations of the instrument being used to take the measurements. It is nearly impossible for the investigator to take the same measurement in the exact same way repeatedly throughout an experiment. There will be random, minute fluctuations with every measurement.

For example, a cell biologist measuring the volume of media in a petri dish uses a pipet to make her measurement. She takes three measurements, and they range between 4.95 ml and 5.05 ml. Similarly, a chemist takes the weight of a sodium chloride sample three times using the same scale and finds the weights to be 14.2 g, 14.1 g, and 14.4 g. This type of error can be minimized or eliminated by taking more data and increasing the number of observations. Then, a statistical evaluation can determine whether or not the differences in the measurements are meaningful or simply random error.

Systematic errors are fluctuations in the measurements in only one direction and are reproducible. They are caused by a flaw in the device used for measurement or by an error made consistently by the investigator. These types of errors cannot be determined by statistical analysis. Instead, the experimental design and procedure must be closely examined to determine where the error is being made. For example, the temperature of a chemical solution is measured by an electronic thermometer and the measurements are consistently low because the thermometer has not been calibrated. Another example is that the length of a fossil is consistently off by an inch in all measurements because the investigator is taking the measurement starting at the one-inch mark instead of the zero mark.

Identifying the Strengths and Weaknesses of Different Types of Science Investigations
There are several factors that determine how strong or weak a scientific investigation is. Participants or items must be assigned to a group randomly. There should not be a bias in determining which participants or items are placed in which group. Control groups must be included to account for the many variables that are not being tested by the experiment. The control group should be treated identically to the experimental group but without receiving the experimental treatment. Only one variable should be tested in the experiment, while all other variables are consistent between all of the groups. The results that are collected must be quantifiable in order to be comparable between the groups.

Three types of scientific investigations are descriptive, comparative, and experimental. A **descriptive investigation** includes an experimental design, data collection and analysis, and a conclusion. A **comparative investigation** compares two or more things against each other. It does not include a control group to account for more than the variable being tested. It includes a hypothesis, an experimental design that includes independent and dependent variables, data collection and analysis, and a conclusion. An **experimental investigation** includes a hypothesis, an experimental design that includes independent and dependent variables, control and experimental groups, data collection and analysis, and a conclusion, and is the most thorough and strongest type of investigation.

Using Evidence to Draw Conclusions or Make Predictions

Deciding Whether Conclusions are Supported by Data

Once the results of an investigation are analyzed, a conclusion can be drawn based on the evidence collected. The conclusion is limited in scope to the items or participants tested in the investigation. If an unbiased, representative sample was tested, the conclusion can be broadened to the whole population but not beyond that. To determine whether a conclusion is supported by the data, logic and reasoning skills have to be used to analyze the data and evidence collected in the investigation.

In an investigation determining whether pineapple trees can survive in Florida, the conclusion was made that pineapple trees cannot survive in Florida. Data were collected by recording the amount of rainfall in Florida for three months. Pineapple trees need wet climates to survive and produce pineapples, with an average requirement of five inches of water each month. The data shows that there was one inch of water in May, two inches of water in June, and one inch of water in July. Because pineapple trees need five inches of water per month, the conclusion of the investigation is supported by the data collected, which detail the amount of water that is actually available for pineapple trees in Florida.

Making Conclusions Based on Data

Data can either be qualitative (observations, interviews, or focus groups) or quantitative (measured data, as in the population of a certain country, a person's height, or the depth of the Earth's oceans). In order for students to interpret and analyze the data, they first must understand the information that has been collected. Collaboration with other students and with professors helps students to further comprehend the collected data. Once fully understood and analyzed, the data can now be interpreted. For instance, how has the analysis of this data affected the students' initial assumptions, thoughts, and beliefs? Data interpretation helps to make a student's knowledge more meaningful. Encouraging metacognitive awareness by asking students to think about how this analysis affects the learning process will also help to strengthen a student's overall understanding and make the learning process more meaningful.

Analysis of data must precede interpretation. Data analysis can take many forms. For instance, students may begin to identify various relationships within the data. Certain patterns or trends may come to light that help students better grasp the meaning or the relevance of the results. The opposite could also be true; students may not identify any patterns, trends, or relationships that would lead to further discussion and evaluation. After fully analyzing the data, students can then begin to interpret this information, which places them in a better position to make informed decisions. Interpreting the data means acquiring a greater understanding of the results of the data. For example, the analysis of collected data on temperature patterns throughout the year and around the globe might reveal that there is a pattern of increasingly hotter temperatures. The interpretation of this data may then result in an individual's greater understanding of global warming.

When students learn to interpret qualitative data in a social studies or historical context, or quantitative data collected in a scientific experiment, their knowledge base expands. The process of data collection, analysis, and interpretation helps to develop an appreciation for how knowledge is attained and why it is critically important to challenge all hypotheses in an attempt to continue to further our collective understanding of the world.

Making Predictions Based on Data

Predictions can have a greater scope than the conclusions of the experiment. They can use the data collected along with limited conclusions that were made to apply a conclusion to a larger population or similar populations that were not tested in the original investigation. Predictions can include a hypothesis that predicts the outcome if the same experiment was completed on a different set of items or a different population.

Let's say an investigation is done to look at hair growth with a certain vitamin in yellow Labradors. It was concluded that the vitamin does not help with hair growth. The scientist could then predict how the same vitamin would affect chocolate Labradors or other breeds of similar dogs. Similarly, thinking about the pineapple trees again, if they do not grow well in Florida because of their need for an abundance of rainfall every month, the scientist could predict how other tropical trees would grow in Florida. Knowing that banana plants require a similar amount of rainfall, it could be predicted that they would not grow well in Florida. However, orange trees only require one inch of water a month, so it could be predicted that they would grow well in Florida.

Science Theories and Processes

Science theories are explanations about different aspects of the natural world. They are based on observations and experiments that have been done to obtain evidence about a certain phenomenon, so they are not random guesses. *Scientific processes* are used to investigate the theories and determine whether the theories hold true or false. The scientific process starts with asking a question, formulating a hypothesis about the question, establishing a method for collecting data, and then collecting and analyzing the data. All of these parts put together help to answer the original question and test the scientific theory.

There have been many scientific theories developed in history. In the 1770s, Antoine Lavoisier proposed the oxygen theory of combustion and figured out that oxygen was the element that was combining with other substances to make them burn, which opposed the theory that every combustible substance contained phlogiston (a fire-like element) within it. Einstein developed the theory of special relativity in 1905 and found that space and time are interwoven into one continuum, known as space-time. Thus, two events could happen at the same time for one observer but at different times for another observer.

A more tangible example is that a group of doctors wants to determine whether or not there are more complications with women who deliver their babies at home compared with women who deliver their babies in the hospital. They believe that there are more complications with delivery at home, which would be the scientific theory. The doctors may have formulated this theory based on the patients they have seen and information they received from other patients and doctors, but they have not investigated it themselves, and so do not have any of their own evidence or data. Once the theory is developed, they need to follow an appropriate scientific process to answer their question.

They choose two hundred patients at random to enter their investigation after making sure the patients all have similar healthy pregnancies. Half of the patients have chosen to deliver their babies at home and half have chosen to deliver at the hospital. They count the number of complications that occur with the deliveries both at home and in the hospital. Comparing these numbers, the doctors can see which number is higher and where more complications happen. Ten of the women who delivered at home had complications, whereas five of the women who delivered at the hospital had complications. The doctors conclude that there is a greater chance of complications with deliveries that occur at home, which answers their question and proves their scientific theory to be true.

Using Numbers and Graphics in Science

Science Formulas and Statistics

Applying Science Formulas

There are many standard formulas used to make calculations on scientific data. They can be used on many different sets and types of data to determine the same outcome measure. It is important to critically read the passage to figure out what information is given and what information is missing and being asked for. This will allow you to figure out what formula needs to be applied. For example, converting a temperature measurement from Celsius to Fahrenheit always uses the formula:

$$°F = °C \times (9/5) + 32$$

A mother is trying to decide whether or not to send her son to school with a jacket. If the weather is below 65 °F, she will send him with a jacket. If it is higher than 65 °F, he does not need to wear a jacket. She checks the temperature outside and it says 20 °C. Plugging this in to the equation, she finds that it is 68 °F. The same equation can be used to determine temperature in Fahrenheit in many other situations as well, such as determining whether certain chemicals change the freezing point of water, if the weather outside is cold enough to produce snow, or if the oven is hot enough to bake a cake.

Using Statistics to Describe Science Data

Statistics involves the analysis of quantitative data to determine something about a whole population using a representative sample. It not only includes calculations of the numbers collected, but also considers how the numbers were collected and other important factors about the experimental design. There are many different statistical tests that can be used depending on how the investigation was set up and the data were collected. Statistical tests help to interpret scientific data, which helps determine whether or not the original question was answered and if the hypothesis holds true or not.

The term "significant" is often used to explain the results of a statistical test. If the test shows that the differences seen between the experimental groups are consistent enough to also apply to the larger population, the difference is described as significant. Sometimes differences are seen between experimental groups but they are too small or not consistent enough to hold true through the statistical test. In this case, the investigator would not be able to say that the differences seen in the experimental sample can also be seen in the population as a whole.

Probability and Sampling in Science

Determining the Probability or Likelihood of Something Happening

Probability is the chance, or likelihood, of an event occurring. With enough information, the probability of any event can be calculated. The likelihood of one single event happening is called **simple probability** and can be calculated as follows:

Simple probability = (# of favorable outcomes) / (# of total possible outcomes)

Favorable outcomes are the events that you want to determine the probability of. Probability can be expressed as a decimal, fraction, or using the term "___ out of ___." For example, let's say you want to determine the probability of rolling a number 4 on a die. There is only one number 4 out of six total numbers. You can divide one (the number of 4s on the die) by six (the total number of numbers on the die) and find that the probability of rolling a number 4 on a die is 0.17 or 1/6 or 1 out of 6. **Compound probability** is more complex than simple probability and would allow you to calculate the probability of rolling a number 4 on a die if you rolled more than one time. It determines the likelihood of multiple events happening and is calculated by multiplying the probabilities of each event happening by each other.

Compound probability = (Probability of event #1) x (Probability of event #2) x and so on...

Using a Sample to Answer Science Questions

Often when trying to answer a scientific question, it is not possible to collect data from a whole population or all items in question. Scientific investigations use a sample of the population to collect evidence and data and then draw a conclusion that can be applied to the whole population. The sample size must be large enough to give a true representation of the population and is based on the acceptable margin of error and the confidence level, both of which are statistical calculations. Stratified sampling can also help create a representative sample population. The whole population is divided into several homogeneous subpopulations and then a portion of each subpopulation is chosen to be represented in the sample population.

Here is an example of the use of sampling to determine a characteristic about the whole population. Scientist A wants to determine how many high school students in Rochester, NY carry backpacks that weigh more than ten pounds. It would take a lot of time to weigh the backpacks of the thousands of high school students in the city and would also be a lot of data to analyze. So, Scientist A decides to take a sample of the high school student population and weigh only their backpacks to answer her question. She randomly chooses 250 students from high schools throughout the city to participate in her study and weighs their backpacks. She finds that two hundred of her sample students have backpacks that weigh more than ten pounds. Applying her conclusion to the whole population, she can say that eighty percent of the high school students in Rochester, NY carry backpacks that weigh more than ten pounds.

Using Counting to Solve Science Problems

Many scientific investigations involve the quantification of a sample. In order to quantify the portion of a sample that has a certain feature, that portion must be counted. Comparing different portions of the sample with each other can reveal different facets of the population or sample being studied. The objects of a study can be arranged according to shape, size, color, or any other feature, and the feature chosen will affect the proportion of the sample that is counted. **Permutations** allow for the calculation of the number of different ways a certain number of objects can be arranged. For example, if there are four seats a table, the number of permutations tells us how many different seating arrangements are possible to fill the seats with four different people. Where n represents the number of positions to fill, there would be n choices of people for the first seat, $(n-1)$ choices for the second seat, $(n-2)$ choices for the third seat, and 1 choice for the fourth seat. Permutations are calculated by multiplying these together, giving us the following formula:

$$_nP_n = n \times (n-1) \times (n-2) \times (n-3) \times 1$$

Plugging in $n = 4$ for this case, there are 24 different possible seating arrangements.

Presenting Science Information Using Numbers, Symbols, and Graphics

Using Graphics to Display Science Information

Graphics are a useful tool for presenting data in a visual, rather than a descriptive, manner. Tables, charts, and graphs are some of the ways that data can be presented through graphics. These tools often make it easier to make comparisons between different experimental groups and to visualize when large changes are happening during the experimental process.

For a table, the experimental groups are listed in the header or first row, and the independent variable values are listed in the first column. The rest of the table is filled in with data collected about the dependent variable. Here is an example of a table with data about bacterial growth in the presence of different antibiotics:

	Number of Bacteria (thousands)		
	Day 0	Day 7	Day 14
Control	0	10	40
Antibiotic A	0	2	4
Antibiotic B	0	5	15

A bar graph can be used to display the same data:

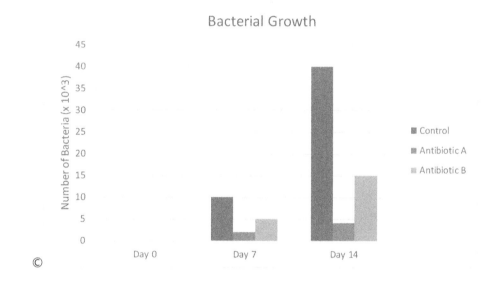

A line graph can also be used to show the same data:

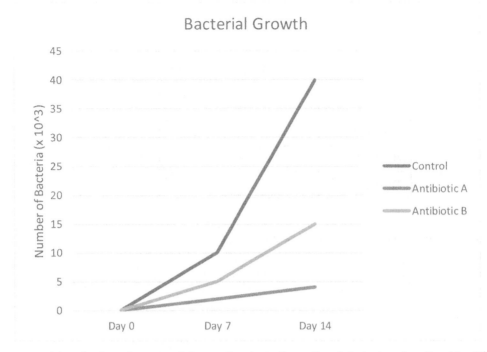

Bacterial Growth

Each of these graphics displays the same information but allows the data to be visualized in different ways. Pie charts are a useful tool for representing different portions of a population:

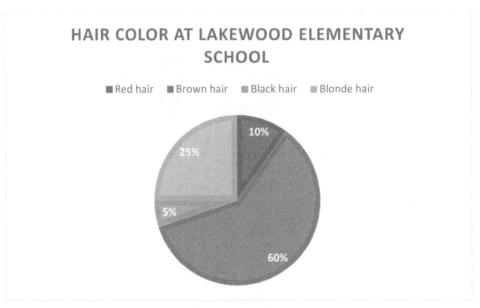

HAIR COLOR AT LAKEWOOD ELEMENTARY SCHOOL

■ Red hair   ■ Brown hair   ■ Black hair   ■ Blonde hair

Using Numbers or Symbols to Display Science Information
Scientific Notation
Scientific notation is a system used to represent numbers that are very large or very small. Sometimes, numbers are way too big or small to be written out with multiple zeros behind them or in decimal form, so scientific notation is used as a way to express these numbers in a simpler way.

140

Scientific notation takes the decimal notation and turns it into scientific notation, like the table below:

Decimal Notation	Scientific Notation
5	5×10^0
500	5×10^2
10,000,000	1×10^7
8,000,000,000	8×10^9
-55,000	-5.5×10^4
.00001	10^{-5}

In scientific notation, the decimal is placed after the first digit and all the remaining numbers are dropped. For example, 5 becomes "5.0×10^0." This equation is raised to the zero power because there are no zeros behind the number "5." Always put the decimal after the first number. Let's say we have the number 125,000. We would write this using scientific notation as follows: 1.25×10^5, because to move the decimal from behind "1" to behind "125,000" takes five counts, so we put the exponent "5" behind the "10." As you can see in the table above, the number ".00001" is too cumbersome to be written out each time for an equation, so we would want to say that it is "10^{-5}." If we count from the place behind the decimal point to the number "1," we see that we go backwards 5 places. Thus, the "-5" in the scientific notation form represents 5 places to the right of the decimal.

Converting Within and Between Standard and Metric Systems
Recall that the metric system has base units of meter for length, kilogram for mass, and liter for liquid volume. This system expands to three places above the base unit and three places below. These places correspond with prefixes with a base of 10. The following table shows the conversions:

kilo-	hecto-	deka-	base	deci-	centi-	milli-
1,000 times the base	100 times the base	10 times the base		1/10 times the base	1/100 times the base	1/1000 times the base

To convert between units within the metric system, values with a base ten can be multiplied. The decimal can also be moved in the direction of the new unit by the same number of zeros on the number. For example, 3 meters is equivalent to 0.003 kilometers. The decimal moved three places (the same number of zeros for kilo-) to the left (the same direction from base to kilo-). Three meters is also equivalent to 3,000 millimeters. The decimal is moved three places to the right because the prefix milli- is three places to the right of the base unit.

The English Standard system used in the United States has a base unit of foot for length, pound for weight, and gallon for liquid volume. These conversions aren't as easy as the metric system because they aren't a base ten model. The following table shows the conversions within this system:

Length	Weight	Capacity
1 foot (ft) = 12 inches (in) 1 yard (yd) = 3 feet 1 mile (mi) = 5280 feet 1 mile = 1760 yards	1 pound (lb) = 16 ounces (oz) 1 ton = 2000 pounds	1 tablespoon (tbsp) = 3 teaspoons (tsp) 1 cup (c) = 16 tablespoons 1 cup = 8 fluid ounces (oz) 1 pint (pt) = 2 cups 1 quart (qt) = 2 pints 1 gallon (gal) = 4 quarts

When converting within the English Standard system, most calculations include a conversion to the base unit and then another to the desired unit. For example, take the following problem: 3 qt = ___ c. There is no straight conversion from quarts to cups, so the first conversion is from quarts to pints. There are 2 pints in 1 quart, so there are 6 pints in 3 quarts. This conversion can be solved as a proportion:

$$\frac{3 \text{ qt}}{x} = \frac{1 \text{ qt}}{2 \text{ pt}}$$

It can also be observed as a ratio 2:1, expanded to 6:3. Then the 6 pints must be converted to cups. The ratio of pints to cups is 1:2, so the expanded ratio is 6:12. For 6 pints, the measurement is 12 cups. This problem can also be set up as one set of fractions to cancel out units. It begins with the given information and cancels out matching units on top and bottom to yield the answer. Consider the following expression:

$$\frac{3 \text{ qt}}{1} \times \frac{2 \text{ pt}}{1 \text{ qt}} \times \frac{2 \text{ c}}{1 \text{ pt}}$$

It's set up so that units on the top and bottom cancel each other out:

$$\frac{3 \ \cancel{\text{qt}}}{1} \times \frac{2 \ \cancel{\text{pt}}}{1 \ \cancel{\text{qt}}} \times \frac{2 \text{ c}}{1 \ \cancel{\text{pt}}}$$

The numbers can be calculated as 3 × 2 × 2 on the top and 1 on the bottom. It still yields an answer of 12 cups.

This process of setting up fractions and canceling out matching units can be used to convert between standard and metric systems. A few common equivalent conversions are 2.54 cm = 1 in, 3.28 ft = 1 m, and 2.205 lb = 1 kg. Writing these as fractions allows them to be used in conversions. For the fill-in-the-blank problem 5 m = ___ ft, an expression using conversions starts with the expression $\frac{5 \text{ m}}{1} \times \frac{3.28 \text{ ft}}{1 \text{ m}}$, where the units of meters will cancel each other out and the final unit is feet. Calculating the numbers yields 16.4 feet. This problem only required two fractions. Others may require longer expressions, but the underlying rule stays the same. When there's a unit on the top of the fraction that's the same as the unit on the bottom, then they cancel each other out. Using this logic and the conversions given above, many units can be converted between and within the different systems.

Temperature Scales
Science utilizes three primary temperature scales. The temperature scale most often used in the United States is the Fahrenheit (F) scale. The Fahrenheit scale uses key markers based on the measurements of

the freezing (32 °F) and boiling (212 °F) points of water. In the United States, when taking a person's temperature with a thermometer, the Fahrenheit scale is used to represent this information. The human body registers an average temperature of 98.6 °F.

Another temperature scale commonly used in science is the Celsius (C) scale (also called *centigrade* because the overall scale is divided into one hundred parts). The Celsius scale marks the temperature for water freezing at 0 °C and boiling at 100 °C. The average temperature of the human body registers at 37 °C. Most countries in the world use the Celsius scale for everyday temperature measurements.

For scientists to easily communicate information regarding temperature, an overall standard temperature scale was agreed upon. This scale is the Kelvin (K) scale. Named for Lord Kelvin, who conducted research in thermodynamics, the Kelvin scale contains the largest range of temperatures to facilitate any possible readings.

The Kelvin scale is the accepted measurement by the International System of Units (from the French *Système international d'unités*), or SI, for temperature. The Kelvin scale is employed in thermodynamics, and its reading for 0 is the basis for absolute zero. This scale is rarely used for measuring temperatures in the medical field.

The conversions between the temperature scales are as follows:

Degrees Fahrenheit to Degrees Celsius:

$$^0C = \frac{5}{9}(^0F - 32)$$

Degrees Celsius to Degrees Fahrenheit:

$$^0F = \frac{9}{5}(^\circ C) + 32$$

Degrees Celsius to Kelvin:

$$K = {}^0C + 273.15$$

For example, if a patient has a temperature of 38 °C, what would this be on the Fahrenheit scale?

Solution:

First, select the correct conversion equation from the list above.

$$^0F = \frac{9}{5}(^\circ C) + 32$$

Next, plug in the known value for °C, 38.

$$^0F = \frac{9}{5}(38) + 32$$

Finally, calculate the desired value for °F.

$$°F = \frac{9}{5}(38) + 32$$

$$°F = 100.4°F$$

For example, what would the temperature 52 °C be on the Kelvin scale?

First, select the correct conversion equation from the list above.

$$K = °C + 273.15$$

Next, plug in the known value for °C, 52.

$$K = 52 + 273.15$$

Finally, calculate the desired value for K.

$$K = 325.15 \ K$$

Social Studies

Reading for Meaning in Social Studies

Main Ideas and Details in Social Studies Readings

Determining the Main Ideas

Determining the main idea of a social studies text is much like determining the main idea of any other text. The **main idea** is the author's central message; it is the chief argument that pervades the *entire* text. The main idea of each paragraph in the text is typically different from the overall main idea. However, the main ideas of the paragraphs typically reinforce the main idea of the entire text. When searching for the main idea of a social studies text, one must first ask whether the main idea is being conveyed explicitly or implicitly. **Explicit main ideas** are directly expressed to the audience. **Implicit main ideas** are implied by the language and literary devices of the text. Explicit main ideas are likely to be found either at the beginning of the text (typically within the first sentence of the introductory paragraph) or in the thesis of the text (which, more often than not, can be found near the conclusion of the introductory paragraph). Most historical court cases, for example, have explicit main ideas—their rulings are delivered in direct language. When the main idea is not explicitly stated, it is more difficult to identify. Implicit main ideas must be deduced or decoded by gathering together the facts, arguments, and images hinted at in the text and drawing information from these clues. Some historical speeches, poems, songs, and political cartoons have implicit main ideas. They are likely to incorporate indirect literary devices to express their main points.

Whether the text has an implicit or explicit main idea, every student should begin with one overarching question: "What point is the author trying to make in this text?"

This overarching question should be followed by several smaller questions:

1. **Who**?—Who wrote the text and who (if anybody) is the text describing?

2. **What**?—What time frame is being discussed by the text? What historical context is it implicitly representative of? What context is it explicitly discussing?

3. **Where**?—Where was the text written? Which region, culture, or place is its main focus?

4. **When**?—When was the text written? Is the date explicitly referenced? Or is it implicitly implied?

5. **Why**?—Why did the author write this text? Is there a good reason or explanation for its existence?

6. **How**?—How is the theory or method of interpretation presented to the audience? Likewise, how did the text even originate?

If you can successfully answer the overarching question and the majority of the underlying questions, then you have successfully pinpointed the main idea of a text.

Now, let's cross-analyze a text with an explicit main idea and a text with an implicit main idea.

DOCUMENT A: EXPLICIT MAIN IDEA SOCIAL STUDIES TEXT

"(2) No person acting under color of law shall—"(A) in determining whether any individual is qualified under State law or laws to vote in any Federal election, apply any standard, practice, or procedure different from the standards, practices, or procedures applied under such law or laws to other individuals within the same county, parish, or similar political subdivision who have been found by State officials to be qualified to vote;

"(B) deny the right of any individual to vote in any Federal election because of an error or omission on any record or paper relating to any application, registration, or other act requisite to voting, if such error or omission is not material in determining whether such individual is qualified under State law to vote in such election."

Voting Rights Act of 1965

DOCUMENT B: IMPLICIT MAIN IDEA SOCIAL STUDIES TEXT

I speak tonight for the dignity of man and the destiny of Democracy. I urge every member of both parties, Americans of all religions and of all colors, from every section of this country, to join me in that cause.

At times, history and fate meet at a single time in a single place to shape a turning point in man's unending search for freedom. So it was at Lexington and Concord. So it was a century ago at Appomattox. So it was last week in Selma, Alabama. There, long-suffering men and women peacefully protested the denial of their rights as Americans. Many of them were brutally assaulted. One good man—a man of God—was killed.

There is no cause for pride in what has happened in Selma. There is no cause for self-satisfaction in the long denial of equal rights of millions of Americans. But there is cause for hope and for faith in our Democracy in what is happening here tonight. For the cries of pain and the hymns and protests of oppressed people have summoned into convocation all the majesty of this great government—the government of the greatest nation on earth. Our mission is at once the oldest and the most basic of this country—to right wrong, to do justice, to serve man. In our time we have come to live with the moments of great crises. Our lives have been marked with debate about great issues, issues of war and peace, issues of prosperity and depression.

But rarely in any time does an issue lay bare the secret heart of America itself. Rarely are we met with a challenge, not to our growth or abundance, or our welfare or our security, but rather to the values and the purposes and the meaning of our beloved nation. The issue of equal rights for American Negroes is such an issue. And should we defeat every enemy, and should we double our wealth and conquer the stars, and still be unequal to this issue, then we will have failed as a people and as a nation. For, with a country as with a person, "what is a man profited if he shall gain the whole world, and lose his own soul?"

There is no Negro problem. There is no Southern problem. There is no Northern problem. There is only an American problem.

And we are met here tonight as Americans—not as Democrats or Republicans; we're met here as Americans to solve that problem. This was the first nation in the history of the world to be founded with a purpose.

The great phrases of that purpose still sound in every American heart, North and South: "All men are created equal." "Government by consent of the governed." "Give me liberty or give me death." And those are not just clever words, and those are not just empty theories. In their name Americans have fought and died for two centuries and tonight around the world they stand there as guardians of our liberty risking their lives. Those words are promised to every citizen that he shall share in the dignity of man. This dignity cannot be found in a man's possessions. It cannot be found in his power or in his position. It really rests on his right to be treated as a man equal in opportunity to all others. It says that he shall share in freedom. He shall choose his leaders, educate his children, provide for his family according to his ability and his merits as a human being.

President Lyndon B. Johnson. "We Shall Overcome" Speech (March 15, 1965)

Both documents discuss American citizens' response to the civil rights movement in different ways. Document A is and excerpt from the Voting Rights Act of 1965. It discusses an explicit legal response to the abrogation of citizens' rights to vote. This act makes legal provisions that protect people's voting rights. Document B is an excerpt from a famous speech made by President Lyndon B. Johnson before Voting Rights Act of 1965 was ratified. President Johnson makes an implicit references to the civil rights movement in the United States by referring to Selma, Alabama, site of a key civil rights march. He ties the civil rights movement to other key points in US history (Lexington and Concord, as well as Appomattox). In referring to these events, he is implicitly tying Dr. Martin Luther King's civil rights marches in Alabama to the broader search for freedom and the longer march for democracy in US history. In doing this, he is implicitly claiming racism and the voting disenfranchisement of African-Americans are barriers to democracy. He is also implicitly stating that the division between North and South is one that must end. Together, these texts show how main ideas can be either explicit (the Voting Rights Act of 1965) or implicit (the "We Shall Overcome" Speech).

Using Details to Make Inferences or Claims
Whether their main idea is explicit or implicit, all texts employ details that enable the audience to make inferences. Let's once again use the example of Lyndon B. Johnson's "We Shall Overcome" speech to gain a better understanding of how details build up claims.

Take a look at some of the details highlighted in bold below.

DOCUMENT C: EXAMPLES OF DETAILS IN TEXTS

(1) I speak tonight for the dignity of man and the destiny of Democracy. I urge every member of both parties, Americans of all religions and of all colors, from every section of this country, to join me in that cause.

(2) At times, history and fate meet at a single time in a single place to shape a turning point in man's unending search for freedom. So it was at Lexington and Concord. So it was a century ago at Appomattox. So it was last week in Selma, Alabama. There, long-suffering men and women peacefully protested the denial of their rights as Americans. **(3) Many of them were brutally assaulted. One good man—a man of God—was killed.**

There is no cause for pride in what has happened in Selma. There is no cause for self-satisfaction in the long denial of equal rights of millions of Americans. But there is cause for hope and for faith in our Democracy in what is happening here tonight. For the cries of pain and the hymns and protests of oppressed people have summoned into convocation all the majesty of this great government—the government of the greatest nation on earth. Our mission is at once the oldest and the most basic of this country—to right wrong, to do justice, to serve man. In our time we have come to live with the moments of great crises. **(4) Our lives have been marked with debate about great issues, issues of war and peace, issues of prosperity and depression.**

(5) But rarely in any time does an issue lay bare the secret heart of America itself. Rarely are we met with a challenge, not to our growth or abundance, or our welfare or our security, but rather to the values and the purposes and the meaning of our beloved nation. The issue of equal rights for American Negroes is such an issue. And should we defeat every enemy, and should we double our wealth and conquer the stars, and still be unequal to this issue, then we will have failed as a people and as a nation. For, with a country as with a person, **(6) "But what is a man profited if he shall gain the whole world, and lose his own soul?"**

(7) There is no Negro problem. There is no Southern problem. There is no Northern problem. There is only an American problem.

And we are met here tonight as Americans—not as Democrats or Republicans; we're met here as Americans to solve that problem. **(8) This was the first nation in the history of the world to be founded with a purpose.**

(9) The great phrases of that purpose still sound in every American heart, North and South: "All men are created equal." "Government by consent of the governed." "Give me liberty or give me death." And those are not just clever words, and those are not just empty theories. In their name Americans have fought and died for two centuries and tonight around the world they stand there as guardians of our liberty risking their lives. Those words are promised to every citizen that he shall share in the dignity of man. This dignity cannot be found in a man's possessions. It cannot be found in his power or in his position. It really rests on his right to be treated as a man equal in opportunity to all others. It says that he shall share in freedom. He shall choose his leaders, educate his children, provide for his family according to his ability and his merits as a human being.

President Lyndon B. Johnson. "We Shall Overcome" Speech (March 15, 1965)

These details form the philosophical fabric of the excerpt; they combine to form evidence that can be used to make claims. Details in texts can take the form of facts, opinions, references or citations, or literary devices. The sentences and phrases in bold above are examples of these kinds of details. The first bold sentence—*I speak tonight for the dignity of man and the destiny of Democracy*—emphasizes American democracy, tying it to the dignity of humanity. The phrase partially personifies democracy by providing it with something typically afforded to people: a destiny.

In the second bold section—*At times, history and fate meet at a single time in a single place to shape a turning point in man's unending search for freedom. So it was at Lexington and Concord. So it was a century ago at Appomattox. So it was last week in Selma, Alabama*—the author is using another literary device to provide details: allusion. President Johnson is using allusions to important historical events that championed freedom and unity in the United States to support his claim that Selma is yet another wave of increasing, broadening democracy.

At times, Johnson steps away from the literary devices in order to cite the facts of the matter, as he does in the third bold section, where he reports that "Many of [the protesters at Selma] were brutally assaulted. One good man—a man of God—was killed."

Allusion, which seems to be the kind of detail most used in this speech, also emerges in the fourth bold section, where Johnson discusses debates over abolition, wars (which includes the Civil War over slavery and abolition, among other wars), and the Great Depression. These allusions place history on the side of the president and on the side of the Selma protesters.

The fifth bold section once again employs personification, humanizing America by giving it a beating heart, one that is plagued by racial hatred and injustice. In order to challenge the ethics of the audience,

Johnson even cites the Bible in the sixth bold section: "But what is a man profited if he shall gain the whole world, and lose his own soul?" This is a quotation from Matthew 16:26 in the New Testament, which contrasts the emptiness of material gain with the fullness of spiritual peace.

Sometimes authors use syntactical techniques such as repetition to make their claims known. Johnson does so in this passage: "There is no Negro problem. There is no Southern problem. There is no Northern problem. There is only an American problem." The repetition of the phrase "there is no [x]problem" emphasizes the problem is shared by all Americans rather than limited to one segment or region.

To bolster his arguments, Johnson sometimes uses hyperbolic statements like the one presented in the eighth bold section: "This was the first nation in the history of the world to be founded with a purpose."

The final bold section returns to allusion, noting the historical tensions between North and South, and then it concludes with two citations, one from the Declaration of Independence and one on from Patrick Henry's famous "Give Me Liberty or Give Me Death" revolutionary speech. These are just a few examples of the ways in which in-text details can be used to make inferences and claims. The sections below show how details in maps, graphs, charts, tables, photographs, and political cartoons can also be used to make claims.

The biggest take-away, however, is that these details are essentially useless without historical background knowledge. Background knowledge allows us to make more logical inferences.

Social Studies Vocabulary

The chart on the next page presents some key people, places, events, documents, eras, and terms commonly encountered in social studies content.

People	Places	Events or Documents	Eras	General Terms
Constantine: the Roman Emperor who famously shifted power to Byzantium (Constantinople) and converted the Roman Empire to Christianity	**Thirteen Colonies:** the first colonies of England in North America that eventually became the first states of the United States of America in 1776; included the Massachusetts Bay Colony, New Hampshire, Connecticut, Rhode Island, New York, Pennsylvania, New Jersey, Delaware, Maryland, Virginia North Carolina, South Carolina, and Georgia	**American Revolution:** the American revolt against the British Crown, which culminated in American colonists ratifying the Declaration of Independence and eventually declaring victory over Great Britain in the American War for Independence; famous for establishing rights to life, liberty, and the pursuit of happiness; has been honored and replicated by ensuing revolutions across the globe	**Antiquity or Ancient Times:** the time prior to the Middle Ages that witnessed the First Agricultural Revolution and the rise of the world's first great civilizations	**Evidence:** the concrete qualitative and quantitative data, found in primary and secondary sources, that allow historians to make arguments
Julius Caesar: a famous Roman dictator who was assassinated; known for consolidating power in a triumvirate and later a dictatorship; famous for transitioning Rome from a Republic to an Empire	**Nazi Germany:** an expansive but brief German empire built under the leadership of Nazi dictator Adolf Hitler in the 1930s and 1940s; known for the Holocaust and its belief in Aryan supremacy	**Declaration of Independence:** document signed on July 4, 1776 that declared the Thirteen Colonies independence from King George III of Great Britain; formed the United States of America	**Medieval or Middle Ages:** the era that followed the fall of the Roman Empire; famous for its reliance on religious doctrine and feudal practices	**Timeline:** a tool historians use to place sequences of events in chronological order

People	Places	Events or Documents	Eras	General Terms
Adolf Hitler: the fascist leader of Nazi Germany; famous for starting World War II and carrying out the Holocaust (a genocide) of the Jews in Europe	**Pearl Harbor:** an infamous US naval base that was bombed by Imperial Japan on December 7, 1941, during World War II; famously described as a "day that will live in infamy," the bombing brought the US into World War II	**Articles of Confederation:** The first governmental document of the United States that gave states more power than the federal government; famous for its failures and its leading the US into a critical period before the ratification of the Constitution	**Early Modern Era:** an era, roughly between the 1400s and 1800s, that witnessed the rise of modern nation-states as a result of the combined forces of Exploration and Enlightenment	**Primary Source:** a first-hand source, such as a diary, journal, or war recollection, that is closer in proximity to a particular historical era or event than a secondary source
Charlemagne: a Frankish king who tried to revive the Roman Empire as the "Holy Roman Empire" but only succeeded in uniting a portion of Western Europe for a short time period	**New England:** the northernmost colonies, located in the modern-day Northeast of the US; included the Massachusetts Bay Colony (Massachusetts and Maine), New Hampshire, Connecticut, and Rhode Island; known for its fishing and trade economy and its revolutionary spirit during the War for Independence	**US Constitution:** the current governing document of the United States of America, this document was written in the late 1780s in response to the failures of the Articles of Confederation; famous for granting the US government more federal power	**Age of Exploration:** the era between the 1400s and 1700s in which European colonies voyaged across the oceans and colonize places like the New World; famous for creating institutions of global power and colonization that persisted well into the 20th century	**Secondary Source:** a source that uses primary sources to create a second-hand account of a historical event or era or person; often written long after the event or era or person's life in question

People	Places	Events or Documents	Eras	General Terms
Galileo: a famous scientist who invented a telescope, supported Copernicus's heliocentric theory, and was declared a heretic by the Roman Catholic Church	**Middle Colonies** the Mid-Atlantic colonies sandwiched between New England and the Southern Colonies, which included Pennsylvania, New York, New Jersey, and Delaware; known for their relative religious tolerance	**Magna Carta:** a famous government document that extended greater rights to English citizens in 1215; served as a philosophical foundation for later government and rights documents	**Renaissance:** an era in European art and science between the 1400s and 1700s that witnessed the rise of humanism; famous for producing such artists as Michelangelo and Leonardo da Vinci	**Revolution:** a word that describes a sweeping political, economic, or cultural transformation; based on the spatial metaphor of turning upside down or turning in a circle, thereby changing the reigning order
Henry VIII: an English king who was excommunicated by the Roman Catholic Church in the early 1500s and formed a new Protestant Church: the Church of England (or Anglican Church)	**Southern Colonies:** the predominantly agricultural southern-most colonies of the original Thirteen Colonies, which included Maryland, Virginia, North Carolina, South Carolina, and Georgia; known for its loyalist tendencies during the American Revolution	**War of 1812:** sometimes described as a "Second American Revolution," this war eventually ended in a treaty between the US and Great Britain, but not before the British invaded Washington, D.C. and burned down the original White House	**Reformation:** a religious transformation that began in the 1500s; famous for its attacks on Roman Catholic corruption and the prominence of such Protestant religious leaders as John Calvin and Martin Luther	**Colonization:** a process by which one country settles another country and makes it its economic vassal state; was first widely and aggressively used as a political and economic system during the Age of Exploration and continued into the 20th century

People	Places	Events or Documents	Eras	General Terms
William Shakespeare: a 16th century English poet and playwright who, during the Renaissance, created some the most famous plays in Western Civilization; his plays include *Romeo and Juliet* and *Hamlet*	**Ancient Rome:** an ancient republic that emerged from the ashes of ancient Hellenistic culture, spanning, at its peak, from the British Isles to the Middle East/North Africa; it is known for its impressive architecture and trade routes, republican government, and eventual conversion to Christianity	**Mexican-American War:** an expansionist American War in the 1840s that led to the Mexican Cession according to the Treaty of Guadalupe-Hidalgo; famous for its Manifest Destiny attitudes (the belief that the US had the God-given right to expand from the Atlantic to the Pacific)	**Scientific Revolution:** a sweeping paradigmatic transformation in scientific knowledge and technology that emerged from the human-centered focus of the Renaissance; famous for spawning the Enlightenment and contributing to Western mathematics and science	**Civilization:** a word used to describe complex governing societies that have strong food sources, economic systems, cultures and arts, and infrastructures
George Washington: a famous general of the Seven Years' War and the American Revolution who became the first president of the United States of America; he set the precedent for a two-term presidency	**Ancient Greece:** an ancient civilization that emerged near the modern-day Balkans region of the Mediterranean; it is known for developing Hellenistic culture, democracy, and traditional liberal arts education	**Treaty of Guadalupe-Hidalgo:** the treaty that ended the Mexican-American war; it led to the Mexican Cession, which witnessed Mexico ceding about one third of its total territory in the American West in 1848	**Enlightenment:** an era of humanist awakening that witnessed a wave of revolutions (the American Revolution) and a newfound interest in inalienable human rights, and logic, reason, math, and science over tradition, religion, and superstition	**Research:** the name given to the analysis and evaluation of historical facts and opinions
Abraham Lincoln: the 16th president of the United States, who helped repair the Union during the Civil War and ratified the Emancipation	**Mesopotamia:** an ancient civilization that emerged between the Tigris and Euphrates Rivers near modern-day Iraq; it is known for its development of	**Civil War:** a civil conflict, mainly fought over slavery, between two segments of the United States: the North (Union) and the South (Confederacy);	**Early Republic:** the name given to the era in United States history spanning the Declaration of Independence (1776) and the Market Revolution	**Culture:** a blanket term for the broader sociological variables that form our identities and societies (such as Black

People	Places	Events or Documents	Eras	General Terms
Proclamation, which freed slaves; after the Civil War he was assassinated by John Wilkes Booth	agriculture in the Fertile Crescent	lasted from 1861 to 1865 and was one of the bloodiest wars in US history	(early 1800s); known for its focus on making the US a reputable nation among other nations in the globe	culture or American culture)
Thomas Jefferson: the author of the Declaration of Independence, and the third president of the United States of America; he famously created the Jeffersonian-Republican platform, which focused on states' rights and agrarian republicanism	**Ancient Egypt:** an ancient civilization located on the Upper Nile River in Africa; known for its contributions to Western Civilization and the architectural genius of its pyramids	**World War I:** a war that lasted from 1914 to 1918 and that was set off by the assassination of Austro-Hungarian Archduke Franz Ferdinand; during it the Central Powers (Germany, Austria-Hungary, Bulgaria and the Ottoman Empire) fought against the Allied Powers (Great Britain, France, Russia, Italy, Romania, Japan and the United States)	**Antebellum Period:** the period before the Civil War in the 1860s; known for its sectional tensions and disputes over slavery	**Economics:** the name given to the study of overarching, superstructural financial forces such as trade, commerce, and the stock market
Robert E. Lee: a famous Confederate general who eventually conceded to the Union at the Appomattox Courthouse at the end of the Civil War; known for his military strategy and prowess	**Atlantic World:** a region of the globe spanning the Atlantic Ocean and uniting the continents of North America, South America, Africa, and Europe; became the epicenter of triangular trade during the Age of Exploration	**World War II:** a war in which the Axis powers (primarily Germany, Italy, and Japan) fought against the Allies (primarily France, Russia, the US, and Great Britain) during the 1940s as Germany, Italy and Japan sought to expand their empires; the US victory in World	**Postbellum Period:** the period following the Civil War; known for failed attempts at establishing a more racially equitable during Reconstruction and the greed of the Gilded Age	**Society:** a blanket term used to describe all social institutions and relations created by human beings

People	Places	Events or Documents	Eras	General Terms
		War II launched the country into a period of superpower economic prosperity		
John F. Kennedy: a president who was infamously assassinated in Dallas, Texas, in the early 1960s; he was the only Roman Catholic president in US history	**Middle East:** the epicenter of the Islamic world, with sites holy to Judaism and Christianity as well; characterized by its unique position between Central and South Asia, Europe, and Africa; its largest land mass is the Arabian Peninsula (modern-day Saudi Arabia)	**Vietnam War:** a controversial American armed conflict in Southeast Asia in the 1960s that famously divided the country into the anti-war, countercultural "doves" and the pro-war "hawks"	**Westward Expansion:** an era of intensified American expansion into the trans-Appalachian region of North America; driven by Manifest Destiny, it eventually paved the way to a coast-to-coast nation	**Geography:** the study and mapping of the earth's physical features; it helps students understand history as a spatial phenomenon
Theodore Roosevelt: a late 19th and early 20th century political figure who became a Progressive US President; famous for his masculine, athletic persona and his conservation efforts (the National Park system)	**Caribbean:** a region of the globe dominated by a series of tropical islands in the Caribbean Sea, Gulf of Mexico, and Middle Atlantic; Caribbean islands include Jamaica, the Bahamas, Puerto Rico, Haiti, the Dominican Republic, the US Virgin Islands	**September 11th or 9/11:** an infamous terrorist attack on the United States of America that witnessed the collapse of the twin towers of the World Trade Center in New York City; inaugurated the "War on Terror" under the leadership of President George W. Bush	**Progressive Era:** an era in US history that lasted from the late 1890s to roughly 1920; known as an "Age of Reform" that famously passed Prohibition and Women's Suffrage in the US	**Paradigm:** a broad, overarching pattern of thought or culture that reigns during a particular historical era or within a particular academic discipline or profession; paradigms are thought to succeed one another in what are "paradigm shifts"

People	Places	Events or Documents	Eras	General Terms
Dr. Martin Luther King, Jr.: a famous Southern Baptist civil rights leader who fought for racial desegregation and equality for African-Americans in the 1950s and 1960s; he was assassinated in 1968	**Latin America:** a region of the Southern Hemisphere known for its Latin or Latino culture as a result of Spanish and Portuguese colonization in the New World; includes Central America, South America, and the Caribbean	**Bill of Rights:** the first ten amendments of the US Constitution, which makes the Constitution a living governmental document; created as a result of political compromises between federalists and anti-federalists during the Constitutional Convention	**Cold War:** a post-WWII era of increased tensions between the United States and the Soviet Union, lasting from the late 1940s to the early 1990s	**Liberty:** an Enlightenment value of autonomy and independence, lack of subordination to outside rulers; it sparked the American Revolution and continues to spark civil rights struggles across the globe

How Authors Use Language in Social Studies

When a reader is evaluating the point of view or purpose of a historical text or piece of art, he or she must be aware of how language is used to convey the author's message(s). Authors and cartoonists or painters can incorporate concrete facts or strategic imagery into a text or work of art to help establish their point of view or purpose. Language exists in both text-based forms and visual art forms. Language can be used to convey facts, opinions, references or citations, or literary devices. The example of Dr. Martin Luther King's "I Have a Dream" Speech has already helped exhibit the power, complexity, and utility of language. There are countless social studies sources that employ the power of language. Consider this 1962 speech by John F. Kennedy as another example:

... **(1) William Bradford, speaking in 1630 of the founding of the Plymouth Bay Colony, said that all great and honorable actions are accompanied with great difficulties, and both must be enterprised and overcome with answerable courage.**

If this **(2) capsule history of our progress** teaches us anything, it is that man, in his quest for knowledge and progress, is determined and cannot be deterred. The exploration of space will go ahead, whether we join in it or not, and it is one of the great adventures of all time, and no nation which expects to be the leader of other nations can expect to stay behind in the race for space.

Those who came before us made certain that this country **(3) rode the first waves of the industrial revolutions, the first waves of modern invention, and the first wave of nuclear power, and this generation does not intend to founder in the backwash of the coming age of space.** We mean to be a part of it—we mean to lead it. For the eyes of the world now look into space, to the moon and to the planets beyond, and we have vowed that we shall not see it governed by a hostile flag of conquest,

---

but by a banner of freedom and peace. **(4) We have vowed that we shall not see space filled with weapons of mass destruction, but with instruments of knowledge and understanding.**

Yet the vows of this Nation can only be fulfilled if we in this Nation are first, and, therefore, we intend to be first. In short, our leadership in science and in industry, our hopes for peace and security, our obligations to ourselves as well as others, all require us to make this effort, to solve these mysteries, to solve them for the good of all men, and to become the world's leading space-faring nation....

...**(5) We choose to go to the moon. We choose to go to the moon in this decade and do the other things**, not because they are easy, but because they are hard, because that goal will serve to organize and measure the best of our energies and skills, because that challenge is one that we are willing to accept, one we are unwilling to postpone, and one which we intend to win, and the others, too...

President John F. Kennedy, "We Choose to Go to the Moon" Speech, September 12, 1962

Readers should notice how John F. Kennedy (JFK) uses language—including facts, opinions, references, and literary devices such as imagery—to establish his point of view and purpose. In the first highlighted passage JFK uses a historical reference; he refers to the Plymouth Colony in order to connect the early history of the United States to what he hopes will be its future: flights to the moon. He uses this reference to historical fact to stir the emotions of the audience and to highlight Americans' destiny as courageous explorers.

In the second highlighted passage, JFK uses the phrase "capsule history of progress" to help the audience recall the history he has just summarized. In the third highlighted passage, JFK uses the image of progress acting like a wave; this is a play on words, a double meaning. The waves refer to eras of sweeping change in history, but also actual waves, which conjure up notions of seafaring discovery. When he says that "this nation does not intend to founder in the backwash of the coming space age," he is again employing aquatic imagery, encouraging people to move forward instead of falling back. The fourth highlighted passage sets forth an opinion: JFK declares that he does not want weapons of mass destruction (a reference to the Cold War) planted in space. Finally, the repetition of "we choose to go to the moon" in the final paragraph drives the point home, making it a collective endeavor rather than an individual one.

Language—in the form of facts, opinions, references, and literary devices—is used in this speech to make one major point: JFK believes America should be the first nation to land on the moon.

Fact versus Opinion

A **fact** is a statement that is true empirically or an event that has actually occurred in reality, and can be proven or supported by evidence; it is generally objective. In contrast, an **opinion** is subjective, representing something that someone believes rather than something that exists in the absolute. People's individual understandings, feelings, and perspectives contribute to variations in opinion. Though facts are typically objective in nature, in some instances, a statement of fact may be both factual and yet also subjective. For example, emotions are individual subjective experiences. If an individual says that they feel happy or sad, the feeling is subjective, but the statement is factual; hence, it is a subjective fact. In contrast, if one person tells another that the other is feeling happy or sad—whether this is true or not— that is an assumption or an opinion.

Claims and Evidence in Social Studies

Determining Whether a Claim Is or Is Not Supported by Evidence

There are three major ways to determine whether a claim in social studies is or is not supported by evidence: 1) cross-referencing the claim with information provided in-text, 2) cross-referencing the claim with information provided in *other* texts, 3) cross-referencing the claim with information generally accepted as fact, according to our background knowledge.

For example, someone could write:

Coach Smith's basketball team did not improve during his years as head coach.

There are three ways you can evaluate this statement.

1. Cross-referencing the claim with information provided in-text. The writer could then list the win/loss record of the team during his time as coach:

Year	Coach	Wins	Losses	Winning Percentage
2015	Taylor	14	6	70%
2016	Smith	13	7	65%
2017	Smith	11	9	55%
2018	Smith	11	9	55%

The 2015 season is the year before Smith became the head coach. In his first year (2016) as coach, the team's record decreased to a 65% winning percentage. It decreased again in 2017 to 55% and remained the same in 2018. By providing information in the article, you can see if the data supports the writer's claim. Of course, the writer could have made up the information, so you must evaluate if the source seems trustworthy. You can also verify the data by using a secondary source.

2. Cross-referencing the claim with information provided in *other* texts. To do this, you must find another article, book, or source for your information. Perhaps the wins and losses of the basketball teams are recorded on the basketball league's website or in other news articles.

3. Cross-referencing the claim with information generally accepted as fact, according to our background knowledge. In other cases, you may follow the basketball team so closely that you know from memory that this is true.

Sometimes you know from your own sources and knowledge that something is correct or incorrect. For example, if someone said that the U.S. Senate had 100 members, you may not have to verify it if you remember from school or previous learning that it has 100 members.

Comparing Information that Differs Between Sources

A **primary source** is a piece of original work. This can include books, musical compositions, recordings, movies, works of visual art (paintings, drawings, photographs), jewelry, pottery, clothing, furniture, and other artifacts. Within books, primary sources may be of any genre. Whether nonfiction based on actual events or a fictional creation, the primary source relates the author's firsthand view of some specific event, phenomenon, character, place, process, ideas, field of study or discipline, or other subject matter. Whereas primary sources are original treatments of their subjects, **secondary sources** are a step removed from the original subjects; they analyze and interpret primary sources. These include journal articles, newspaper or magazine articles, works of literary criticism, political commentaries, and academic textbooks.

In the field of history, primary sources frequently include documents that were created around the same time period that they were describing, and most often produced by someone who had direct experience or knowledge of the subject matter. In contrast, secondary sources present the ideas and viewpoints of other authors about the primary sources; in history, for example, these can include books and other written works about the particular historical periods or eras in which the primary sources were produced. Primary sources pertinent in history include diaries, letters, statistics, government information, and original journal articles and books. In literature, a primary source might be a literary novel, a poem or book of poems, or a play. Secondary sources addressing primary sources may be criticism, dissertations, theses, and journal articles. **Tertiary sources,** typically reference works referring to primary and secondary sources, include encyclopedias, bibliographies, handbooks, abstracts, and periodical indexes.

In scientific fields, when scientists conduct laboratory experiments to answer specific research questions and test hypotheses, lab reports and reports of research results constitute examples of primary sources. When researchers produce statistics to support or refute hypotheses, those statistics are primary sources. When a scientist is studying some subject longitudinally or conducting a case study, they may keep a journal or diary. For example, Charles Darwin kept diaries of extensive notes on his studies during sea voyages on the *Beagle*, visits to the Galápagos Islands, etc.; Jean Piaget kept journals of observational notes for case studies of children's learning behaviors. Many scientists, particularly in past centuries, shared and discussed discoveries, questions, and ideas with colleagues through letters, which also constitute primary sources. When a scientist seeks to replicate another's experiment, the reported results, analysis, and commentary on the original work is a secondary source, as is a student's dissertation if it analyzes or discusses others' work rather than reporting original research or ideas.

Analyzing Historical Events and Arguments

Making Inferences

Essentially, inferences are educated guesses based on the presented evidence and the reader's background knowledge that help form conclusions in the absence of one being directly stated. Inferences should be conclusions based off of sound evidence and reasoning.

Connections Between Different Social Studies Elements

Analyzing Cause-and-Effect Relationships
As mentioned previously, a common fallacy is incorrectly classifying correlation as causality. When this occurs, two events that may or may not be associated with one another are said to be linked such that one was the cause or reason for the other, which is considered the effect. To be a true "cause-and-effect" relationship, one factor or event must occur first (the cause) and be the sole reason (unless others are also listed) that the other occurred (the effect). The cause serves as the initiator of the relationship between the two events.

For example, consider the following argument:

> Last weekend, the local bakery ran out of blueberry muffins and some customers had to select something else instead. This week, the bakery's sales have fallen. Therefore, the blueberry muffin shortage last weekend resulted in fewer sales this week.

In this argument, the author states that the decline in sales this week (the effect) was caused by the shortage of blueberry muffins last weekend. However, there are other viable alternate causes for the decline in sales this week besides the blueberry muffin shortage. Perhaps it is summer and many normal patrons are away this week on vacation, or maybe another local bakery just opened or is running a special sale this week. There might be a large construction project or road work in town near the bakery, deterring customers from navigating the detours or busy roads. It is entirely possible that the decline this week is just a random coincidence and not attributable to any factor other than chance, and that next week, sales will return to normal or even exceed typical sales. Insufficient evidence exists to confidently assert that the blueberry muffin shortage was the sole reason for the decline in sales, thus mistaking correlation for causation.

With that said, when using cause and effect to extrapolate meaning from text, readers must determine the cause when the author only communicates effects. For example, if a description of a child eating an ice cream cone includes details like beads of sweat forming on the child's face and the ice cream dripping down her hand faster than she can lick it off, the reader can infer or conclude it must be hot outside. A useful technique for making such decisions is wording them in "If...then" form, e.g. "If the child is perspiring and the ice cream melting, it may be a hot day." Cause and effect text structures explain why certain events or actions resulted in particular outcomes. For example, an author might describe America's historical large flocks of dodo birds, the fact that gunshots did not startle/frighten dodos, and that because dodos did not flee, settlers killed whole flocks in one hunting session, explaining how the dodo was hunted into extinction.

Describing the Connections Between People, Places, Environments, Processes, and Events
History is different than mathematics and hard sciences because it deals with human phenomena that are difficult to define: people, places, environments, processes, and events. The best way to define these phenomena is to find connections, which are typically extracted from historical sources—which can be both primary and secondary sources—and artifacts. It may be helpful to think of historians as "forensic detectives" who must find clues by analyzing evidence from the past. These clues help to create conceptual connections that show how certain people, places, environments, processes, and events are connected.

Let's take a look at the Industrial Revolution to understand how multiple factors can fit together:

> During the Industrial Revolution, factories moved from using individual labor to machines and assembly lines. As part of this chance, industry began using fossil fuels such as coal and oil over energy sources like water and wood. The industry changes began in Great Britain, and then much of the innovation spread to the United States. The ability to accomplish and create more generated economic growth. This increased prosperity made way for greater food, more stable lives, and health innovations that led to a drop in infant mortality rates and increases in life expectancies. Both of those factors led to extreme population rise.

The chart below shows how these factors affected one another:

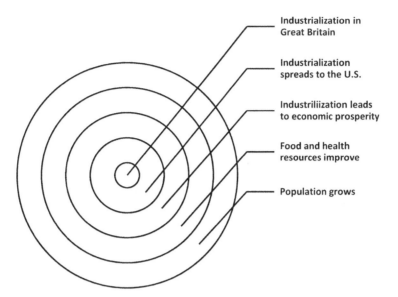

Industrialization in
Great Britain

Industrialization
spreads to the U.S.

Industriliization leads
to economic prosperity

Food and health
resources improve

Population grows

Putting Events in Order and Understanding the Steps in a Process
One of the biggest tasks for any historian or student is placing events in order so that the steps of the historical process can be understood. Historical events are linked by causation (cause and effect) and correlation (associations that are not cause and effect). In order to gain a better understanding of the bigger historical picture, it helps to think about the order in which things happen. In some cases, historians might take a step-by-step look at one event's effects over time; in other cases, historians might try to list several events within one era.

Take a look at the two examples that follow:

EXAMPLE A: ONE EVENT'S EFFECTS OVER TIME: Watergate Scandal

Year	Date	Event
1972	June 17	Watergate Break-in
1973	May 17	Senate hearings begin
	June 25	John Dean begins testifying
	October 20	Saturday Night Massacre (string of events reflecting poorly on Nixon)
1974	July 27	Judiciary Committee votes for impeachment of Nixon
	August 5	Release of "Smoking Gun" tape
	August 8	Nixon announces his resignation
	August 9	Nixon resigns
	August 9	Ford inaugurated as President
	September 8	Ford pardons Nixon
	October 17	Ford testifies before Congress regarding the pardon

European Colonization

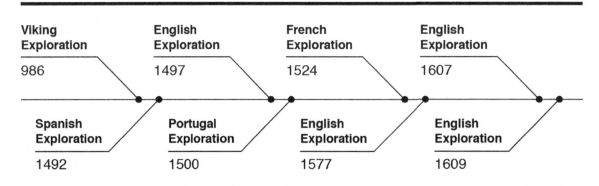

Notice how both timelines place events in order, according to date, and evaluate their impact on history. However, the first timeline looks at the impact of one event—the Watergate Scandal—on a broad span of history, while the second timeline looks at many events and categorizes them under one broad theme—

European colonization. Yet both timelines are great examples of how historians place events in order, according to a process, which in this case, is time.

Analyzing the Relationship of Events, Processes, and/or Ideas

As social science students and scholars create timelines or conceptual maps of events, processes, identities, or ideas, they can evaluate whether earlier events caused later events, simply occurred before these events, or indirectly affected these events. In order to do this, they must contextualize or historicize the events, processes, identities, and ideas they encounter in sources. Likewise, they must compare these sources to other sources—via cross-textual analysis or cross-referencing—in order to validate their relationships between events, processes, identities, and ideas.

Here are some questions that a person can ask in order to illuminate these events:

- Which event, process, or idea came first?

- Are the events, processes, and ideas causally or correlatively connected?

- If the events, processes, or ideas are causally related, then how significant influence did the cause have on the effect?

- If the events, processes, or ideas are correlatively related, then how direct or indirect is the relationship?

- Is there anything that may prove this causal or correlative relation wrong?

- What evidence do you have to support or prove all of the above answers?

Let's attempt this vetting process for analyzing the relationships of two events, processes, or ideas. For this example, let's try to analyze the relationship between the Japanese attack on Pearl Harbor and the United States' entry into World War II.

- Which event, process, or idea came first?
  ANSWER: The attack on Pearl Harbor happened before the United States' entry into World War II.

- Are the events, processes, and ideas causally or correlatively connected?
  ANSWER: These events are causally correlated because one led to another.

- If the events, processes, or ideas are causally related, then how significantly did the preceding event influence the later one?
  ANSWER: There is an extremely significant influence of the attack on the declaration of war.

- If the events, processes, or ideas are correlatively related, then how direct or indirect is the relationship?
  ANSWER: The events are causally related, not correlatively related, so this question is not applicable.

- Is there anything that may prove this causal or correlative relation wrong?
  ANSWER: Some may argue that the entry of the US into WWII was inevitable, but this is a weaker argument in comparison to the more generally accepted causal argument. This argument is speculative in comparison to the causal argument.

- What evidence do you have to support or prove all of the above answers?
  ANSWER: Perhaps the best evidence is President Franklin D. Roosevelt's speech to Congress requesting a declaration of war. See the evidence below for confirmation.

DOCUMENT A: Evidence Supporting the Answers Above

Mr. Vice President, Mr. Speaker, members of the Senate and the House of Representatives:

Yesterday, December 7th, 1941—a date which will live in infamy—the United States of America was suddenly and deliberately attacked by naval and air forces of the Empire of Japan.

The United States was at peace with that nation, and, at the solicitation of Japan, was still in conversation with its government and its Emperor looking toward the maintenance of peace in the Pacific.

Indeed, one hour after Japanese air squadrons had commenced bombing in the American island of Oahu, the Japanese Ambassador to the United States and his colleague delivered to our Secretary of State a formal reply to a recent American message. And, while this reply stated that it seemed useless to continue the existing diplomatic negotiations, it contained no threat or hint of war or of armed attack.

It will be recorded that the distance of Hawaii from Japan makes it obvious that the attack was deliberately planned many days or even weeks ago. During the intervening time the Japanese Government has deliberately sought to deceive the United States by false statements and expressions of hope for continued peace.

The attack yesterday on the Hawaiian Islands has caused severe damage to American naval and military forces. I regret to tell you that very many American lives have been lost. In addition, American ships have been reported torpedoed on the high seas between San Francisco and Honolulu.

Yesterday the Japanese Government also launched an attack against Malaya.
Last night Japanese forces attacked Hong Kong.
Last night Japanese forces attacked Guam.
Last night Japanese forces attacked the Philippine Islands.
Last night the Japanese attacked Wake Island.
And this morning the Japanese attacked Midway Island.

Japan has therefore undertaken a surprise offensive extending throughout the Pacific area. The facts of yesterday and today speak for themselves. The people of the United States have already formed their opinions and well understand the implications to the very life and safety of our nation.

As Commander-in-Chief of the Army and Navy I have directed that all measures be taken for our defense, that always will our whole nation remember the character of the onslaught against us.

> No matter how long it may take us to overcome this premeditated invasion, the American people, in their righteous might, will win through to absolute victory.
>
> I believe that I interpret the will of the Congress and of the people when I assert that we will not only defend ourselves to the uttermost but will make it very certain that this form of treachery shall never again endanger us.
>
> Hostilities exist. There is no blinking at the fact that our people, our territory and our interests are in grave danger.
>
> With confidence in our armed forces, with the unbounding determination of our people, we will gain the inevitable triumph. So help us God.
>
> I ask that the Congress declare that since the unprovoked and dastardly attack by Japan on Sunday, December 7th, 1941, a state of war has existed between the United States and the Japanese Empire.
>
> *-President Franklin D. Roosevelt—December 8, 1941*

Cross-referencing evidence is crucial to ensuring that the perceived connections do, indeed, exist. In this particular instance, the evidence supporting a causal relationship is very clear: "I ask that the Congress declare that since the unprovoked and dastardly attack by Japan on Sunday, December 7th, 1941, a state of war has existed between the United States and the Japanese Empire." There are, however, times when the causal relation might not be as clear or the evidence might not be as firm. This forces historians to comb through thousands of documents to find the right clues to validate perceived connections. Regardless, this process of questioning can be used to vet and validate relationships.

The Effect of Different Social Studies Concepts on an Argument or Point Of View

Analyzing How Events and Situations Shape the Author's Point of View

There are many things we human beings can overcome; history is not one of these things. We can make history, shape history, and analyze history. We can even write history or change our views of history. But we cannot undo what happened in the past. Put simply, *our historical context inevitably shapes our points of view.* We cannot escape—sociologically speaking—many of the events and situations that shape our character. Thus, when you are analyzing primary and secondary sources in social studies, keep in mind that the authors of these sources are always influenced, some more than others, by their historical context.

Thus, whenever we consider an author's point of view, we must also ask the following questions to help us historicize or contextualize their opinions:

Can we validate the exact date or era of the source?

> Certain events and certain eras live within particular paradigms in history. A paradigm is worldview or overarching pattern of thought that underlies the theories and methodology of a particular field of knowledge, in this case, history. These paradigms help us understand people's ideas and people's actions.

Are there any direct or indirect references to other historical events, moments, ideas, or figures?

If you do not know the exact date or era of the source, you may be able to approximate it by identifying direct or indirect references to other historical events, moments, ideas, or figures.

Are there any implicit or explicit citations in the source that hint at the intellectual lineage of the author?

Most ideas or points of view do not emerge in a historical vacuum. It is human inclination to pilfer ideas from others. We all owe much of our intellect to specific books, authors, or mentors. Sometimes these books, authors, and mentors are alluded to in a text or piece of art.

If there is not a reference to time, is there a reference to place or culture?

We are all shaped by nationalism, regionalism, localism, and culturalism. References to place or culture can provide clues to how events and situations shape an author's point of view.

Evaluating Whether the Author's Evidence is Factual, Relevant, and Sufficient

There are three major ways students and scholars can evaluate whether an author's evidence is factual, relevant, and sufficient.

1. **Background Knowledge and Fact-Checking**: In order to evaluate an author's evidence, every student or scholar must have a broad background knowledge. In social studies, students or scholars must be able to reference events, facts, primary sources, secondary sources, theories, and historiographies. This background knowledge allows students and scholars to fact-check particular statements. Much like a journalist fact-checks the statements of their interviewees, a student or scholar of social studies must fact-check the statements of a source.
2. **Cross-Referencing Sources**: Sometimes the truth becomes muddled in history. One event can have millions of sources, which begs the question "What is Truth?" The best way to figure out the truth is to cross-reference and triangulate sources, paying close attention to common statements, phrases, perspectives, and themes.
3. **Deconstructing Biases**: Every source has biases that need to be deconstructed. These biases can either hide the truth, and, in some cases, they illuminate the truth. For examples, biases may help historians historicize or contextualize a certain source because every context or era has its own set of paradigmatic biases.

Making Judgments About How Different Ideas Impact the Author's Argument

In order to make judgments about how different ideas impact the author's argument, you must first follow the same pattern as evaluating sources. However, there are some slight nuances to this approach and one additional step in the process.

1. **Background Knowledge and Fact-Checking**: In order to evaluate an author's arguments, every student or scholar must some background knowledge. In social studies, students or scholars must be able to reference events, facts, primary sources, secondary sources, theories, and historiographies. This background knowledge allows students and scholars to fact-check particular statements. Much like a journalist fact-checks the statements of their interviewees, a student or scholar of social studies must fact-check the statements of a source. This process allows you to validate an author's claims.

2. **Cross-Referencing Sources**: Who else is making these arguments? Do you trust the person or people making these arguments? Are they a reliable source? Do the author's arguments have some semblance of collective validity'? In order to validate an argument, it helps to either a) discover the source of the argument, or b) understand the academic lineage of an argument. This can be accomplished through cross-referencing sources.

3. **Examining Biases**: While every argument has biases, the best arguments are as objective as possible. How objective or subjective is the argument? The only way you can answer this question is by examining the biases of an argument. The biases can be illuminated through background knowledge, fact-checking, and cross-referencing.

4. **Comparing the Arguments to Your Own Worldview**: We all bring our own biases, worldviews, theories, and arguments to the sources we encounter. The sources students and scholars use are filtered through their own perspectives. Thus, since this process is about making judgments rather than simply evaluating, we must be prepared to become aware of our own biases. That is the only way we can address the biases we encounter in other sources.

Identifying Bias and Propaganda in Social Studies Readings

All sources have a relative amount of bias. However, when the bias of the historical materials verges on coercion, censorship, or indoctrination, then the source becomes what we call **propaganda.** Propaganda is different than bias. Bias can be conscious or unconscious. But propaganda is always conscious and strategic. Propaganda can best be described as an attempt at swaying the beliefs or opinions of a target audience. Some people may refer to propaganda as brainwashing. The best way to identify propaganda is to: a) be aware of what is fact and fiction in history (background knowledge), b) be aware of the ways in which truth can be manipulated for certain causes, and c) be conscious of intensified historical prejudices.

Biases
Biases usually occur when someone allows their personal preferences or ideologies to interfere with what should be an objective decision. In personal situations, someone is biased towards someone if they favor them in an unfair way. In academic writing, being biased in your sources means leaving out objective information that would turn the argument one way or the other. The evidence of bias in academic writing makes the text less credible, so be sure to present all viewpoints when writing, not just your own, so to avoid coming off as biased. Being objective when presenting information or dealing with people usually allows the person to gain more credibility.

Stereotypes
Stereotypes are preconceived notions that place a particular rule or characteristics on an entire group of people. Stereotypes are usually offensive to the group they refer to or allies of that group, and often have negative connotations. The reinforcement of stereotypes isn't always obvious. Sometimes stereotypes can be very subtle and are still widely used in order for people to understand categories within the world. For example, saying that women are more emotional and intuitive than men is a stereotype, although this is still an assumption used by many in order to understand the differences between one another.

Using Numbers and Graphics in Social Studies

Using Data Presented in Visual Form, Including Maps, Charts, Graphs, and Tables

Making Sense of Information that is Presented in Different Ways

To make arguments, social studies fields—history, political science, sociology, geography, philosophy, law, and criminal justice—make use of both qualitative data (words, quotes, passages) and quantitative data (numbers and statistics). This qualitative and quantitative information can be presented in a variety of ways. In most cases, the qualitative and quantitative information will be highlighted in in-text references or quotations. Take a look at the two examples below to gain a better understanding of what these in-text references look like.

EXAMPLE A: QUALITATIVE INFORMATION REFERENCE

According to [Émile] Durkheim, such a society produces, in many of its members, **psychological states characterized by a sense of purposelessness, emotional emptiness and despair.**

EXAMPLE B: QUANTITATIVE INFORMATION REFERENCE

According to a report released by the Milwaukee Police Department in 2015, **the median homicide rate per 100,000 residents was 23, but District 5 had a murder rate of 55.**

Analyzing Information from Maps, Tables, Charts, Photographs, and Political Cartoons
Besides in-text citations, data and information can also be presented in the forms of maps and graphs.

Take a look at how information is represented spatially on the map below by showing the state lines and the areas of each land purchases. The information on this map is not only spatially represented, but also color coded.

Bar Graphs

Bar graphs feature equally spaced, horizontal or vertical rectangular bars representing numerical values. They can show change over time as line graphs do, but unlike line graphs, bar graphs can also show differences and similarities among values at a single point in time. Bar graphs are also helpful for visually representing data from different categories, especially when the horizontal axis displays some value that is not numerical, like basketball players with their heights:

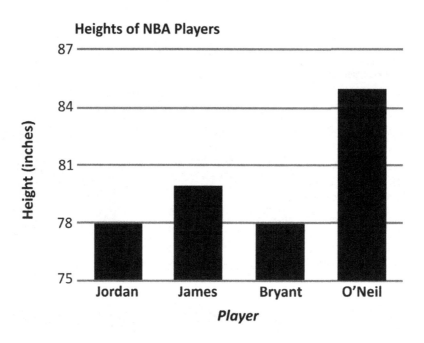

Pie Charts

Pie charts, also called **circle graphs**, are good for representing percentages or proportions of a whole quantity because they represent the whole as a circle or "pie", with the various proportion values shown as "slices" or wedges of the pie. This gives viewers a clear idea of how much of a total each item occupies. For example, biologists may have information that 62% of dogs have brown eyes, 20% have green eyes, and 18% have blue eyes. A pie chart of these distributions would look like this:

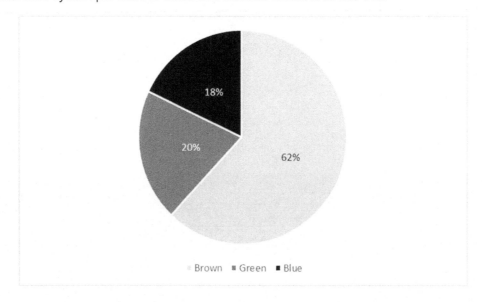

Line Graphs

Line graphs are useful for visually representing data that vary continuously over time, like temperatures over time. The horizontal or *x*-axis shows days of the week; the vertical or *y*-axis shows temperatures. A dot is plotted on the point where each horizontal line intersects each vertical line, and then these dots are connected, forming a line. Line graphs show whether changes in values over time exhibit trends like ascending, descending, flat, or more variable, like going up and down at different times. Take a look at this one below that tracks the high and low temperatures each day for five days.

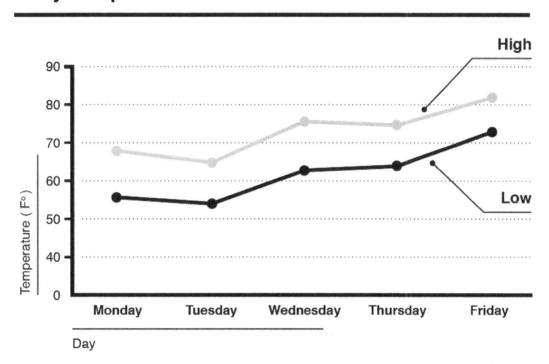

Tables

In some cases, social studies authors will forego in-text citations and graphs, charts, and maps in favor of using a table. Tables, which often look much like a spreadsheet, list some key facts and compare these facts across categories. The simple table below lists the age and gender distribution of residents in a village. These straightforward table comparisons, while simple in presentation, can present students or scholars with a lot of data.

Distribution of the Residents of a Particular Village

	70 or older	69 or younger	Totals
Women	20	40	60
Men	5	35	40
Total	25	75	100

Photographs and political cartoons are different than maps, tables, and charts because they normally employ aesthetic or artistic aspects that are typically lacking in these quantitative platforms for data visualization. In fact, in most photographs or political cartoons, statistics are altogether absent.

Take a look at the political cartoon below to gain a better understanding of how data or evidence is available even in the absence of statistics.

The political cartoon above does not have any numbers (quantitative data) for analysis or interpretation. Nevertheless, it does offer some data. It may even be easier to consider this evidence as historical clues. This particular political cartoon has several clues:

> 1. An oil storage tank is labeled with the name Standard Oil company. This means that the oil storage tank represents that company.

> 2. The tank is shown as an octopus causing terror. This means the cartoonist is depicting the oil company in a negative way.

> 3. The octopuses' limbs are holding onto oil rigs, transportation (ships and railroads), banks, and governmental buildings as if it has control over all of it.

> 4. One of the oil rigs is labeled "Small Oil Co.", showing that it is stifling out competition.

While the clues offer a lot of insight, it is background knowledge that the viewer uses to make inferences about what the image's claims are. Background knowledge of oil companies in the early 1900's and of anti-monopoly laws allows you to use the clues in the political cartoon to understand this cartoon. The cartoon depicts a very large and "evil" Standard Oil Company having control over business, infrastructure, government, and competitors because of its size and power. Public distrust of the size and power of the company eventually led to laws that caused it to break up into smaller companies. Political cartoons, therefore, are best interpreted when one has adequate background knowledge.

Photographs, like political cartoons, offer clues about history. However, they are different from political cartoons because they are somewhat less manipulated than drawings. Political cartoons are created. Photographs are also created, but with less choice; the photographer takes a picture of what shows up in the viewfinder. Both are artifacts of history, but only photographs capture history in "real time."

Take a look at one of these historical snapshots below.

Unlike the political cartoon example, there are no words here, just images. Yet this image is starkly powerful. The photograph depicts a man standing in front of four tanks. He appears to be confronting the tanks, staring at them directly. The tanks have red stars on them (though these are difficult to see) and they appear to be stopped in their tracks by the lone man.

Without even understanding history, one can appreciate the powerful protest in this picture. However, background knowledge allows us to have greater access to the meaning and significance of the photo. This photo, entitled "Tank Man," was taken in Tiananmen Square in China in 1989. The man is trying to agitate for democratic reform by protesting the brute power of the communist regime (symbolized by the red star and tanks).

Thus, the absence of quantitative data in photographs and political cartoons does not necessarily mean that these important historical artifacts are devoid of data. Yet, much like maps, graphs, and charts, these historical tools are best interpreted when one has adequate background knowledge.

Representing Textual Data into Visual Form

Sometimes it helps to represent textual data in visual form in order to reach a broader audience or reinforce a main argument. Visual images engage more audience members; they also strengthen textual data. That is why many students and scholars in social studies use charts, graphs, maps, and tables to support their claims.

Take the following statement about the annual rainfall in Woodriver as an example.

EXAMPLE A: TEXTUAL DATA

Rainfall in Woodriver was high in 2018 but not unprecedented.

This in-text statement is sufficient by itself. It is saying that while rainfall was high in 2018, it has been that high before. Nevertheless, this qualitative claim is ev©en ©more powerful when it is backed by quantitative data in a map, chart, graph, or table.

Take a look at how this type of statement might be repeated or amplified in visual form.
©
EXAMPLE B: VISUAL FORM

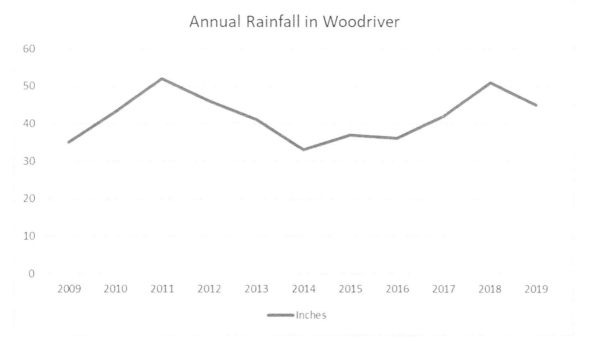

Annual Rainfall in Woodriver

Graphs—or other visual forms like the one above—can help support or reiterate claims by illustrating quantitative examples. Notice how the actual claim—"**Rainfall in Woodriver was high in 2018 but not unprecedented,**"—makes its way into visual form. The graph is accompanied by a title: "Annual Rainfall in Woodriver." The line graph charts rainfall each year in the city of Woodriver from 2009 to 2019, illustrating that, although the rainfall was high in 2018, it was not unprecedented in the city. The year 2011 had even higher rainfall. This visualization of data breathes life into an already powerful statement. In social studies, this type of visualization is often repeated in order to quantify or qualify statements.

Interpreting, Using, and Creating Graphs with Appropriate Labeling, and Using the Data to Predict Trends

All graphs should be interpreted, used, and created with appropriate labels, which include (but are not limited to) the following labels:

The Main Title: The main title offers a brief explanation of what is in the graph. Titles help the audience to understand the main point of a graph.

The Subtitle: The subtitle offers more specific information about the purpose of the graph. Subtitles are brief sentences or phrases that enhance main titles. In some cases, the subtitle can be placed below or beside the main title.

X-Axis and X-Axis Label: Bar graphs and line graphs have an *x*-axis, which runs horizontally (flat). The *x*-axis has quantities representing different categories, statistics, or times that are being compared.

Y-Axis and Y-Axis Label: Bar graphs and line graphs have a *y*-axis, which runs vertically (up and down). The *y*-axis usually measures quantities, typically starting at zero or another designated number.

Take a look at the graph below to get a better understanding of where each label needs to be. This bar graph has the following elements:

> **Title**: What kind of pet do you own?
>
> **Y-Axis Label**: Number of People
>
> **X-Axis Label**: Kind of Pet

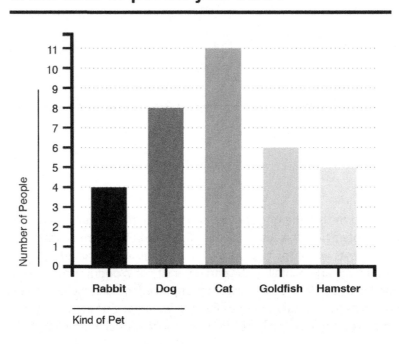

Some graphs also incorporate **keys**, which define colors and symbols. Notice how the graph below uses a key to explain the meaning of each color.

VIOLENT CRIME PER 100,000 POPULATION

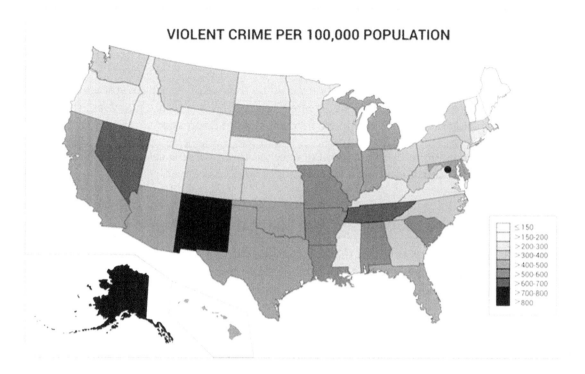

Keys, titles, subtitles, and axes are incorporated in order to both clarify arguments, engage a broader audience of learners, and offer alternative forms of organizing data.

Dependent and Independent Variables

In social studies, students and scholars must constantly analyze variables. Variables in social studies refer to people, places, eras, events, things, or phenomena that students and scholars try to measure or evaluate. In essence, students and scholars are trying to evaluate relationships between certain historical people, places, eras, events, things, or phenomena. In history, students or scholars must constantly be asking, "How are these two variables related?" In answering this question, one can determine which variables are dependent on each other and which variables are independent of each other.

Dependent variables literally *depend* on other variables or factors. As a result, they change accordingly as these other variables or factors change. Take the Japanese attack on Pearl Harbor, for example. Most scholars would agree that this attack forced the US to declare its entry into World War II. Thus, the dependent variable in this case would be the United States' entry into World War II; it was a dependent effect or result of the Japanese attack on Pearl Harbor. In fact, most dependent variables are considered to be effects, while most independent variables are considered to be the root causes in social studies. Thus, the independent variable in this case would be the Japanese attack on Pearl Harbor. It is the variable that accounts for the effect on a dependent variable. In this case, the cause-and-effect relationship is easy to decipher; however, this is not always the case when analyzing history and social studies.

Correlation Versus Causation

When analyzing variables in social studies, it is important to distinguish between correlation and causation. Correlation and causation are not always easy to infer in social studies. Correlation does not imply cause-and-effect; rather, it simply implies that there is an implied (typically a *statistically* implied) connection or relationship between two variables. This connection or relationship does not necessarily clearly designate a cause-and-effect relationship. Correlations can be positive or direct, or they can also be negative or indirect. An example of a correlation would be the rise of fascist governments during the Great Depression. The Great Depression did not create fascist governments, but many fascist governments drew inspiration from the economic hardships of the Depression.

In causation, one phenomenon is the direct result of another; this is called a cause-and-effect relationship. A connection can be called **causation** only if three major conditions are met:

- The cause has to precede the effect in time; the cause happens first. For example, the Japanese attack on Pearl Harbor clearly preceded United States' entry into World War II.

- There has to be empirical evidence that supports the claim of causation; there has to be qualitative or quantitative data that supports the cause-and-effect relationship. For instance, references to the Japanese attack in the United States' declaration of war are support the claim of causation.

- The effect cannot be explained away by other variables. For instance, it is difficult to understand causation when it comes to the Great Depression because there are so many variables that contributed to (or are correlated with) the stock market crash in 1929.

Causation, unlike correlation, is concrete: it proves that *this* led to *that.* Correlation, on the other hand, is much less defined. It implies that this *might have* contributed to that.

Using Statistics in Social Studies

Mean, Median, Mode, and Range of a Data Set
In statistics, measures of central tendency are measures of average. They include the mean, median, mode, and midrange of a data set. The **mean**, otherwise known as the **arithmetic average**, is found by dividing the sum of all data entries by the total number of data points. The **median** is the midpoint of the data points. If there is an odd number of data points, the median is the entry in the middle. If there is an even number of data points, the median is the mean of the two entries in the middle. The **mode** is the data point that occurs most often. Finally, the **midrange** is the mean of the lowest and highest data points. Given the spread of the data, each type of measure has pros and cons. In a **right-skewed distribution**, the bulk of the data falls to the left of the mean. In this situation, the mean is on the right of the median and the mode is on the left of the median. In a **normal distribution,** where the data are evenly distributed on both sides of the mean, the mean, median, and mode are very close to one another. In a **left-skewed distribution**, the bulk of the data falls to the right of the mean. The mean is on the left of the median and the mode is on the right of the median.

Here is an example of each type of distribution:

Left Skew

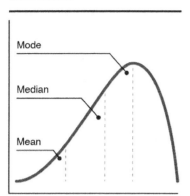

Normal Distribution

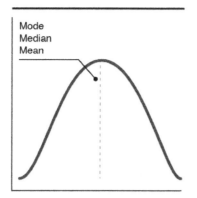

Right Skew

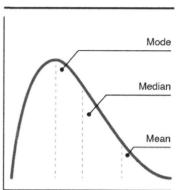

Practice Test #1

Mathematical Reasoning

No Calculator Questions

1. What is the resultant matrix when addition is performed?

$$\begin{bmatrix} 1 & -5 \\ 4 & 2 \end{bmatrix} + \begin{bmatrix} 3 & 2 \\ -2 & 1 \end{bmatrix} = \begin{bmatrix} & \\ & \end{bmatrix}$$

   a. $\begin{bmatrix} 3 & -2 \\ 5 & 0 \end{bmatrix}$

   b. $\begin{bmatrix} 4 & 3 \\ 2 & 3 \end{bmatrix}$

   c. $\begin{bmatrix} 4 & -3 \\ 2 & 3 \end{bmatrix}$

   d. $\begin{bmatrix} 3 & -3 \\ -3 & 3 \end{bmatrix}$

2. What is the domain of $f(x) = 4x^2 + 2x - 1$?
   a. $(0, \infty)$
   b. $(-\infty, 0)$
   c. $(-64, 64)$
   d. $(-\infty, \infty)$

3. What is the domain of the logarithmic function $f(x) = \log_2(x - 2)$?
   a. 2
   b. $(-\infty, \infty)$
   c. $(0, \infty)$
   d. $(2, \infty)$

4. What are the zeros of the following quadratic function?
$$f(x) = 2x^2 - 12x + 16$$
   a. $x = 2$ and $x = 4$
   b. $x = 8$ and $x = 2$
   c. $x = 2$ and $x = 0$
   d. $x = 0$ and $x = 4$

5. What is the equivalent exponential form of the following logarithmic function?
$$f(x) = \log_5(x + 3)$$
   a. $3^y = x$
   b. $5^y = x$
   c. $5^{(x+3)} = y$
   d. $5^y = x + 3$

180

6. What is the range of the polynomial function $f(x) = 2x^2 + 5$?

    a. $(-\infty, \infty)$

    b. $(2, \infty)$

    c. $(0, \infty)$

    d. $[5, \infty)$

7. What is an equivalent logarithmic form of the following exponential function?

$$f(x) = 10^{x+3}$$

    a. $x = 3 + \log y$

    b. $-3 + \log y = x$

    c. $\log(x - 3) = y$

    d. $\log_3 x = y$

8. For which two values of x is $g(x) = 4x + 4$ equal to $g(x) = x^2 + 3x + 2$?

    a. 1, 0

    b. 2, -1

    c. 0, 2

    d. -1, 2

9. Given the linear function $g(x) = \frac{1}{4}x - 2$, which domain value corresponds to a range value of $\frac{1}{8}$?

    a. $\dfrac{17}{2}$

    b. $-\dfrac{63}{32}$

    c. 0

    d. $\dfrac{2}{17}$

10. What are the zeros of the following cubic function?

$$g(x) = x^3 - 2x^2 - 9x + 18$$

    a. 2, 3

    b. 2, 3, -2

    c. 2, 3, -3

    d. 2, -2

11. What is the domain of the following rational function?

$$f(x) = \frac{x^3 + 2x + 1}{2 - x}$$

    a. $(-\infty, -2) \cup (-2, \infty)$

    b. $(-\infty, 2) \cup (2, \infty)$

    c. $(2, \infty)$

    d. $(-2, \infty)$

12. Given the function $f(x) = 4x - 2$, what is the correct form of the simplified difference quotient:

$$\frac{f(x + h) - f(x)}{h}$$

    a. $4x - 1$

    b. $4x$

    c. 4

d. $4x + h$

13. Determine whether the function $g(x) = 7x^3 + 5x - 2$ is odd, even, both even and odd, or neither even nor odd.
    a. Odd
    b. Even
    c. Both
    d. Neither

14. How many kiloliters are in 6 liters?

15. How many centimeters are in 3 feet? (Note: 2.54cm = 1 inch)

16. If $4x - 3 = 5$, then $x =$

Calculator Questions

17. Which of the following is equivalent to $16^{\frac{1}{4}}16^{\frac{1}{2}}$?
    a. 8
    b. 16
    c. 4
    d. 4,096

18. The function $f(t) = \frac{20,000}{1+10e^{-2t}}$ represents the number of people who catch a disease t weeks after its initial outbreak in a population of 20,000 people. How many people initially had the disease at the time of the initial outbreak? Round to the nearest whole number.
    a. 20,000
    b. 1,818
    c. 2,000
    d. 0

19. The mass of the moon is about 7.348×10^{22} kilograms and the mass of Earth is 5.972×10^{24} kilograms. How many times GREATER is Earth's mass than the moon's mass?
    a. 8.127×10^1
    b. 8.127
    c. 812.7
    d. 8.127×10^{-1}

20. Which expression is equivalent to $\sqrt[4]{x^6} - \frac{x}{x^3} + x - 2$?

a. $x^{\frac{3}{2}} - x^2 + x - 2$

b. $x^{\frac{2}{3}} - x^{-2} + x - 2$

c. $x^{\frac{3}{2}} - \frac{1}{x^2} + x - 2$

d. $x^{\frac{2}{3}} - \frac{1}{x^2} + x - 2$

21. How many possible positive zeros does the polynomial function $f(x) = x^4 - 3x^3 + 2x + x - 3$ have?

a. 4

b. 3

c. 2

d. 1

22. What is the solution to the following linear inequality?
$$7 - \frac{4}{5}x < \frac{3}{5}$$

a. $(-\infty, 8)$

b. $(8, \infty)$

c. $[8, \infty)$

d. $(-\infty, 8]$

23. What is an equivalent form of the rational expression $\sqrt{200x^6y^7z^2}$? Assume all variables represent positive real numbers.

a. $20x^3y^3z\sqrt{y}$

b. $10x^3y^3z\sqrt{2y}$

c. $10xyz\sqrt{2xyz}$

d. $10x^2y^3z\sqrt{2z}$

24. How is a transposition of a matrix performed?

a. Multiply each number by negative 1

b. Switch the rows and columns

c. Reverse the order of each row

d. Find the inverse of each number

25. What is the label given to a problem that multiplies a matrix by a constant?

a. Vector multiplication

b. Scalar multiplication

c. Inverse of a matrix

d. Transposition of a matrix

26. What are the two values that always describe a vector?

a. Magnitude and direction

b. Magnitude and length

c. Length and position

d. Direction and position

27. Simplify $\left[\dfrac{2vw^3}{-4w}\right]^2$

a. $\dfrac{v^2w^3}{8}$

b. $\dfrac{v^2w^2}{16}$

c. $\dfrac{v^2w^4}{8}$

d. $\dfrac{v^2w^4}{4}$

28. Which set of matrices represents the following system of equations?

$$\begin{cases} x - 2y + 3z = 7 \\ 2x + y + z = 4 \\ -3x + 2y - 2z = -10 \end{cases}$$

a. $\begin{bmatrix} 1 & -2 & 3 \\ 2 & 3 & 1 \\ -3 & 1 & 2 \end{bmatrix}\begin{bmatrix} x \\ y \\ z \end{bmatrix} = \begin{bmatrix} 7 \\ -4 \\ 10 \end{bmatrix}$

b. $\begin{bmatrix} 1 & 2 & -3 \\ -2 & 1 & 2 \\ 3 & 1 & -2 \end{bmatrix}\begin{bmatrix} x \\ y \\ z \end{bmatrix} = \begin{bmatrix} 7 \\ 4 \\ -10 \end{bmatrix}$

c. $\begin{bmatrix} 1 & -2 & 4 \\ -3 & 1 & 7 \\ 2 & 2 & -10 \end{bmatrix}\begin{bmatrix} 3 \\ 1 \\ -2 \end{bmatrix} = \begin{bmatrix} x \\ y \\ z \end{bmatrix}$

d. $\begin{bmatrix} 1 & -2 & 3 \\ 2 & 1 & 1 \\ -3 & 2 & -2 \end{bmatrix}\begin{bmatrix} x \\ y \\ z \end{bmatrix} = \begin{bmatrix} 7 \\ 4 \\ -10 \end{bmatrix}$

29. Which of the statements below is a statistical question?
    a. What was your grade on the last test?
    b. What were the grades of the students in your class on the last test?
    c. What kind of car do you drive?
    d. What was Sam's time in the marathon?

30. What is the volume of the given figure?

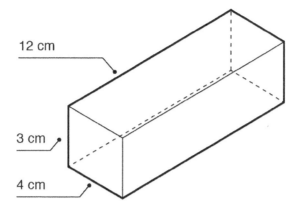

a. 36 cm²
b. 144 cm³
c. 72 cm³
d. 36 cm³

31. What is the perimeter of the following figure?

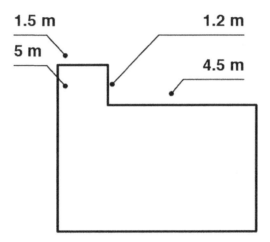

a. 13.4 m
b. 22 m
c. 12.2 m
d. 22.5 m

32. Which shapes could NOT be used to compose a hexagon?

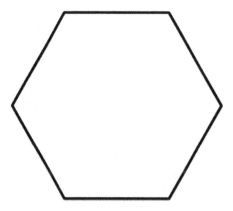

    a. Six triangles
    b. One rectangle and two triangles
    c. Two rectangles
    d. Two trapezoids

33 What is the equation of a straight line through (0, 3) and (4, 8).
    a. $y = \frac{5}{3}x + 3$
    b. $y = 5x + 3$
    c. $y = \frac{5}{4}x + 3$
    d. $y = 4x + 3$

34. What is the solution to the following compound inequality?
$$-14 < 4x + 6 < 18$$

    a. [-5, 3]
    b. [3, 5]
    c. (-5, 3)
    d. [-5, 3)

35. What is the correct result when rationalizing the denominator of $\frac{49}{x-\sqrt{7}}$?
    a. $\frac{49}{x^2-7}$

    b. $\frac{49x+49\sqrt{7}}{x^2-7}$

    c. $\frac{x}{x^2-49}$

    d. $\frac{49}{x}$

36. Which of the following inequalities is equivalent to $3 - \frac{1}{2}x \geq 2$?

    a. $x \geq 2$
    b. $x \leq 2$
    c. $x \geq 1$
    d. $x \leq 1$

37. If $g(x) = x^3 - 3x^2 - 2x + 6$ and $f(x) = 2$, then what is $g(f(x))$?

    a. -26
    b. 6
    c. $2x^3 - 6x^2 - 4x + 12$
    d. -2

38. What is the definition of a factor of the number 36?

    a. A number that can be divided by 36 and have no remainder
    b. A number that 36 can be divided by and have no remainder
    c. A prime number that is multiplied times 36
    d. An even number that is multiplied times 36

39. What are the coordinates of the focus of the parabola $y = -9x^2$?

    a. $(-3, 0)$

    b. $\left(-\frac{1}{36}, 0\right)$

    c. $(0, -3)$

    d. $\left(0, -\frac{1}{36}\right)$

40. A grocery store is selling individual bottles of water, and each bottle contains 750 milliliters of water. If 12 bottles are purchased, what conversion will correctly determine how many liters that customer will take home?

    a. 100 milliliters equals 1 liter
    b. 1,000 milliliters equals 1 liter
    c. 1,000 liters equals 1 milliliter
    d. 10 liters equals 1 milliliter

41. What is the volume of a rectangular prism with the height of 3 centimeters, a width of 5 centimeters, and a depth of 11 centimeters?

    a. 19 cm³
    b. 165 cm³
    c. 225 cm³
    d. 150 cm³

42. What is the volume of a cylinder, in terms of π, with a radius of 5 inches and a height of 10 inches?

    a. 250 π in³
    b. 50 π in³
    c. 100 π in³
    d. 200 π in³

43. What is the solution to the following system of equations?
$$x^2 - 2x + y = 8$$
$$x - y = -2$$
   a. $(-2, 3)$
   b. There is no solution.
   c. $(-2, 0)\ (1, 3)$
   d. $(-2, 0)\ (3, 5)$

44. An equation for the line passing through the origin and the point $(2, 1)$ is
   a. $y = 2x$
   b. $y = \frac{1}{2}x$
   c. $y = x - 2$
   d. $2y = x + 1$

45. A rectangle was formed out of pipe cleaner. Its length was $\frac{1}{2}$ feet and its width was $\frac{11}{2}$ inches. What is its area in square inches?
   a. $\frac{11}{4}$ inch²
   b. $\frac{11}{2}$ inch²
   c. 22 inch²
   d. 33 inch²

46. What type of function is modeled by the values in the following table?

X	f(x)
1	2
2	4
3	8
4	16
5	32

   a. Linear
   b. Exponential
   c. Quadratic
   d. Cubic

Reasoning Through Language Arts

Questions 1–6 are based on the following passage:

Alice was beginning to get very tired of sitting by her sister on the bank, and of having nothing to do: once or twice she had peeped into the book her sister was reading, but it had no pictures or conversations in it, "and what is the use of a book," thought Alice "without pictures or conversation?"

So she was considering in her own mind (as well as she could, for the hot day made her feel very sleepy and stupid), whether the pleasure of making a daisy-chain would be worth the trouble of getting up and picking the daisies, when suddenly a White Rabbit with pink eyes ran close by her.

There was nothing so VERY remarkable in that; nor did Alice think it so VERY much out of the way to hear the Rabbit say to itself, "Oh dear! Oh dear! I shall be late!" (when she thought it over afterwards, it occurred to her that she ought to have wondered at this, but at the time it all seemed quite natural); but when the Rabbit actually TOOK A WATCH OUT OF ITS WAISTCOAT-POCKET, and looked at it, and then hurried on, Alice started to her feet, for it flashed across her mind that she had never before seen a rabbit with either a waistcoat-pocket, or a watch to take out of it, and burning with curiosity, she ran across the field after it, and fortunately was just in time to see it pop down a large rabbit-hole under the hedge.

In another moment down went Alice after it, never once considering how in the world she was to get out again.

The rabbit-hole went straight on like a tunnel for some way, and then dipped suddenly down, so suddenly that Alice had not a moment to think about stopping herself before she found herself falling down a very deep well.

Either the well was very deep, or she fell very slowly, for she had plenty of time as she went down to look about her and to wonder what was going to happen next. First, she tried to look down and make out what she was coming to, but it was too dark to see anything; then she looked at the sides of the well, and noticed that they were filled with cupboards and book-shelves; here and there she saw maps and pictures hung upon pegs. She took down a jar from one of the shelves as she passed; it was labelled "ORANGE MARMALADE," but to her great disappointment it was empty: she did not like to drop the jar for fear of killing somebody, so managed to put it into one of the cupboards as she fell past it.

"Well!" thought Alice to herself, "after such a fall as this, I shall think nothing of tumbling down stairs! How brave they'll all think me at home! Why, I wouldn't say anything about it, even if I fell off the top of the house!" (Which was very likely true.)

Excerpt from <u>Alice's Adventures in Wonderland</u> by Lewis Carroll, 1865

1. Based on this passage, which three words BEST describe Alice?
   a. Patient, sleepy, observant
   b. Whimsical, conformist, doting
   c. Impulsive, restless, curious
   d. Thoughtful, inventive, boisterous

2. Why does Alice follow the white rabbit?
   a. She hopes to catch him and keep him for a pet.
   b. He looks like one of the pictures in her sister's book.
   c. Her sister suggested she find something to do.
   d. She was bored, and she had never seen a rabbit with a waist-coat pocket or a watch.

3. This passage reveals what about the setting:
   a. It establishes the transition between Alice's world and the rabbit's world.
   b. It shows us the world as Alice sees it.
   c. It shows us that the white rabbit is equally at home in both places.
   d. It establishes the characters and where they live.

4. In this passage, Alice:
    a. Reads her sister's book, picks daisies, follows the rabbit, falls down a well
    b. Falls asleep, chases a rabbit, climbs under a hedge, falls down a well
    c. Is bored, chases a rabbit across a field, follows it down a rabbit hole, falls down a very deep well
    d. Talks to her sister about a book, follows the white rabbit down the rabbit hole, eats marmalade and drops the jar, falls for a very long time

5. The passage states: "she did not like to drop the jar for fear of killing somebody, so managed to put it into one of the cupboards as she fell past it." This tells us that Alice is:
    a. Tidy
    b. Considerate
    c. Unsure
    d. Clumsy

6. Based on the passage, the White Rabbit is:
    a. Punctual
    b. Worried
    c. Friendly
    d. Fast

Follow the directions below.

1. Drive from Florida to Georgia.

2. Drive from Georgia to Texas.

3. Drive from Texas to Oklahoma.

4. Drive from Oklahoma to California.

7. Translate the above into general cardinal directions.
    a. Drive west, then south, then north, then west.
    b. Drive north, then west, then north, then south.
    c. Drive north, then west, then north, then west.
    d. Drive south, then west, then north, then south.

8. Based on the label above, this juice contains the most of which two ingredients?
    a. Vitamin C and cornstarch
    b. Corn syrup and Vitamin C
    c. Water and corn syrup
    d. Water and cornstarch

9. On the label above, how many calories would you consume if you drank two bottles of this drink?
    a. 80 calories
    b. 100 calories
    c. 160 calories
    d. 180 calories

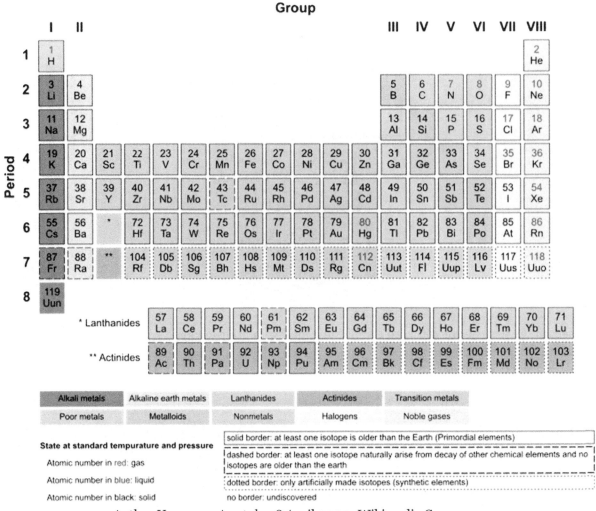

Author Username Armtuk, 28 April 2007, Wikimedia Commons,
https://commons.wikimedia.org/wiki/File:Periodic_Table_Armtuk3.svg

10. Based on the image above, which element is under Group V, Period 6?
    a. Pb
    b. Bi
    c. Uup
    d. At

11. According to the image above, what is the temperature in Fahrenheit?

    a. 17°

    b. 20°

    c. 61°

    d. 62°

12. A reader comes across a word they do not know in the book they are reading, and they need to find out what the word means in order to understand the context of the sentence. Where should the reader look?

    a. Table of contents

    b. Introduction

    c. Index

    d. Glossary

Questions 13–20 are based on the following passage:

> Oliver, being left to himself in the undertaker's shop, set the lamp down on a workman's bench, and gazed timidly about him with a feeling of awe and dread, which many people a good deal older than he will be at no loss to understand. An unfinished coffin on black tressles, which stood in the middle of the shop, looked so gloomy and death-like that a cold tremble came over him, every time his eyes wandered in the direction of the dismal object: from which he almost expected to see some frightful form slowly rear its head, to drive him mad with terror. Against the wall were ranged, in regular array, a long row of elm boards cut in the same shape: looking in the dim light, like high-shouldered ghosts with their hands in their breeches pockets. Coffin-plates, elm-chips, bright-headed nails, and shreds of black cloth, lay scattered on the floor; and the wall behind the counter was ornamented with a lively representation of two mutes in very stiff neckcloths, on duty at a large private door, with a hearse drawn by four black steeds, approaching in the distance. The shop was close and hot. The atmosphere seemed tainted with the smell of coffins. The recess beneath the counter in which his flock mattress was thrust, looked like a grave.

Nor were these the only dismal feelings which depressed Oliver. He was alone in a strange place; and we all know how chilled and desolate the best of us will sometimes feel in such a situation. The boy had no friends to care for, or to care for him. The regret of no recent separation was fresh in his mind; the absence of no loved and well-remembered face sank heavily into his heart.

But his heart was heavy, notwithstanding; and he wished, as he crept into his narrow bed, that that were his coffin, and that he could be lain in a calm and lasting sleep in the churchyard ground, with the tall grass waving gently above his head, and the sound of the old deep bell to soothe him in his sleep.

Oliver was awakened in the morning, by a loud kicking at the outside of the shop-door: which, before he could huddle on his clothes, was repeated, in an angry and impetuous manner, about twenty-five times. When he began to undo the chain, the legs desisted, and a voice began.

"Open the door, will yer?" cried the voice which belonged to the legs which had kicked at the door.

"I will, directly, sir," replied Oliver: undoing the chain, and turning the key.

"I suppose yer the new boy, ain't yer?" said the voice through the key-hole.

"Yes, sir," replied Oliver.

"How old are yer?" inquired the voice.

"Ten, sir," replied Oliver.

"Then I'll whop yer when I get in," said the voice; "you just see if I don't, that's all, my work'us brat!" and having made this obliging promise, the voice began to whistle.

Oliver had been too often subjected to the process to which the very expressive monosyllable just recorded bears reference, to entertain the smallest doubt that the owner of the voice, whoever he might be, would redeem his pledge, most honorably. He drew back the bolts with a trembling hand, and opened the door.

For a second or two, Oliver glanced up the street, and down the street, and over the way: impressed with the belief that the unknown, who had addressed him through the key-hole, had walked a few paces off, to warm himself; for nobody did he see but a big charity-boy, sitting on a post in front of the house, eating a slice of bread and butter: which he cut into wedges, the size of his mouth, with a clasp-knife, and then consumed with great dexterity.

Excerpt from <u>Oliver Twist</u> by Charles Dickens, 1838

13. Based on the passage, how is Oliver feeling?
    a. Excited
    b. Hopeful
    c. Lonely
    d. Challenged

14. Based on context, what does the phrase "charity-boy" likely mean?
    a. Orphan
    b. Fundraiser
    c. Truant
    d. Schoolboy

15. The phrase "...like high-shouldered ghosts with their hands in their breeches pockets" is an example of:
    a. Metaphor
    b. Simile
    c. Symbolism
    d. Onomatopoeia

16. Where does Oliver imagine himself when he lies down to sleep?
    a. Home in bed
    b. Outside camping
    c. A coffin
    d. Beneath the counter

17. In this passage, the following events happen:
    a. Oliver gets dropped off at the shop, the undertaker tells him to sleep, Oliver sleeps in a coffin, the undertaker sends him outside in the morning
    b. Oliver is put to work building coffins, a hearse pulls up, Oliver ventures out to the churchyard, Oliver learns to whistle
    c. Oliver is visited in the night by ghosts, he thinks about his family and friends, someone tells Oliver to open the door, Oliver refuses
    d. Oliver is left alone in the shop, he sleeps and is awakened by someone kicking on the door, the voice threatens him, Oliver learns it's another orphan boy

18. When Oliver imagines himself in a coffin, he also imagines he hears:
    a. Mourners crying
    b. The church bell
    c. Tall grass in the wind
    d. His own breathing

19. The tone of this passage is:
    a. Hopeful
    b. Lighthearted
    c. Empowered
    d. Gloomy

20. The arrival of the other boy at the end of the passage suggests:
    a. Oliver will lose his job to this boy.
    b. Oliver's life might be subject to potential changes.
    c. The boy is the undertaker's son.
    d. The boy will share his bread and butter with Oliver.

The first part of their journey was performed in too melancholy a disposition to be otherwise than tedious and unpleasant. But as they drew toward the end of it, their interest in the appearance of a country which they were to inhabit overcame their dejection, and a view of Barton Valley as they entered it gave them cheerfulness. It was a pleasant fertile spot, well wooded, and rich in pasture. After winding along it for more than a mile, they reached their own house. A small green court was the whole of its demesne in front; and a neat wicket gate admitted them into it.

As a house, Barton Cottage, though small, was comfortable and compact; but as a cottage it was defective, for the building was regular, the roof was tiled, the window shutters were not painted green, nor were the walls covered with honeysuckles. A narrow passage led directly through the house into the garden behind. On each side of the entrance was a sitting room, about sixteen feet square; and beyond them were the offices and the stairs. Four bed-rooms and two garrets formed the rest of the house. It had not been built many years and was in good repair. In comparison of Norland, it was poor and small indeed!—but the tears which recollection called forth as they entered the house were soon dried away. They were cheered by the joy of the servants on their arrival, and each for the sake of the others resolved to appear happy. It was very early in September; the season was fine, and from first seeing the place under the advantage of good weather, they received an impression in its favour which was of material service in recommending it to their lasting approbation.

The situation of the house was good. High hills rose immediately behind, and at no great distance on each side; some of which were open downs, the others cultivated and woody. The village of Barton was chiefly on one of these hills, and formed a pleasant view from the cottage windows. The prospect in front was more extensive; it commanded the whole of the valley, and reached into the country beyond. The hills which surrounded the cottage terminated the valley in that direction; under another name, and in another course, it branched out again between two of the steepest of them.

With the size and furniture of the house Mrs. Dashwood was upon the whole well satisfied; for though her former style of life rendered many additions to the latter indispensable, yet to add and improve was a delight to her; and she had at this time ready money enough to supply all that was wanted of greater elegance to the apartments. "As for the house itself, to be sure," said she, "it is too small for our family, but we will make ourselves tolerably comfortable for the present, as it is too late in the year for improvements. Perhaps in the spring, if I have plenty of money, as I dare say I shall, we may think about building. These parlors are both too small for such parties of our friends as I hope to see often collected here; and I have some thoughts of throwing the passage into one of them with perhaps a part of the other, and so leave the remainder of that other for an entrance; this, with a new drawing room which may be easily added, and a bed-chamber and garret above, will make it a very snug little cottage. I could wish the stairs were handsome. But one must not expect every thing; though I suppose it would be no difficult matter to widen them. I shall see how much I am before-hand with the world in the spring, and we will plan our improvements accordingly."

In the mean time, till all these alterations could be made from the savings of an income of five hundred a-year by a woman who never saved in her life, they were wise enough to be contented with the house as it was; and each of them was busy in arranging their particular concerns, and endeavoring, by placing around them books and other possessions, to form themselves a home.

Marianne's pianoforte was unpacked and properly disposed of; and Elinor's drawings were affixed to the walls of their sitting room.

In such employments as these they were interrupted soon after breakfast the next day by the entrance of their landlord, who called to welcome them to Barton, and to offer them every accommodation from his own house and garden in which theirs might at present be deficient. Sir John Middleton was a good looking man about forty. He had formerly visited at Stanhill, but it was too long for his young cousins to remember him. His countenance was thoroughly good-humoured; and his manners were as friendly as the style of his letter. Their arrival seemed to afford him real satisfaction, and their comfort to be an object of real solicitude to him. He said much of his earnest desire of their living in the most sociable terms with his family, and pressed them so cordially to dine at Barton Park every day till they were better settled at home, that, though his entreaties were carried to a point of perseverance beyond civility, they could not give offence. His kindness was not confined to words; for within an hour after he left them, a large basket full of garden stuff and fruit arrived from the park, which was followed before the end of the day by a present of game.

Excerpt from Chapter 6 of <u>Sense and Sensibility</u> by Jane Austen

21. What is the point of view in this passage?
    a. Third-person omniscient
    b. Second-person
    c. First-person
    d. Third-person objective

22. Which of the following events occurred first?
    a. Sir John Middleton stopped by for a visit.
    b. The servants joyfully cheered for the family.
    c. Mrs. Dashwood discussed improvements to the cottage.
    d. Elinor hung her drawings up in the sitting room.

23. Over the course of the passage, the Dashwoods' attitude shifts. Which statement best describes that shift?
    a. From appreciation of the family's former life of privilege to disdain for the family's new landlord
    b. From confidence in the power of the family's wealth to doubt in the family's ability to survive
    c. From melancholy about leaving Norland to excitement about reaching Barton Cottage in the English countryside
    d. From cheerfulness about the family's expedition to anxiety about the upkeep of such a big home

24. Which of the following is a theme of this passage?
    a. All-conquering love
    b. Power of wealth
    c. Wisdom of experience
    d. Reality vs. expectations

25. At the start of paragraph five, the narrator says, "till all these alterations could be made from the savings of an income of five hundred a-year by a woman who never saved in her life, they were wise enough to be contented with the house as it was." What does the narrator mean?
   a. The family is going through a transition phase.
   b. Mrs. Dashwood needs to obtain meaningful employment.
   c. The family is going through a growth phase.
   d. The Dashwood children need to be concerned about the future.

Questions 26–30 are based on the following passage:

The causes of the increased efficiency given to labor by the division of employments are some of them too familiar to require specification; but it is worthwhile to attempt a complete enumeration of them. By Adam Smith they are reduced to three: "First, the increase of dexterity in every particular workman; secondly, the saving of the time which is commonly lost in passing from one species of work to another; and, lastly, the invention of a great number of machines which facilitate and abridge labor, and enable one man to do the work of many."

(1.) Of these, the increase of dexterity of the individual workman is the most obvious and universal. It does not follow that because a thing has been done oftener it will be done better. That depends on the intelligence of the workman, and on the degree in which his mind works along with his hands. But it will be done more easily. This is as true of mental operations as of bodily. Even a child, after much practice, sums up a column of figures with a rapidity which resembles intuition. The act of speaking any language, of reading fluently, of playing music at sight, are cases as remarkable as they are familiar. Among bodily acts, dancing, gymnastic exercises, ease and brilliancy of execution on a musical instrument, are examples of the rapidity and facility acquired by repetition. In simpler manual operations the effect is, of course, still sooner produced.

(2.) The second advantage enumerated by Adam Smith as arising from the division of labor is one on which I can not help thinking that more stress is laid by him and others than it deserves. To do full justice to his opinion, I will quote his own exposition of it: "It is impossible to pass very quickly from one kind of work to another, that is carried on in a different place, and with quite different tools. A country weaver, who cultivates a small farm, must lose a good deal of time in passing from his loom to the field, and from the field to his loom. When the two trades can be carried on in the same workhouse, the loss of time is no doubt much less. It is even in this case, however, very considerable. A man commonly saunters a little in turning his hand from one sort of employment to another." I am very far from implying that these considerations are of no weight; but I think there are counter-considerations which are overlooked. If one kind of muscular or mental labor is different from another, for that very reason it is to some extent a rest from that other; and if the greatest vigor is not at once obtained in the second occupation, neither could the first have been indefinitely prolonged without some relaxation of energy. It is a matter of common experience that a change of occupation will often afford relief where complete repose would otherwise be necessary, and that a person can work many more hours without fatigue at a succession of occupations, than if confined during the whole time to one. Different occupations employ different muscles, or different energies of the mind, some of which rest and are refreshed while others work. Bodily labor itself rests from mental, and conversely. The variety itself has an invigorating effect on what, for want of a more philosophical appellation, we must term the animal spirits—so important to the efficiency of all work not mechanical, and not unimportant even to that.

(3.) The third advantage attributed by Adam Smith to the division of labor is, to a certain extent, real. Inventions tending to save labor in a particular operation are more likely to occur to any one in proportion as his thoughts are intensely directed to that occupation, and continually employed upon it.

This also can not be wholly true. "The founder of the cotton manufacture was a barber. The inventor of the power-loom was a clergyman. A farmer devised the application of the screw-propeller. A fancy-goods shopkeeper is one of the most enterprising experimentalists in agriculture. The most remarkable architectural design of our day has been furnished by a gardener. The first person who supplied London with water was a goldsmith. The first extensive maker of English roads was a blind man, bred to no trade. The father of English inland navigation was a duke, and his engineer was a millwright. The first great builder of iron bridges was a stone-mason, and the greatest railway engineer commenced his life as a colliery engineer."

(4.) The greatest advantage (next to the dexterity of the workmen) derived from the minute division of labor which takes place in modern manufacturing industry, is one not mentioned by Adam Smith, but to which attention has been drawn by Mr. Babbage: the more economical distribution of labor by classing the work-people according to their capacity.

Excerpt from the Advantages of Division of Labor section in Chapter VI of Principles of Political Economy by John Stuart Mill

26. Which of the following statements would the author agree is an advantage associated with the division of labor proposed by Adam Smith?
    a. Dexterity increases as employees complete repeated tasks
    b. Repetition results in increased monotony for employees
    c. Greater interdependence forms in the production process
    d. Unemployment increases as workers are replaced by machines

27. In paragraph five, why does the author mention that the "first great builder of iron bridges was a stone-mason"?
    a. To provide an example of how a person can do anything that he or she sets his or her mind to
    b. To demonstrate that Adam Smith's third advantage associated with the division of labor is not entirely true
    c. To explain that individuals are not always employed in professions for which they have received schooling
    d. To state that it is possible for individuals to have more than one career during their working life

28. Which statement best expresses the passage's main idea?
    a. The effect of specialization of division of labor
    b. Disproving the economic principles of Adam Smith
    c. Advantages associated with the division of labor
    d. Basic principles of macroeconomics

29. Which advantage associated with the division of labor does the author say is the second most important one?
    a. Increase of dexterity in employees
    b. Savings of time by staying on one type of work
    c. Invention of machines to assist with manual labor
    d. Classification of employees by their abilities

30. In paragraph two of this passage, what does the word "bodily" mean?
    a. Tangible
    b. Animal
    c. Organic
    d. Spiritual

Questions 31–35 are based on the following passage:

Tendon sheaths have the same structure and function as the synovial membranes of joints and are liable to the same diseases. Apart from the tendon sheaths displayed in anatomical dissections, there is a loose peritendinous and perimuscular cellular tissue that is subject to the same pathological conditions as the tendon sheaths proper.

Tenosynovitis. The toxic or infective agent is conveyed to the tendon sheaths through the blood-stream, as in the gouty, gonorrheal, and tuberculous varieties, or is introduced directly through a wound, as in the common pyogenic form of tenosynovitis.

Tenosynovitis crepitans: In the simple or traumatic form of tenosynovitis, although the most prominent etiological factor is a strain or overuse of the tendon, there would appear to be some other, probably a toxic, factor in its production; otherwise the affliction would be much more common than it is: only a small proportion of those who strain or overuse their tendons become the subjects of tenosynovitis. The opposed surfaces of the tendon and its sheath are covered with fibrinous lymph, so that there is friction when they move on one another.

The *clinical features* are pain on movement, tenderness on pressure over the affected tendon, and a sensation of crepitation or friction when the tendon is moved in its sheath. The crepitation may be soft like the friction of snow, or may resemble the creaking of new leather—"saddle-back creaking." There may be swelling in the long axis of the tendon, and redness and edema of the skin. If there is an effusion of fluid into the sheath, the swelling is more marked and crepitation is absent. There is little tendency to the formation of adhesions.

In the upper extremity, the sheath of the long tendon of the biceps may be affected, but the condition is most common in the tendons about the wrist, particularly in the extensors of the thumb, and it is most frequently met with in those who follow occupations which involve prolonged use or excessive straining of these tendons—for example, washerwomen or riveters. It also occurs as a result of excessive piano-playing, fencing, or rowing.

At the ankle it affects the peronei, the extensor digitorum longus, or the tibialis anterior. It is most often met with in relation to the tendo-calcaneus—*Achillo-dynia*—and results from the pressure of ill-fitting boots or from the excessive use and strain of the tendon in cycling, walking, or dancing. There is pain in raising the heel from the ground, and creaking can be felt on palpation.

The *treatment* consists in putting the affected tendon at rest, and with this object a splint may be helpful; the usual remedies for inflammation are indicated: Bier's hyperemia, lead and opium fomentations, and ichthyol and glycerin. The affliction readily subsides under treatment, but is liable to relapse on a repetition of the exciting cause.

Gouty tenosynovitis: A deposit of urate of soda beneath the endothelial covering of tendons or of that lining their sheaths is commonly met with in gouty subjects. The accumulation of urates may result in the formation of visible nodular swellings, varying in size from a pea to a cherry, attached to the tendon and moving with it. They may be merely unsightly, or they may interfere with the use of the tendon. Recurrent attacks of inflammation are prone to occur. We have removed such gouty masses with satisfactory results.

Suppurative tenosynovitis: This form usually follows upon infected wounds of the fingers—especially of the thumb or little finger—and is a frequent sequel to whitlow; it may also follow amputation of a finger. Once the infection has gained access to the sheath, it tends to spread, and may reach the palm or even the forearm, being then associated with cellulitis. In moderately acute cases the tendon and its sheath become covered with granulations, which subsequently lead to the formation of adhesions; while in more acute cases the tendon sloughs. The pus may burst into the cellular tissue outside the sheath, and the suppuration is liable to spread to neighbouring sheaths or to adjacent bones or joints—for example, those of the wrist.

The *treatment* consists in inducing hyperemia and making small incisions for the escape of pus. The site of incision is determined by the point of greatest tenderness on pressure. After the inflammation has subsided, active and passive movements are employed to prevent the formation of adhesions between the tendon and its sheath. If the tendon sloughs, the dead portion should be cut away, as its separation is extremely slow and is attended with prolonged suppuration.

Excerpt from the Muscles, Tendons, and Tendon Sheaths section in Chapter XVIII of <u>Manual of Surgery</u> by Alexis Thomson, F.R.C.S.ED. and Alexander Miles, F.R.C.S. ED.

31. Which statement best expresses the main idea of the passage?
    a. A discussion of the causes, symptoms, and treatments associated with various types of tenosynovitis
    b. The similarities that exist between tendon sheaths and the synovial membranes of joints
    c. An exploration of sports and professions that may be responsible for injuries to tendons
    d. Differences in how tenosynovitis displays in injuries of the wrist versus injuries of the ankle

32. According to the passage, which of the following is a treatment for tenosynovitis crepitans of the ankle?
    a. Depositing urate of soda under the endothelial tendon covering
    b. Placing the tendon at rest, possibly utilizing a splinting device
    c. Incorporating active and passive movements to prevent tendon adhesions
    d. Inducing hyperemia and making small incisions to release pus

33. Which statement is supported by the information in the third paragraph?
    a. Tenosynovitis crepitans directly results from the overuse of a tendon.
    b. Edema and redness of the skin will return a diagnosis of tenosynovitis crepitans.
    c. Since few overuse injuries result in tenosynovitis crepitans, a toxic agent may be at play.
    d. A dancer who strains a tendon will undoubtedly suffer from tenosynovitis crepitans.

34. According to the author, a patient having which of the following clinical features would be presenting with a case of Tenosynovitis crepitans?
    a. Visible nodular swellings
    b. Cellulitis infections
    c. Finger wound infections
    d. Sensation of friction

35. The authors' point of view in this passage can be described as which of the following?
   a. First-person
   b. Second-person
   c. Third-person limited omniscient
   d. Third-person objective

36. Ethicist: Artificial intelligence is rapidly approaching consciousness. What began as simple algorithms that locate and regurgitate information is now capable of independently drawing conclusions. Unfortunately, the free market is responsible for the advances in artificial intelligence, and private companies aren't incentivized to align artificial intelligence with humanity's goals. Without any guidance, artificial intelligence will adopt humanity's worst impulses and mirror the Internet's most violent worldviews.

The ethicist would most likely agree with which one of the following?
   a. The Internet is negatively impacting society.
   b. Unregulated artificial intelligence is a threat to humanity.
   c. Humanity's goals should always be prioritized over technological advancements.
   d. Artificial intelligence should be limited to simple algorithms.

37. Businessman: My cardinal rule is to only invest in privately-held small businesses that exclusively sell tangible goods and have no debt.

Which one of the following is the best investment opportunity according to the businessman's cardinal rule?
   a. Jose owns his own grocery store. He's looking for a partner, because he fell behind on his mortgage and owes the bank three months' worth of payments.
   b. Elizabeth is seeking a partner with business expertise to help expand her standalone store that sells niche board games. The store isn't currently profitable, but it's never been in debt.
   c. A family-owned accounting firm with no outstanding debts is looking for its first outside investor. The firm has turned a profit every year since it opened
   d. A multinational corporation is selling high-yield bonds for the first time.

Questions 38–44 are based on the following passage:

The first and most universal change effected in milk is its souring. So universal is this phenomenon that it is generally regarded as an inevitable change which can not be avoided, and, as already pointed out, has in the past been regarded as a normal property of milk. To-day, however, the phenomenon is well understood. It is due to the action of certain of the milk bacteria upon the milk sugar which converts it into lactic acid, and this acid gives the sour taste and curdles the milk. After this acid is produced in small quantity its presence proves deleterious to the growth of the bacteria, and further bacterial growth is checked. After souring, therefore, the milk for some time does not ordinarily undergo any further changes.

Milk souring has been commonly regarded as a single phenomenon, alike in all cases. When it was first studied by bacteriologists it was thought to be due in all cases to a single species of micro-organism which was discovered to be commonly present and named *Bacillus acidi lactici.* This bacterium has certainly the power of souring milk rapidly, and is found to be very common in dairies in Europe. As soon as bacteriologists turned their attention more closely to the subject it was found that the spontaneous souring of milk was not always caused by the same species of bacterium. Instead of finding this *Bacillus acidi lactici* always present, they found that quite a number of different species of bacteria have the power of souring milk, and are found in different specimens of soured milk. The number of species of bacteria which have been found to sour milk has increased until something over a hundred are known to have this power. These different species do not affect the milk in the same way. All produce some acid, but they differ in the kind and the amount of acid, and especially in the other changes which are effected at the same time that the milk is soured, so that the resulting soured milk is quite variable. In spite of this variety, however, the most recent work tends to show that the majority of cases of spontaneous souring of milk are produced by bacteria which, though somewhat variable, probably constitute a single species, and are identical with the *Bacillus acidi lactici*. This species, found common in the dairies of Europe, according to recent investigations occurs in this country as well. We may say, then, that while there are many species of bacteria infesting the dairy which can sour the milk, there is one which is more common and more universally found than others, and this is the ordinary cause of milk souring.

When we study more carefully the effect upon the milk of the different species of bacteria found in the dairy, we find that there is a great variety of changes which they produce when they are allowed to grow in milk. The dairyman experiences many troubles with his milk. It sometimes curdles without becoming acid. Sometimes it becomes bitter, or acquires an unpleasant "tainted" taste, or, again, a "soapy" taste. Occasionally a dairyman finds his milk becoming slimy, instead of souring and curdling in the normal fashion. At such times, after a number of hours, the milk becomes so slimy that it can be drawn into long threads. Such an infection proves very troublesome, for many a time it persists in spite of all attempts made to remedy it. Again, in other cases the milk will turn blue, acquiring about the time it becomes sour a beautiful sky-blue colour. Or it may become red, or occasionally yellow. All of these troubles the dairyman owes to the presence in his milk of unusual species of bacteria which grow there abundantly.

Excerpt from The Story of Germ Life by Herbert William Conn

38. The word *deleterious* in the first paragraph can be best interpreted as referring to which one of the following?
   a. Amicable
   b. Smoldering
   c. Luminous
   d. Ruinous

39. Which of the following best explains how the passage is organized?
   a. The author begins by presenting the effects of a phenomenon, then explains the process of this phenomenon, and then ends by giving the history of the study of this phenomenon.
   b. The author begins by explaining a process or phenomenon, then gives the history of the study of this phenomenon, then ends by presenting the effects of this phenomenon.
   c. The author begins by giving the history of the study of a certain phenomenon, then explains the process of this phenomenon, then ends by presenting the effects of this phenomenon.
   d. The author begins by giving a broad definition of a subject, then presents more specific cases of the subject, then ends by contrasting two different viewpoints on the subject.

40. What is the primary purpose of the passage?
    a. To inform the reader of the phenomenon, investigation, and consequences of milk souring.
    b. To persuade the reader that milk souring is due to *Bacillus acidi lactici*, found commonly in the dairies of Europe.
    c. To describe the accounts and findings of researchers studying the phenomenon of milk souring.
    d. To discount the former researchers' opinions on milk souring and bring light to new investigations.

41. What does the author say about the ordinary cause of milk souring?
    a. Milk souring is caused mostly by a species of bacteria called *Bacillus acidi lactici*, although former research asserted that it was caused by a variety of bacteria.
    b. The ordinary cause of milk souring is unknown to current researchers, although former researchers thought it was due to a species of bacteria called *Bacillus acidi lactici*.
    c. Milk souring is caused mostly by a species of bacteria identical to that of *Bacillus acidi lactici*, although there are a variety of other bacteria that cause milk souring as well.
    d. The ordinary cause of milk souring will sometimes curdle without becoming acidic, though sometimes it will turn colors other than white, or have strange smells or tastes.

42. The author of the passage would most likely agree most with which of the following?
    a. Milk researchers in the past have been incompetent and have sent us on a wild goose chase when determining what causes milk souring.
    b. Dairymen are considered more expert in the field of milk souring than milk researchers.
    c. The study of milk souring has improved throughout the years, as we now understand more of what causes milk souring and what happens afterward.
    d. Any type of bacteria will turn milk sour, so it's best to keep milk in an airtight container while it is being used.

43. Given the author's account of the consequences of milk souring, which of the following is most closely analogous to the author's description of what happens after milk becomes slimy?
    a. The chemical change that occurs when a firework explodes
    b. A rainstorm that overwaters a succulent plant
    c. Mercury inside of a thermometer that leaks out
    d. A child who swallows flea medication

44. What type of paragraph would most likely come after the third?
    a. A paragraph depicting the general effects of bacteria on milk.
    b. A paragraph explaining a broad history of what researchers have found in regard to milk souring.
    c. A paragraph outlining the properties of milk souring and the way in which it occurs.
    d. A paragraph showing the ways bacteria infiltrate milk and ways to avoid this infiltration.

For the next question, select the choice you think best fits the underlined part of the sentence. If the original is the best answer choice, then choose Choice A.

45. <u>Play baseball, swimming, and dancing</u> are three of Hannah's favorite ways to be active.
    a. Play baseball, swimming, and dancing
    b. Playing baseball; swimming; dancing;
    c. Playing baseball, to swim and to dance,
    d. Playing baseball, swimming, and dancing

Extended Response

Analyze the argument in the two speeches. Develop an argument where you explain which argument is better supported and why. Be sure to present specific evidence from both passages to support the argument. Note that you shouldn't talk about the argument you most agree with; rather, you should talk about which argument has better support and why.

Topic: Are smart cars part of the future?

Passage 1

When we look at the modern passenger vehicle, we see an increase in safety systems to assist drivers, so much so that one can clearly see we are well on our way to driverless cars. From forward collision and late departure warnings to lane centering and parking assists, significantly more vehicles are performing tasks for the driver as we make the shift.

This shift demonstrates how vehicles are already making drivers and our roads safer. Further, The National Highway Traffic Safety Association reports that over 90 percent of accidents are caused by human error. In fact, according to their data, 37,133 people were killed in motor vehicle accidents in 2017. Ten percent of those deaths were attributable to distracted driving, another nine thousand to speeding, and ten thousand to alcohol related incidents. All of these deaths would be preventable with autonomous vehicles.

Similarly, we can improve traffic safety on our highways when we include freight trucks and 18-wheelers in this scenario. Systems currently being tested can reportedly see farther and often react more quickly than a human driver.

If we remove the primary cause of vehicular deaths, human error, we can ensure that not only our drivers are safer, but so are our roadways.

Passage 2

Despite the increase of driverless vehicles currently being developed, tested, and "driven" on our roadways, the technology is not yet prepared to be introduced wide scale. With the inclusion of enhanced safety features on passenger cars, we're already making significant impact on accidents resulting from human error.

Perhaps the most important element to consider is that drivers have one thing computers don't, and that's experience. We learn from driving how to handle a variety of roadway hazards that computers are just unprepared for, such as weather. How will a computer react to hydroplaning? To icy road conditions? Right now, for many driverless cars, these remain an issue. Weather conditions can change in an instant. Are these vehicles able to monitor changing conditions, or conditions visible to a human driver, and adjust in time? Right now, the answer is inconclusive.

And, it's not just weather they're unprepared for. Right now, automated 18-wheelers have no way to respond to emergency vehicles. More specifically, they do not yet know how to be "pulled over," nor do we know if they can avoid emergency vehicles on the shoulder, which most drivers avoid. While driverless cars may be the future, we're not there quite yet.

Science

1. Scientist A is interested in determining the density of three different rocks that are the same size. Density is calculated by dividing the volume of an object by its mass. She hypothesizes that heavier rocks are denser. Which would be a step in her experimental process that would help her test her hypothesis?
    a. Weighing each rock
    b. Observing the color of the rocks
    c. Tracing the shape of each rock
    d. Dropping each rock to see if it breaks

2. Which organelle is the control center of a cell?
    a. Endoplasmic reticulum
    b. Nucleus
    c. Vacuole
    d. Ribosome

3. What is the process that converts carbon dioxide and water to glucose energy in plants.
    a. Photosynthesis
    b. Glycolysis
    c. Mitosis
    d. Krebs Cycle

4. Mitosis is a type of cellular reproduction. At the end of mitosis, what type and how many cells are produced?
    a. Four non-identical cells
    b. Two non-identical cells
    c. Eight identical cells
    d. Two identical cells

Questions 5 and 6 refer to the image below. The following image depicts a Punnett square of light grey and dark grey bean plants being cross-bred. The light grey phenotype (Y) is dominant over the dark grey phenotype (y).

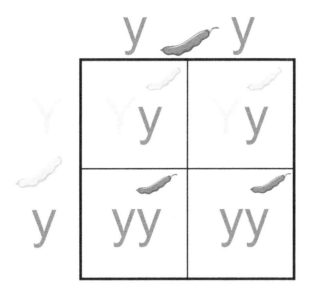

5. What is the predicted fraction offspring for dark grey bean plants?
    a. 1/4
    b. 1/2
    c. 3/4
    d. 1

6. **Short Answer.**
If a mutation occurred that made the light grey and dark grey phenotypes co-dominant, what color beans would the plants produce that have a genetic makeup of Yy? Explain why.

7. **Fill in the blank.**
Many fatal diseases are now preventable through the administration of vaccines. Some viruses always have the same genetic makeup so the vaccine that is administered is standard from year to year. Other viruses, such as the flu, have different prominent strains from year to year. This means that the _____ produced by the body against last year's vaccine may not work against this year's strain.

8. Which of the following steps does NOT contribute to speciation?

    a. A group of individuals develops reproductive traits that no longer allow for reproduction with the original species.

    b. A group of individuals develops new traits that adapt to their environment.

    c. A group of individuals completely loses the ability to reproduce.

    d. A group of individuals becomes isolated from their original species.

9. **Drag and drop question.**

Drag the word "produced" to the part of the diagram where urine is produced by the body. Drag the word "exits" to the part of the diagram where urine exits the body.

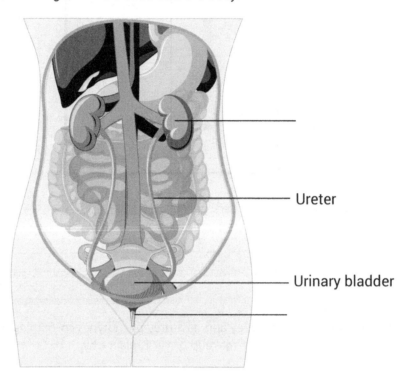

Ureter

Urinary bladder

10. Which part of a flower is considered the male reproductive part of a flower?

    a. Filament

    b. Ovary

    c. Stigma

    d. Style

11. In taxonomy, which category is subclassified next into phylum?

    a. Species

    b. Order

    c. Class

    d. Kingdom

Questions 12 and 13 refer to the following passage and table.

Sarah needs a blood transfusion. Her blood type is O. Five of her family members have volunteered to donate blood for her.

Family Member	Blood Type	Rh factor
Bill	B	Positive
Charlotte	A	Negative
Lucy	AB	Negative
Kevin	O	Negative
Simone	AB	Positive

12. Which family member can Sarah accept blood from?
    a. Simone
    b. Kevin
    c. Lucy
    d. Bill

13. Sarah's cousins Lucy and Charlotte are both pregnant. Both of their babies are Rh positive so they need a blood transfusion with Rh positive blood. Who can receive Rh positive type A blood?
    a. Neither
    b. Only Lucy
    c. Both
    d. Only Charlotte

14. Which is an example of the human body attempting to restore homeostasis?
    a. Voluntary movement of skeletal muscles
    b. Nerves sending signals to the brain
    c. The heart pumping blood throughout the body
    d. Sweating to reduce internal body temperature

15. Which is NOT a characteristic of Protozoa?
    a. Multi-cellular
    b. Unicellular
    c. Has independent locomotive ability
    d. Reproduces asexually

16. During which stage of mitosis do the duplicated chromosomes align at the equator of the cell?
    a. Prophase
    b. Telophase
    c. Metaphase
    d. Anaphase

17. **Short answer.**

Describe the four steps of the Scientific Method.

18. Which of Mendel's Laws of Inheritance states that the alleles of a gene are passed onto an offspring in a random manner, not associated with the alleles of other genes?
    a. The Law of Dominance
    b. The Law of Independent Assortment
    c. The Law of Segregation
    d. The Law of Codominance

19. Looking at the graph below, which type of cell has the fastest rate of replication?

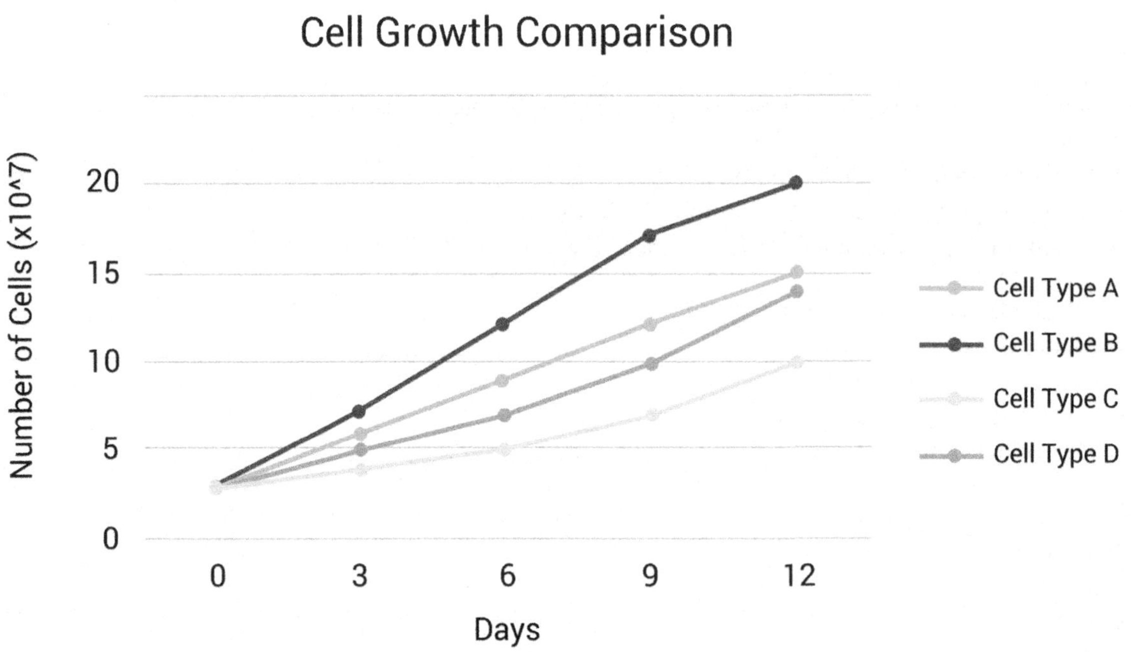

Cell Growth Comparison

    a. Cell Type A
    b. Cell Type B
    c. Cell Type C
    d. Cell Type D

20. Which part of the skeletal system attaches muscles to bones?
   a. Cartilage
   b. Veins
   c. Ligaments
   d. Tendons

21. **Short answer.**
Explain what a radioisotope's half-life is and why it is important for dating of fossils.

22. **Fill in the blank.**
Label the layers of the Earth.

   - Crust
   - Mantle
   - Outer Core
   - Inner Core

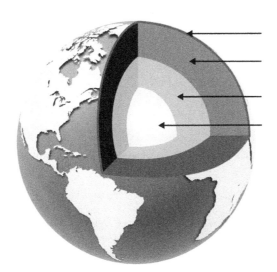

23. The strength of an earthquake is measured by the amplitude of the seismic waves they create. Which is NOT a type of seismic wave?
   a. Secondary
   b. Surface
   c. Tsunami
   d. Primary

24. Which of the following is a source of renewable energy?
   a. Wind
   b. Oil
   c. Coal
   d. Natural Gas

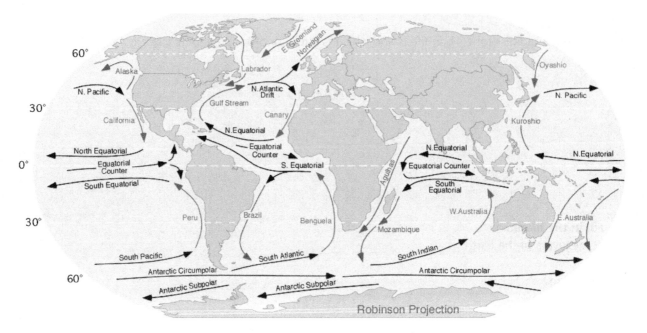

25. The map above depicts how the currents of the ocean move. Which is an accurate explanation of these directional movements?
   a. Water has to squeeze through the small spaces between certain continents
   b. There is more rainfall near the Equator and less rainfall at the poles
   c. Winds always blow from east to west
   d. Warm water rises, whereas cold water sinks

26. **Short answer.**
Weather refers to the conditions of the atmosphere at a given time whereas climate refers to the atmospheric conditions over hundreds or thousands of years. Give an example of both weather and climate in Alaska.

27. What causes people to experience day and night on Earth?
   a. Earth's orbit around the sun
   b. Earth's rotation on its axis
   c. The tilt of the Earth
   d. The position of the moon with respect to the Earth

28. Which is a planet that neighbors Jupiter?
   a. Mercury
   b. Earth
   c. Saturn
   d. Neptune

29. Which is an example of a physical change in matter?
   a. Water freezing to become ice
   b. Rusting of a rake left outside
   c. A yellow liquid combining with a blue liquid to make a green liquid
   d. Burning of fossil fuels

30. **Short answer.**
Sam is comparing two different metals: Metal A and Metal B. He finds that Metal A has a higher melting point, greater malleability, and lower thermal conductivity than Metal B. Explain what these chemical properties told him about the two metals he was testing.

31. What is different about the atomic orbitals of an atom?
   a. The number of neutrons they hold
   b. Nothing, they are all the same
   c. The number of protons they hold
   d. The number of electrons they hold

Questions 32 and 33 refer to the following information and image.

The periodic table of elements is organized according to the atomic number of each element, as well as according to the chemical and physical properties of the elements. From left to right, the elements move from metals to non-metals. The elements in each column have the same number of electrons in their valence shell.

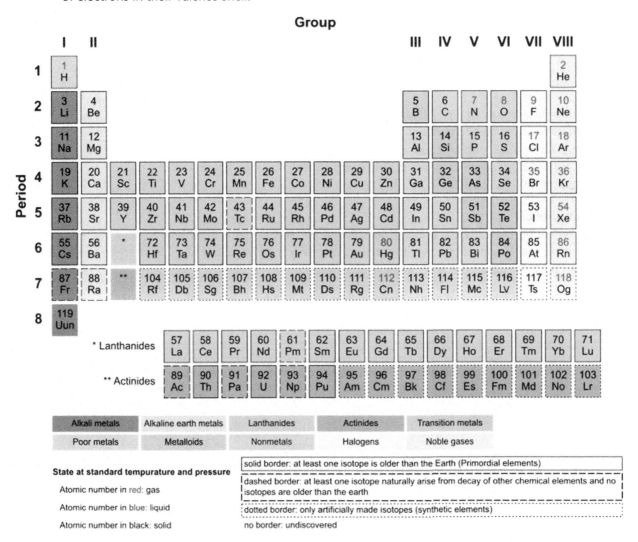

Author Username Armtuk, 28 April 2007, Wikimedia Commons,
https://commons.wikimedia.org/wiki/File:Periodic_Table_Armtuk3.svg

32. What is the atomic number of Nickel?
   a. 7
   b. 10
   c. 28
   d. 11

33. **Short answer.**
Name three differences between Potassium and Krypton.

34. **Circle the correct answer.**
The combustion of methane releases heat, making it an _____ reaction.

$$CH4(g) + 2\,O2(g) \rightarrow CO2(g) + 2\,H2O(g)$$

- Endothermic
- Exothermic

35. Acceleration describes the change in velocity of an object over time. If a car starts out moving 20 miles per hour and ends moving 45 miles per hour after 15 minutes, what is its rate of acceleration?
  a. 60 miles/hr^2
  b. 100 miles/hr^2
  c. 45 miles/hr^2
  d. 25 miles/hr^2

36. Which scenario describes an object with potential energy?
  a. A person singing
  b. A marble rolling down a hill
  c. A car driving on a highway
  d. A charged battery

37. Which is NOT a way in which heat can be transferred from one object to another?
  a. Conduction
  b. Convection
  c. Velocity
  d. Radiation

Social Studies

The following is an excerpt from the Emancipation Proclamation:

> Now, therefore I, Abraham Lincoln, President of the United States, by virtue of the power in me vested as Commander-in-Chief, of the Army and Navy of the United States in time of actual armed rebellion against the authority and government of the United States, and as a fit and necessary war measure for suppressing said rebellion, do, on this first day of January, in the year of our Lord one thousand eight hundred and sixty-three, and in accordance with my purpose so to do publicly proclaimed for the full period of one hundred days, from the day first above mentioned, order and designate as the States and parts of States wherein the people thereof respectively, are this day in rebellion against the United States...

Statement by President Abraham Lincoln, 1863

1. Which government entity was most responsible for overseeing the implementation of the policy outlined in the excerpt?
   a. Executive branch
   B. Judicial branch
   c. House of Representatives
   d. Senate

2. Which statement most accurately summarizes President Lincoln's justification for taking this executive action?
   a. President Lincoln pursued emancipation to remedy the past injustices of slavery.
   b. President Lincoln believed emancipation would deprive the Confederacy of tax revenue, undermining its war effort.
   c. President Lincoln argued emancipation was a moral response to the rebellion against the United States.
   d. President Lincoln characterized emancipation as a necessary and appropriate wartime measure.

3. Which statement best explains why the federal government decided to feature a woman in this poster?
    a. Women received the right to vote immediately before the United States entered World War II .
    b. Approximately 16 million men enlisted in the United States military during World War II.
    c. Traditional household roles declined at the beginning of the twentieth century.
    d. Rosie the Riveter was a feminist symbol and cultural icon.

Electoral Results for the Presidential Election of 1876				
Candidate	Political Party	Popular Vote Count	Popular Vote Percentage	Electoral Vote
Samuel J. Tilden	Democratic	4,286,808	50.92%	184
Rutherford B. Hayes	Republican	4,034,142	47.92%	185
Peter Cooper	Greenback	83,726	0.99%	0
Green Clay Smith	Prohibition	6,945	0.08%	0

4. Which statement about the presidential election of 1876 is true, according to the table?
    a. The Democratic candidate won the presidential election of 1876.
    b. Rutherford B. Hayes won a higher percentage of the popular vote than Samuel J. Tilden.
    c. Green Clay Smith received more popular votes than Peter Cooper.
    d. No candidate received both the majority of the popular vote and the majority of electoral votes.

5. Which candidate won the presidential election of 1876?
    a. Rutherford B. Hayes
    b. Peter Cooper
    c. Samuel J. Tilden
    d. Green Clay Smith

Questions 6 and 7 are based on the following statements:

> When efficiency in production increases, costs per unit of production decrease. When efficiency in production decreases, costs per unit of production increase. When the total amount of production increases, efficiency in production increases.

6. The statements most directly refer to which economic principle?
    a. Comparative advantage
    b. Economies of scale
    c. Law of demand
    d. Law of supply

7. Which statement is true based on the statements above?
    a. If a widget manufacturer introduces more efficient methods of production, production costs of widgets will most likely increase.
    b. If a bank adopts software to more efficiently process claims, costs per claim will most likely remain stagnant.
    c. If a restaurant buys more chicken to meet soaring demand for its chicken entrées, the costs for producing chicken entrées will most likely decrease.
    d. If a farmer experiments with new methods of planting crops, the total amount of crops harvested will most likely increase.

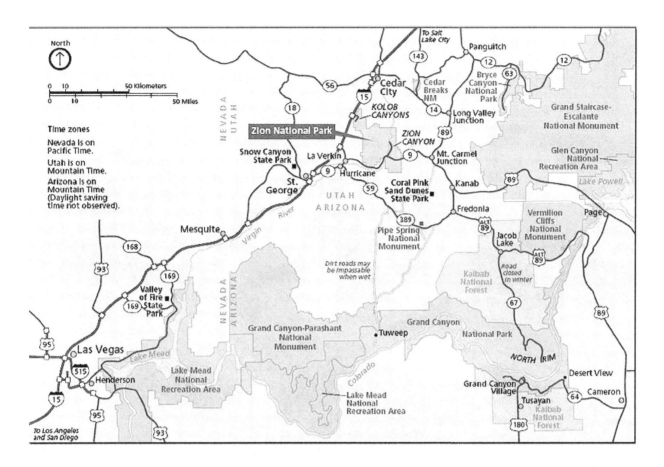

8. Which publicly owned land is in Utah?
    a. Kaibab National Forest
    b. Lake Mead National Recreation Area
    c. Vermillion Cliffs National Monument
    d. Zion National Park

9. Which road crosses through the Grand Staircase-Escalante National Monument?
    a. U.S. Route 93
    b. U.S. Route 180
    c. Utah State Route 12
    d. Utah State Route 18

10. Which statement describes the primary benefit of the production method depicted in the photograph?

    a. The production method required workers to have significantly more skills.

    b. The production method allowed for more customization in the final product.

    c. The production method both increased total production and dramatically reduced costs.

    d. The production method enhanced the quality, reliability, and durability of Ford automobiles.

U.S. natural gas net imports fall to a record low in 2016

billion cubic feet

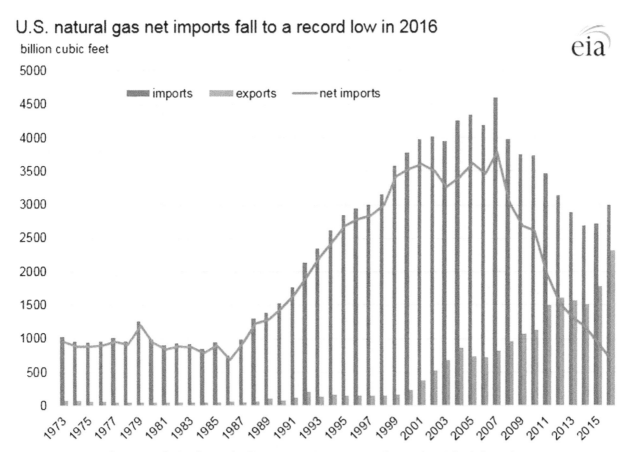

Natural Gas Trade in the United States, U.S. Energy Information Administration, 2017

11. Net imports of natural gas were the lowest in which year?
   a. 1979
   b. 1995
   c. 2003
   d. 2011

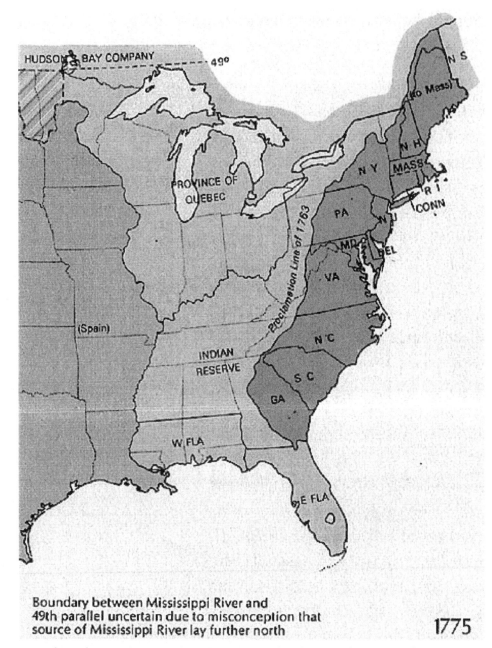

Boundary between Mississippi River and
49th parallel uncertain due to misconception that
source of Mississippi River lay further north

1775

Map of North American Territory in 1775, National Atlas of the United States, 1970

12. Which statement explains why the Eastern Seaboard and Midwest are shaded differently?
   a. Great Britain controlled most of the Eastern Seaboard, while Spain controlled all of the Midwest.
   b. The United States controlled all of the Eastern Seaboard, while Great Britain controlled much of the Midwest.
   c. The Midwest represented a buffer area between the English and Spanish colonial powers.
   d. Great Britain originally colonized most of the Eastern Seaboard and later annexed parts of the Midwest.

13. Which event occurred immediately after the Proclamation Line of 1763 was introduced?
    a. American colonists felt betrayed and sought more independence.
    b. American Indian tribes revolted over the creation of an Indian Reserve.
    c. Spain funded expeditions to build more northern settlements and preserve its territorial claims.
    d. France militarized its colonial borders to prevent further British encroachment.

The following is the full text of the Tenth Amendment to the United States Constitution.

> The powers not delegated to the United States by the Constitution, nor prohibited by it to the states, are reserved to the states respectively, or to the people.

14. The Tenth Amendment directly reinforces which of the following political concepts?
    a. Citizenship
    b. Equality
    c. Federalism
    d. Sovereignty

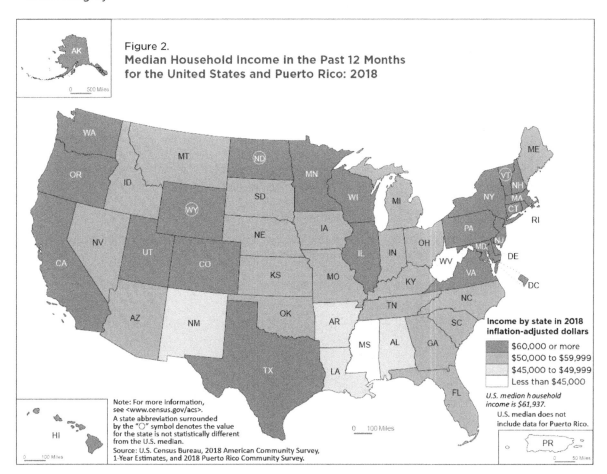

Median Household Income in the United States (2018), United States Censur Bureau

15. Which state had the highest median household income?
    a. Arizona
    b. Colorado
    c. Maine
    d. West Virginia

Table 1. Shares of total expenditures for selected categories, the United States, United Kingdom, and Japan, 2009 and 2014

Categories	U.S.		U.K.		Japan	
	2009	2014	2009	2014	2009	2014
Food	14.6	13.9	19.9	16.7	21.8	23.2
Food at home	8.6	8.2	13.6	11.2	17.2	18.5
Food away from home	6.0	5.7	6.3	5.5	4.6	4.8
Housing	26.3	26.6	27.1	28.5	22.2	23.7
Transportation	17.5	18.7	15.2	16.2	9.8	10.0
Automobile purchases	6.1	6.8	5.1	5.4	2.2	1.6
Public transportation	1.1	1.2	2.5	3.5	2.3	2.5
All other transportation	10.3	10.7	7.6	7.3	5.3	5.9
Healthcare	7.2	8.8	1.4	1.5	4.3	4.3
Clothing	4.0	3.7	5.5	5.1	4.2	4.1
Education	2.4	2.5	1.8	2.1	4.0	3.6
Culture, entertainment, and recreation	6.5	5.8	15.1	14.9	10.6	9.8
Alcoholic beverages and tobacco products	1.9	1.7	4.8	6.3	1.6	1.6
Other categories[1]	19.7	18.4	9.1	8.7	21.6	19.6

[1] Other categories include miscellaneous expenditure shares and also categories that are unique to a particular country. Since these categories were different for each country, they were not used as a direct comparison for this article.

Source: U.S. Bureau of Labor Statistics, Office of National Statistics (U.K.), and Statistics Bureau (Japan).

Comparison of Consumer Spending Habits Calculated as a Percentage of Total Expenditures, Brett Creech, U.S. Bureau of Labor Statistics (2017)

16. Compared to Americans, what did British consumers spend a higher percentage of their income on in 2009?
   a. Alcoholic beverages and tobacco products
   b. Automobile purchases
   c. Education
   d. Transportation

17. Between 2009 and 2014, the percentage of total expenditures increased in all three countries for which of the following categories?
   a. Clothing
   b. Culture, entertainment, and recreation
   c. Healthcare
   d. Housing

Federal Government		
Executive	**Judicial**	**Legislative**
Enforce laws	Interpret laws	Pass laws
Appointment powers	Settle legal disputes	Regulate interstate commerce
Command armed forces	Punish criminal violations	Raise government revenue

18. Which branch oversees how a corporation distributes its products nationwide?
    a. Executive
    b. Judicial
    c. Legislative
    d. None

19. Which branch is responsible for passing and enforcing laws related to conducting elections?
    a. Executive
    b. Judicial
    c. Legislative
    d. None

Berlin Wall, Central Intelligence Agency (1961)

20. What was the Berlin Wall's primary function?

    a. The Berlin Wall created four zones controlled by the victorious Allied Powers.

    b. The Berlin Wall prevented defections from East Germany and protected the Soviet Union's territory.

    c. The Berlin Wall symbolized the beginning of Germany's reconstruction after World War II.

    d. The Berlin Wall prevented the American military from resupplying the zone of territory it controlled.

To summarize, we hold that when an individual is taken into custody or otherwise deprived of his freedom by the authorities in any significant way and is subjected to questioning, the privilege against self-incrimination is jeopardized....

He must be warned prior to any questioning that he has the right to remain silent, that anything he says can be used against him in a court of law, that he has the right to the presence of an attorney, and that if he cannot afford an attorney one will be appointed for him prior to any questioning if he so desires. Opportunity to exercise these rights must be afforded to him throughout the interrogation. After such warnings have been given, and such opportunity afforded him, the individual may knowingly and intelligently waive these rights and agree to answer questions or make a statement. But unless and until such warnings and waiver are demonstrated by the prosecution at trial, no evidence obtained as a result of interrogation can be used against him.

Excerpt from the Supreme Court case of <u>Miranda v. Arizona</u> (1966)

21. Which statement best expresses the meaning of "self-incrimination" as used in the excerpt?
    a. Self-incrimination is when someone exposes themselves to criminal liability.
    b. Self-incrimination is when someone intentionally exposes themself as a criminal.
    c. Self-incrimination is when a prosecutor begins evaluating whether to bring criminal charges.
    d. Self-incrimination is when someone acts to protect themselves from being implicated in a crime.

22. In the excerpt, which warning did the Supreme Court NOT require the police to provide before any questioning?
    a. Consequences of speaking
    b. Consequences of waiving right to counsel
    c. Right to counsel
    d. Right to remain silent

Figure 2. Reported Voting Rates by Race and Hispanic Origin: 1980-2016
(Percentage)

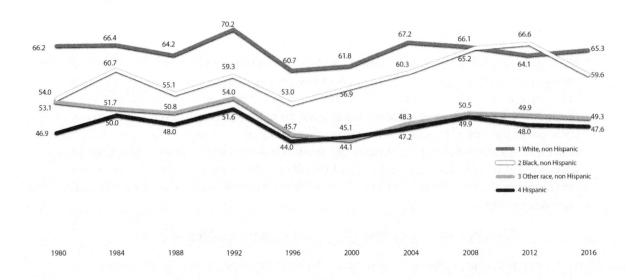

Comparison of Voting Rates in Presidential Elections (1980–2016), United States Census Bureau (2017),

23. From 2012 to 2016, the voting rate declined the most for which group?
    a. Black, non-Hispanic
    b. Hispanic
    c. Other race, non-Hispanic
    d. White, non-Hispanic

24. In which years was voting rate for non-Hispanic black voters the highest?
    a. 2004
    b. 2008
    c. 2012
    d. 2016

Figure 3. Private industry employment, by healthcare industry group, 2001–14 annual averages

Period	All industries	Total healthcare	Offices of physicians, except mental health	General medical and surgical hospitals	Skilled nursing care facilities	Home healthcare services
			Total United States			
2001	109,304,802	11,118,844	1,872,377	3,790,580	1,539,554	635,348
2002	107,577,281	11,482,083	1,940,147	3,881,448	1,567,825	675,660
2003	107,065,553	11,761,194	1,964,069	3,964,761	1,566,775	723,604
2004	108,490,066	11,994,435	2,009,660	4,006,455	1,573,256	770,673
2005	110,611,016	12,247,774	2,049,554	4,054,590	1,570,772	818,867
2006	112,718,858	12,536,713	2,102,235	4,115,486	1,575,603	867,351
2007	114,012,221	12,875,889	2,160,695	4,195,600	1,595,593	915,024
2008	113,188,643	13,228,745	2,209,350	4,294,587	1,616,656	957,911
2009	106,947,104	13,494,501	2,235,673	4,339,758	1,641,479	1,027,612
2010	106,201,232	13,728,075	2,272,233	4,335,639	1,654,706	1,082,731
2011	108,184,795	13,972,516	2,300,225	4,363,863	1,667,523	1,140,402
2012	110,645,869	14,220,690	2,342,075	4,409,965	1,658,761	1,186,670
2013	112,958,334	14,430,444	2,381,270	4,437,590	1,650,979	1,223,952
2014	115,568,686	14,630,002	2,416,054	4,434,972	1,649,686	1,256,831

Michael L. Dolfman, Matthew Insco, and Richard J. Holden, "Healthcare jobs and the Great Recession

25. Which of the following is true with the data provided from the years 2008 to 2009?
    a. Employment increased in overall industries yet decreased in the healthcare industry.
    b. Employment decreased in overall industries yet increased in the healthcare industry.
    c. Employment decreased in skilled nursing care facilities.
    d. Employment increased by 10% in offices of physicians.

Review the following timeline about the American Civil War:

- November 6, 1860: Abraham Lincoln wins the presidential election of 1860.

- February 9, 1861: Southern states establish the Confederate States of America.

- April 12, 1861: Battle of Fort Sumter marks the beginning of the Civil War.

- September 17, 1862: Union forces survive the bloodiest day of the war, achieving victory at the Battle of Antietam.

- September 22, 1862: President Lincoln issues the Emancipation Proclamation.

- July 3, 1863: Union forces win the Battle of Gettysburg.

- March 2,1864: President Lincoln gives Ulysses S. Grant command over all Union forces.

- May 7, 1864: General William T. Sherman launches the Atlanta Campaign.

- November 8, 1864: President Lincoln wins the presidential election of 1864.

- April 9, 1865: Confederate General Robert E. Lee surrenders to Grant at Appomattox Court House, VA.

26. The Battle of Antietam occurred after which event?
    a. Atlanta Campaign
    b. Battle of Fort Sumter
    c. Battle of Gettysburg
    d. Emancipation Proclamation

27. Which two events occurred closest together in time?
    a. Battle of Gettysburg and Battle of Antietam
    b. Establishment of Confederacy and Emancipation Proclamation
    c. Presidential election of 1864 and Lee's surrender
    d. Presidential election of 1860 and Atlanta Campaign

A POOR WAY TO PUT OUT A FIRE.

—Talburt in the Toledo *News-Bee.*

A Poor Way to Put Out a Fire, H.M. Talburt, <u>The Literary Digest</u> (1920),

28. Which war is the political cartoon about?
    a. Korean War
    b. Vietnam War
    c. World War I
    d. World War II

Question 29 is based on the following excerpt:

> We the People of the United States, in Order to form a more perfect Union, establish Justice, insure domestic Tranquility, provide for the common defence, promote the general Welfare, and secure the Blessings of Liberty to ourselves and our Posterity, do ordain and establish this Constitution for the United States of America.

Excerpt from the United States Constitution (1789)

29. Which statement best explains the meaning of a "more perfect Union" as used in the excerpt?
    a. The United States Constitution featured more checks and balances.
    b. The United States Constitution outlawed secession, ensuring that states would be bound to the new system of government.
    c. The United States Constitution required states to cede all claims of sovereignty.
    d. The United States Constitution improved the Articles of Confederation by empowering the federal government.

30. Which intellectual movement most influenced the Framers of the United States Constitution?
    a. Enlightenment
    b. Existentialism
    c. Romanticism
    d. Renaissance

Map of Southwest Asia, Central Intelligence Agency (2020)

31. Which city is located on the Ganges River?
    a. Amritsar
    b. Kanpur
    c. Kolkata
    d. Srinagar

32. The Himalayas function as a natural boundary between which two countries?
    a. China and Nepal
    b. India and Bangladesh
    c. Nepal and India
    d. Pakistan and India

Martin Luther King Jr. Delivers "I Have a Dream" Speech, Rowland Scherman
National Archives and Records Administration (1963)

33. Which landmark piece of legislation did Martin Luther King Jr. advocate for most?
    a. Equal Pay Act of 1963
    b. Food Stamp Act of 1964
    c. Public Works and Economic Development Act of 1965
    d. Voting Rights Act of 1965

34. Martin Luther King Jr. delivered his "I Have a Dream" speech during which protest?
    a. March for Our Lives
    b. March on Washington for Jobs and Freedom
    c. Million Man March
    d. Moratorium to End the War in Vietnam

12 Interior Region Names Based on Watersheds

Regional Watersheds, Department of the Interior (2018)

35. A portion of the Rio Grande watershed is located in which watershed?
    a. California-Great Basin
    b. Lower Colorado Basin
    c. Missouri Basin
    d. Upper Colorado Basin

36. Which watershed borders Canada?
    a. Arkansas-Rio Grande-Texas-Gulf
    b. Mississippi Basin
    c. North Atlantic-Appalachian
    d. South Atlantic Gulf

Answer Explanations #1

Mathematical Reasoning

No Calculator Questions

1. C: The correct answer is Choice *C* because matrices are added by adding the corresponding numbers in the two given matrices. For matrices to be added, they must have the same number of rows and columns. These two matrices have two rows and two columns. The first row, first column numbers are 1 and 3, and adding them together yields a total of 4. The second row, first column has numbers 4 and –2, which yields a sum of 2. The first row, second column has –5 and 2, which yields a sum of –3. The second row, second column have 2 and 1, which yield a sum of 3. The resulting matrix after addition is:

$$\begin{bmatrix} 4 & -3 \\ 2 & 3 \end{bmatrix}$$

2. D: The function given is a polynomial function. Anything can be plugged into a polynomial function to get an output. Therefore, its domain is all real numbers, which is expressed in interval notation as $(-\infty, \infty)$.

3. D: The argument of a logarithmic function has to be greater than or equal to zero. Basically, one cannot take the logarithm of a negative number or 0. Therefore, to find the domain, set the argument greater than 0 and solve the inequality. This results in $x - 2 > 0$, or $x > 2$. Therefore, in order to obtain an output of the function, the number plugged into the function must be greater than 2. This domain is represented as $(2, \infty)$.

4. A: The zeros of a polynomial function are the *x*-values where the graph crosses the *x*-axis, or where $y = 0$. Therefore, set $y = 0$ and solve the polynomial equation. This quadratic can be solved using factoring, as follows:

$$0 = 2x^2 - 12x + 16 = 2(x^2 - 6x + 8) = 2(x - 4)(x - 2).$$

Setting both factors equal to 0 results in the two solutions $x = 4$ and $x = 2$, which are the zeros of the original function.

5. D: Given a logarithmic function $f(x) = \log_b x$, its base is *b* and it can be written as $y = \log_b x$ with equivalent exponential equation $b^y = x$. In this problem, the base is 5, so its equivalent exponential equation is:

$$5^y = x + 3$$

6. D: This is a parabola that opens up, as the coefficient on the x^2 term is positive. The smallest number in its range occurs when plugging 0 into the function $f(0) = 5$. Any other output is a number larger than 5, even when a positive number is plugged in. When a negative number gets plugged into the function, the output is positive, and same with a positive number. Therefore, the domain is written as $[5, \infty)$ in interval notation.

7. B: The equivalent equation form of the function is $y = 10^{x+3}$, with base 10. The expression can be written in logarithmic form as:

$$\log_{10} y = x + 3$$

A logarithm with base 10 is known as the common logarithm and is written without the base. Adding -3 to both sides results in:

$$-3 + \log y = x$$

8. D: First set the functions equal to one another, resulting in:

$$x^2 + 3x + 2 = 4x + 4$$

This is a quadratic equation, so the equivalent equation in standard form is:

$$x^2 - x - 2 = 0$$

This equation can be solved by factoring into:

$$(x - 2)(x + 1) = 0$$

Setting both factors equal to zero results in $x = 2$ and $x = -1$.

9. A: The range value is given, and this is the output of the function. Therefore, the function must be set equal to $\frac{1}{8}$ and solved for x. Thus, $\frac{1}{8} = \frac{1}{4}x - 2$ needs to be solved. The fractions can be cleared by multiplying times the LCD 8. This results in $1 = 2x - 16$. Add 16 to both sides and divide by 2 to obtain $x = \frac{17}{2}$.

10. C: To find the zeros, set the function equal to 0 and factor the polynomial. Because there are four terms, it should be factored by grouping. Factor a common factor out of the first set of two terms, and then find a shared binomial factor in the second set of two terms. This results in:

$$x^2(x - 2) - 9(x - 2) = 0$$

The binomial can then be factored out of each set to get:

$$(x^2 - 9)(x - 2) = 0$$

This can be factored further as:

$$(x + 3)(x - 3)(x - 2) = 0$$

Setting each factor equal to zero and solving results in the three zeros -3, 3, and 2.

11. B: Given a rational function, the expression in the denominator can never be equal to 0. To find the domain, set the denominator equal to 0 and solve for x. This results in $2 - x = 0$, and its solution is $x = 2$. This value needs to be excluded from the set of all real numbers, and therefore the domain written in interval notation is $(-\infty, 2) \cup (2, \infty)$.

12. C: Plugging the function into the formula results in:

$$\frac{4(x+h) - 2 - (4x - 2)}{h}$$

which is simplified to:

$$\frac{4x + 4h - 2 - 4x + 2}{h} = \frac{4h}{h} = h$$

This value is also equal to the derivative of the given function. The derivative of a linear function is its slope.

13. D: To determine whether a function is even or odd, plug $-x$ into the function. If the result is $f(x)$ the function is even, and if the result is $-f(x)$ the function is odd.

$$g(-x) = 7(-x)^3 + 5(-x) - 2 = -7x^3 - 5x - 2$$

This function is neither $f(x)$ nor $-f(x)$, so the given function is neither even nor odd.

14. 0.006. There are 0.006 kiloliters in 6 liters because 1 liter=0.001kiloliters. The conversion comes from the chart where the prefix kilo is found three places to the left of the base unit.

15. 91.44. The conversion between feet and centimeters requires a middle term. As there are 2.54 centimeters in 1 inch, the conversion between inches and feet must be found. As there are 12 inches in a foot, the fractions can be set up as follows:

$$3\ feet \times \frac{12\ inches}{1\ foot} \times \frac{2.54\ cm}{1\ inch}$$

The feet and inches cancel out to leave only centimeters for the answer. The numbers are calculated across the top and bottom to yield:

$$\frac{3 \times 12 \times 2.54}{1 \times 1} = 91.44$$

The number and units used together form the answer of 91.44 cm.

16. 2. Add 3 to both sides to get $4x = 8$. Then divide both sides by 4 to get $x = 2$.

Calculator Questions

17. A: The corresponding expression written using common denominators of the exponents is $16^{\frac{1}{4}}16^{\frac{2}{4}}$, and then the expression is written as $(16 \times 16^2)^{\frac{1}{4}}$. This can be written in radical notation as:

$$\sqrt[4]{16^3} = \sqrt[4]{4{,}096} = 8$$

18. B: The time of the initial outbreak corresponds to $t = 0$. Therefore, 0 must be plugged into the function. This results in $\frac{20{,}000}{1+10e^0} = \frac{20{,}000}{1+10} = \frac{20{,}000}{11} = 1{,}818.182$, which rounds to 1,818. Therefore, there were 1,818 people in the population that initially had the disease.

19. A: Division can be used to solve this problem. The division necessary is:

$$\frac{5.972 \times 10^{24}}{7.348 \times 10^{22}}$$

To compute this division, divide the constants first then use algebraic laws of exponents to divide the exponential expression. This results in about 0.8127×10^2, which written in scientific notation is 8.127×10^1.

20. C: By switching from a radical expression to rational exponents:

$$\sqrt[4]{x^6} = x^{\frac{6}{4}} = x^{\frac{3}{2}}$$

Also, properties of exponents can be used to simplify $\frac{x}{x^3}$ into:

$$x^{1-3} = x^{-2} = \frac{1}{x^2}$$

The other terms can be left alone, resulting in an equivalent expression:

$$x^{\frac{3}{2}} - \frac{1}{x^2} + x - 2$$

21. B: Using Descartes' Rule of Signs, count the number of sign changes in coefficients in the polynomial. This results in the number of possible positive zeros. The coefficients are 1, -3, 2, 1, and -3, so the sign changes from 1 to -3, -3 to 2, and 1 to -3, a total of 3 times. Therefore, there are at most 3 positive zeros.

22. B: The goal is to first isolate the variable. The fractions can easily be cleared by multiplying the entire inequality by 5, resulting in $35 - 4x < 3$. Then, subtract 35 from both sides and divide by -4. This results in $x > 8$. Notice the inequality symbol has been flipped because both sides were divided by a negative number. The solution set, all real numbers greater than 8, is written in interval notation as $(8, \infty)$. A parenthesis shows that 8 is not included in the solution set.

23. B: The expression under the radical can be factored into perfect squares as such:

$$\sqrt{100 \times 2 \times x^6 \times y^6 \times y \times z^2}$$

Using rational exponents, this is the same as

$$(100 \times 2 \times x^6 \times y^6 \times y \times z^2)^{\frac{1}{2}}$$

$$100^{\frac{1}{2}} \times 2^{\frac{1}{2}} \times x^{\frac{6}{2}} \times y^{\frac{6}{2}} \times y^{\frac{1}{2}} \times z^{\frac{2}{2}}$$

$$\sqrt{100} \times \sqrt{2} \times x^3 \times \sqrt{y} \times z$$

$$10x^3 y^3 z \sqrt{2y}$$

24. B: The correct choice is *B* because the definition of transposing a matrix says that the rows and columns should be switched. For example, the matrix $\begin{bmatrix} 3 & 4 \\ 2 & 5 \\ 1 & 6 \end{bmatrix}$ can be transposed into $\begin{bmatrix} 3 & 2 & 1 \\ 4 & 5 & 6 \end{bmatrix}$. Notice that the first row, 3 and 4, becomes the first column. The second row, 2 and 5, becomes the second column. This is an example of transposing a matrix.

25. B: The correct answer is Choice *B* because multiplying a matrix by a constant is called scalar multiplication. A scalar is a constant number, which means the only thing it changes about a matrix is its magnitude. For a given matrix, $\begin{bmatrix} 3 & 4 \\ 6 & 5 \end{bmatrix}$, scalar multiplication can be applied by multiplying by 2, which yields the matrix $\begin{bmatrix} 6 & 8 \\ 12 & 10 \end{bmatrix}$. Notice that the dimensions of the matrix did not change, just the magnitude of the numbers.

26. A: The vector is described as having both magnitude and direction. The magnitude is the size of the vector and the direction is the path along with which the force is being applied. The second answer choice has magnitude and length, which are essentially the same. The third and fourth answer choices include length and position, but position is not part of the description of a vector.

27. D: Choice *D* is the simplified expression by using the order of operations. The expression $\left[\frac{2vw^3}{-4w} \right]^2$ can be simplified by first distributing the exponent to each term inside the brackets. The expression then becomes $\frac{4v^2w^6}{16w^2}$. By simplifying the fraction at the front and cancelling out the common variables on top and bottom, the expression becomes $\frac{v^2w^4}{4}$.

28. D: The correct matrix to describe the given system of equations is the Choice *D* because it has values that correspond to the coefficients in the right order. The top row corresponds to the coefficient in the first equation, the second row corresponds to the coefficients in the second equation, and the third row corresponds to the coefficients in the third equation. The second matrix (Choice *B*) is filled with the three variables in the system. One thing to also look for is the sign on the numbers, to make sure the signs are correct from the equation to the matrix.

29. B: This is a statistical question because to determine this answer one would need to collect data from each person in the class, and it is expected the answers would vary. The other answers do not require data to be collected from multiple sources, therefore the answers will not vary.

30. B: The volume of a rectangular prism is found by multiplying the length by the width by the height. This formula yields an answer of 144 cubic units. The answer must be in cubic units because volume involves all three dimensions.

31. B: The perimeter is found by adding the length of all the exterior sides. When the given dimensions are added, the perimeter is 22 meters. The equation to find the perimeter can be:

$$P = 5 + 1.5 + 1.2 + 4.5 + 3.8 + 6 = 22$$

The last two dimensions can be found by subtracting 1.2 from 5, and adding 1.5 and 4.5, respectively.

32. C: A hexagon can be formed by any combination of the given shapes except for two rectangles. There are no two rectangles that can make up a hexagon.

33. C: One form for an equation of a straight line is $y = mx + b$, where the m is the slope and the b is the y-intercept. To first solve for slope, the change in y over the change in x, the following equation is used

$$\frac{8-3}{4-0} = \frac{5}{4}$$

When this slope is substituted into the equation and one other point is used, the value for the y-intercept can be found. The equation becomes:

$$8 = \frac{5}{4}(4) + b$$

Solving for b yields a value of 3, so the linear equation becomes:

$$y = \frac{5}{4}x + 3$$

34. C: To solve a compound inequality, the variable must be isolated in the middle. Therefore, subtract 6 from all three parts of the inequality, and then divide all three parts by 4. This results in $-5 < x < 3$. The corresponding interval notation is $(-5, 3)$.

35. B: To rationalize the denominator, the conjugate of the denominator must be multiplied by both the numerator and the denominator of the expression. This will result in a denominator without radicals. The conjugate of the denominator is $x + \sqrt{7}$. Multiplying this by both the numerator and denominator results in:

$$\frac{49(x + \sqrt{7})}{(x - \sqrt{7})(x + \sqrt{7})} = \frac{49(x + \sqrt{7})}{x^2 - 7}$$

36. B: To simplify this inequality, subtract 3 from both sides to get $-\frac{1}{2}x \geq -1$. Then, multiply both sides by -2 (remembering this flips the direction of the inequality) to get $x \leq 2$.

37. D: This problem involves a composition function, where one function is plugged into the other function. In this case, the $f(x)$ function is plugged into the $g(x)$ function for each x-value. The composition equation becomes:

$$g\big(f(x)\big) = 2^3 - 3(2^2) - 2(2) + 6$$

Simplifying the equation gives the answer:

$$g\big(f(x)\big) = 8 - 3(4) - 2(2) + 6$$
$$8 - 12 - 4 + 6 = -2$$

38. B: A factor of 36 is any number that can be divided into 36 and have no remainder. $36 = 36 \times 1, 18 \times 2, 9 \times 4,$ and 6×6. Therefore, it has 7 unique factors: 36, 18, 9, 6, 4, 2, and 1.

39. D: A parabola of the form $y = \frac{1}{4f}x^2$ has a focus $(0, f)$. Because $y = -9x^2$, set $-9 = \frac{1}{4f}$. Solving this equation for f results in $f = -\frac{1}{36}$. Therefore, the coordinates of the focus are $\left(0, -\frac{1}{36}\right)$.

40. B: $12 \times 750 = 9,000$. Therefore, there are 9,000 milliliters of water, which must be converted to liters. 1,000 milliliters equals 1 liter; therefore, 9 liters of water are purchased.

41. B: The volume of a rectangular prism is the $length \times width \times height$, and $3cm \times 5cm \times 11cm$ is 165 cm³. Choice A is not the correct answer because that is $3cm + 5cm + 11cm$. Choice C is not the correct answer because that is 15^2. Choice D is not the correct answer because that is:

$$3cm \times 5cm \times 10cm$$

42. A: The volume of a cylinder is $\pi r^2 h$, and $\pi \times 5^2 \times 10$ is $250\,\pi\ in^3$. Choice B is not the correct answer because that is $5^2 \times 2\pi$. Choice C is not the correct answer since that is $5in \times 10\pi$. Choice D is not the correct answer because that is $10^2 \times 2in$.

43. D: This system of equations involves one quadratic function and one linear function, as seen from the degree of each equation. One way to solve this is through substitution. Solving for y in the second equation yields $y = x + 2$. Plugging this equation in for the y of the quadratic equation yields:

$$x^2 - 2x + x + 2 = 8$$

Simplifying the equation, it becomes $x^2 - x + 2 = 8$. Setting this equal to zero and factoring, it becomes:

$$x^2 - x - 6 = 0 = (x - 3)(x + 2)$$

Solving these two factors for x gives the zeros $x = 3, -2$. To find the y-value for the point, each number can be plugged in to either original equation. Solving each one for y yields the points $(3, 5)$ and $(-2, 0)$.

44. B: The slope will be given by $\frac{1-0}{2-0} = \frac{1}{2}$. The y-intercept will be 0, since it passes through the origin. Using slope-intercept form, the equation for this line is $y = \frac{1}{2}x$.

45. D: Area = length x width. The answer must be in square inches, so all values must be converted to inches. $\frac{1}{2}$ ft is equal to 6 inches. Therefore, the area of the rectangle is equal to:

$$6 \times \frac{11}{2} = \frac{66}{2} = 33 \text{ square inches}$$

46. B: The table shows values that are increasing exponentially. The differences between the inputs are the same, while the differences in the outputs are changing by a factor of 2. The values in the table can be modeled by the equation $f(x) = 2^x$.

Reasoning Through Language Arts

1. C: Alice's behaviors in the passage suggest that she is all three of the characteristics named in Choice C, whereas the other answers include only one or two characteristics of Alice's character. Choice A suggests she is patient, Choice B suggests she is conformist or doting, and Choice D suggests she is thoughtful and boisterous, none of which fit based on the text.

2. D: Choice A is partially correct as we learn Alice is bored in the very first sentences of the passage. However, Choice D is more complete as we learn later in the passage that in addition to being bored, Alice has never seen anything like the rabbit.

3. A: The passage establishes the world Alice is in and the transition, via the well, into the new world as she passes the shelves and other items. Because it is third person narration, we do not see as Alice sees (Choice *B*), nor do we learn much about the white rabbit here, other than his being late (Choice *C*). Though we do get introduced to two main characters, we do not know yet where they live (Choice *D*).

4. C: Choices *A*, *B*, and *D* all include actions Alice does not complete or participate in. Therefore, Choice *C* is the correct choice.

5. B: Because the text reads "for fear of killing someone," it suggests that she is considering others who might be at the end/bottom of the well.

6. B: Though it's possible that the rabbit is normally punctual (Choice *A*), we don't know that from this passage. However, his one line of dialogue suggests that he is worried he will be late. We cannot tell from this passage if he is friendly (Choice *C*), or if he is fast (Choice *D*).

7. C: From Florida to Georgia is north. From Georgia to Texas is west. From Texas to Oklahoma is north. From Oklahoma to California is west.

8. C: Water and corn syrup. The list of ingredients on a food or drink label go from the most common ingredient found to the least common ingredient found. Water and corn syrup are both at the very top of the list. Cornstarch is found in the ingredients, but it is not at the top of the list.

9. C: 160 calories. You would consume 160 calories because one serving of this drink has 80 calories. If you double 80 calories, that gives you 160 calories. The other answer choices are incorrect.

10. B: Bi. The element listed under Group V, Period 6 in the table is Bi. The other elements are close by, but they do not fall under this group and period.

11. D: 62°. Choice *A* is the Celsius temperature, not Fahrenheit. Choices *B* and *C* are incorrect, although Choice *C* is just one degree off.

12. D: Glossary. A glossary is a section in a book that provides brief definitions/explanations for words that the reader may not know. Choice *A* is incorrect because a table of contents shows where each section of the book is located. Choice *B* is incorrect because the introduction is usually a chapter that introduces the book about to be read. Choice *C* is incorrect because an index is usually a list of alphabetical references at the end of a book that a reader can look up to get more information.

13. C: Though the circumstances Oliver is in are challenging, this is not the overwhelming feeling (Choice *D*) nor is he excited (Choice *A*). Nothing in the passage suggests hope (Choice *B*). However, in the second paragraph, Oliver's loneliness is clear (Choice *C*).

14. A: While an alternate meaning could be someone raising money for charity, that the boy is eating in the street, and threatens to beat Oliver up, makes Choice *B* unlikely. We do not know if the boy attends school, so whether he is a schoolboy (Choice *C*) or truant (Choice *D*) is unclear. However, as noted, the boy is eating in the street and speaks "street slang," and Oliver is able to identify him that way, suggesting his appearance matches that of an orphan (Choice *A*).

15. B: A simile (Choice *B*) is a comparison using the words *like* or *as*. In this case, the elm boards are compared to high-shouldered ghosts using the word *like*. Metaphors draw comparisons as well, but do not use *like* or *as* (Choice *A*). Symbolism (Choice *C*) is not a comparison; it uses images or symbols to represent something else. Onomatopoeia (Choice *D*) is when a word sounds like the sound it connotes, like "pop" or "fizz."

16. C: In the third full paragraph, Oliver clearly imagines himself in a coffin, in the churchyard (Choice *C*). Oliver does not have a home (Choice *A*), and though he imagines himself outside, he is not camping (Choice *B*). Finally, his bed literally is under the counter, so he need not imagine this (Choice *D*).

17. D: In Choice *A*, we never meet the undertaker, so none of these things happens. In Choice *B*, Oliver imagines some of these things (sees the coffin parts, imagines the hearse, etc.). In Choice *C*, someone knocking/kicking on the door is the only event that happens in this passage. All of the events in Choice *D* happen.

18. B: Oliver mentions that the sound of the old bell would soothe him to sleep, so the correct answer is Choice *B*. There is no mention of mourners (Choice *A*) or his breathing (Choice *D*). While he mentions the grass, there is no indication of the sound it makes (Choice *C*).

19. D: While there is some hope at the end with the arrival of a possible friend (Choice *A*), greater focus is on the gloomy setting, the feelings of isolation and loneliness, and the absence of love and friends or family (Choice *D*). Because Oliver has been brought to the undertaker and feels stuck, the tone is not empowered (Choice *C*), and the gravity of the situation does not lend itself to lighthearted (Choice *B*).

20. B: Quite a bit suggests things are going to change (Choice *B*) including: this boy doesn't seem to answer to anyone, has access to food, is free to wander on his own, and has confidence (and a knife), none of which Oliver has. Further, his arrival, which is startling (a kick at the door), is an entrance that demands attention, so this in itself foreshadows something big is coming. The boy mocks the fact that Oliver is working, so it's unlikely that he will take his job (Choice *A*). If the boy were the undertaker's son, he likely wouldn't need to knock and may, in fact, already work for the undertaker (Choice *C*). That the boy is eating the bread and butter, dexterously, suggests he's also eating it quickly and not likely to share with anyone (Choice *D*).

21. A: The point of view of the narrator of the passage can best be described as third-person omniscient. The narrator refers to the characters in the story by third-person pronouns: *he, it,* and *they.* The narrator also comes across as "all-knowing" (omniscient) by relating the information and feelings about all the characters (instead of just those of a single character). Second-person point of view would incorporate the second-person pronoun, *you.* First-person point of view would incorporate first-person pronouns: *I* and *we.* Finally, the third-person objective point of view would also refer to the characters in the story by third-person pronouns *he, it,* and *they.* However, the narrator would stay detached, only telling the story and not expressing what the characters feel and think.

22. B: The chronological order of these four events in the passage is as follows:

1. The servants joyfully cheered for the family.
2. Mrs. Dashwood discussed improvements to the cottage.
3. Elinor hung her drawings up in the sitting room.
4. Sir John Middleton stopped by for a visit.

23. C: In the course of this passage, the Dashwoods' attitude shifts. The narrator initially describes the family's melancholy disposition at the start of their journey from Norland. However, the narrator then describes their spirits beginning to lift and becoming more upbeat as they make their way through the scenic English countryside to their new home at Barton Cottage. Although the remaining answer choices may contain some partial truths (appreciation for the family's former privileged life and fear about the family's expedition), they do not accurately depict the narrator's theme throughout the entire passage. For example, the family did not express disdain for their new landlord or doubt their future ability to survive.

24. D: Reality versus expectations is a theme of this passage. Although Mrs. Dashwood expects her new financial circumstances to be trying, and the family expects to miss their former place of residence for quite some time, the Dashwoods seem to begin to adapt well with the help of Sir John Middleton, their generous new landlord. A theme of love conquers all is typically used in literature when a character overcomes an obstacle due to his or her love for someone. A theme of power of wealth is used in literature to show that money either accomplishes things or is the root of evil. Finally, a theme of wisdom of experience is typically used in literature to show that improved judgment comes with age.

25. A: The statement at the start of paragraph five in this passage signifies the family is going through a transition. The Dashwoods find themselves in a time of great transition as they learn to accept their family's demotion in social standing and their reduced income. The statement was not meant to signify that Mrs. Dashwood needs to obtain meaningful employment, that the family is going through a growth phase, or that the Dashwood children need to be concerned about the future.

26. A: The author would agree that an advantage associated with the division of labor proposed by Adam Smith is that dexterity increases as employees complete repeated tasks. This answer can be inferred from the second paragraph. The remaining three answer choices are known disadvantages related to the division of labor concept.

27. B: The author mentions that the "first great builder of iron bridges was a stone-mason" in paragraph five in order to demonstrate how the third advantage associated with the division of labor proposed by Adam Smith is not entirely true. Even though the author agrees with Adam Smith that the invention of machines helps employees to save time, he does not agree that employees who are intimately involved in a daily routine of work come up with the ideas for those inventions (e.g., "the inventor of the power-loom was a clergyman," not a weaver). The remaining answer choices are not addressed by the author in the passage.

28. C: The passage's main idea is best expressed by the answer choice "advantages associated with the division of labor." The passage does not focus solely on the effect of specialization of labor, which consists of increasing productivity by dividing up larger tasks into smaller tasks to be completed by workers with specialized skills. The passage is not concerned with disproving Adam Smith's economic principles. Finally, macroeconomics deals with the larger economy as a whole and topics such as interest rates and gross domestic product. Thus, macroeconomics is not discussed in this passage.

29. D: In paragraph six, the author lists classification of employees by their abilities as the second most important advantage associated with the division of labor. The author states that this is "the greatest advantage (next to the dexterity of the workmen)," which is not mentioned as an advantage by Adam Smith. The remaining three answer choices are listed in the passage as advantages associated with the division of labor. However, they are not listed by the author as being the second most important advantage.

30. A: The author uses the word "bodily" in the second paragraph to mean tangible. The author is referring to the hands-on acts of dancing, gymnastics, and playing musical instruments. "Animal" as an adjective relates to the natural state or primal nature of something, while "organic" similarly refers to something natural. These words are similar to "bodily" but do not fit the author's meaning in this passage, so Choices *B* and *C* are incorrect. "Spiritual" is an antonym to "bodily," so Choice *D* is incorrect.

31. A: The main focus of the passage is a discussion of the causes, symptoms, and treatments associated with the various types of tenosynovitis. The remaining answer choices are all mentioned throughout the passage; however, they are separate, subordinate topics, not the main idea of the passage.

32. B: Placing the tendon at rest and possibly utilizing a splinting device is mentioned in the passage as a treatment for tenosynovitis crepitans of the ankle. Depositing urate of soda under the endothelial tendon covering is used to treat gouty tenosynovitis. Inducing hyperemia and making small incisions to release pus, as well as incorporating active and passive movements to prevent tendon adhesions, are both mentioned in the passage as treatments for suppurative tenosynovitis.

33. C: The statement is supported by the information found in the third paragraph, since the author states that only a small proportion of individuals who overuse their tendons are afflicted with the condition. This means that something else, such as the toxic factor that is mentioned, must be the reason.

34. D: According to the author, a patient having a sensation of crepitation or friction when the tendon is moved in its sheath would be presenting with a case of tenosynovitis crepitans. Visible nodular swellings are a clinical feature of gouty tenosynovitis. Finally, cellulitis and finger wound infections are both clinical features of suppurative tenosynovitis.

35. A: The authors use first-person point of view in this passage, which can be determined by the use of the first-person pronoun, *we,* in paragraph eight. Second-person point of view would involve the authors utilizing the second-person pronoun, *you.* Third-person limited omniscient view would involve the authors using third-person pronouns (*he, she, it,* or *they*), and the authors would tell the entire story through the eyes of a single character. Finally, third-person objective point of view would also involve the authors using the third-person pronouns: *he, she, it,* or *they.* However, the authors would stay detached and would not refer to themselves with the pronouns "I" or "we."

36. B: Choice *B* correctly identifies a statement that the ethicist would agree with. The ethicist is troubled by the lack of incentives for private companies to regulate artificial intelligence. According to the ethicist's conclusion, artificial intelligence is adopting humanity's worst impulses and violent worldviews, and thereby threatening humanity. Choice *A* is incorrect. The ethicist might agree that artificial intelligence is negatively impacting society, but the Internet is barely mentioned in this argument. Choice *C* is incorrect. Although the ethicist would likely agree with this general sentiment, Choice *C* is too broad. Choice *D* is incorrect. The ethicist doesn't argue for limiting artificial intelligence to simple algorithms.

37. B: Choice *B* correctly identifies the best investment opportunity. The cardinal rule has three requirements—privately held small business, sells tangible goods, and no debt. Elizabeth's store is a privately held small business (standalone and owned by her), it sells tangible goods (board games), and it has no debt. The lack of profitability is irrelevant, acting as a red herring. The cardinal rule doesn't mention it, presumably since the businessman thinks he can increase profitability as long as the business meets those three requirements. Choice *A* is incorrect. Jose's grocery store owes the bank three months' worth of mortgage payments, so it has debt, violating the cardinal rule. Choice *C* is incorrect. The accounting firm violates the cardinal rule, because it does not sell a tangible good. Choice *D* is incorrect. A multinational corporation is not a small business, so it violates the cardinal rule.

38. D: The word *deleterious* can be best interpreted as referring to the word *ruinous.* The first paragraph attempts to explain the process of milk souring, so the "acid" would probably prove "ruinous" to the growth of bacteria and cause souring. Choice *A, amicable,* means friendly, so this does not make sense in context. Choice *B, smoldering,* means to boil or simmer, so this is also incorrect. Choice *C, luminous,* has positive connotations and doesn't make sense in the context of the passage. Luminous means shining or brilliant.

39. B: The author begins by explaining a process or phenomenon, then gives the history of the study of this phenomenon, then ends by presenting the effects of this phenomenon. The author explains the process of souring in the first paragraph by informing the reader that "it is due to the action of certain of the milk bacteria upon the milk sugar which converts it into lactic acid, and this acid gives the sour taste and curdles the milk." In the second paragraph, we see how the phenomenon of milk souring was viewed when it was "first studied," and then we proceed to gain insight into "recent investigations" toward the end of the paragraph. Finally, the passage ends by presenting the effects of the phenomenon of milk souring. We see the milk curdling, becoming bitter, tasting soapy, turning blue, or becoming thread-like. All of the other answer choices are incorrect.

40: A: To inform the reader of the phenomenon, investigation, and consequences of milk souring. Choice *B* is incorrect because the passage states that *Bacillus acidi lactici* is not the only cause of milk souring. Choice *C* is incorrect because, although the author mentions the findings of researchers, the main purpose of the text does not seek to describe their accounts and findings, as we are not even told the names of any of the researchers. Choice *D* is tricky. We do see the author present us with new findings in contrast to the first cases studied by researchers. However, this information is only in the second paragraph, so it is not the primary purpose of the *entire passage*.

41. C: Milk souring is caused mostly by a species of bacteria identical to that of *Bacillus acidi lactici* although there are a variety of other bacteria that cause milk souring as well. Choice *A* is incorrect because it contradicts the assertion that the souring is still caused by a variety of bacteria. Choice *B* is incorrect because the ordinary cause of milk souring *is known* to current researchers. Choice *D* is incorrect because this names mostly the effects of milk souring, not the cause.

42. C: The study of milk souring has improved throughout the years, as we now understand more of what causes milk souring and what happens afterward. None of the choices here are explicitly stated, so we have to rely on our ability to make inferences. Choice *A* is incorrect because there is no indication from the author that milk researchers in the past have been incompetent—only that recent research has done a better job of studying the phenomenon of milk souring. Choice *B* is incorrect because the author refers to dairymen in relation to the effects of milk souring and their "troubles" surrounding milk souring, and does not compare them to milk researchers. Choice *D* is incorrect because we are told in the second paragraph that only certain types of bacteria are able to sour milk. Choice *C* is the best answer choice here because although the author does not directly state that the study of milk souring has improved, we can see this might be true due to the comparison of old studies to newer studies, and the fact that the newer studies are being used as a reference in the passage.

43. A: The chemical change that occurs when a firework explodes. The author tells us that after milk becomes slimy, "it persists in spite of all attempts made to remedy it," which means the milk has gone through a chemical change. It has changed its state from milk to sour milk by changing its odor, color, and material. After a firework explodes, there is nothing one can do to change the substance of a firework back to its original form—the original substance is turned into sound and light. Choice *B* is incorrect because, although the rain overwatered the plant, it's possible that the plant is able to recover from this. Choice *C* is incorrect because although Mercury leaking out may be dangerous, the actual substance itself stays the same and does not alter into something else. Choice *D* is incorrect; this situation is not analogous to the alteration of a substance.

44. D: A paragraph showing the ways bacteria infiltrate milk and ways to avoid this infiltration. Choices *A, B,* and *C* are incorrect because these are already represented in the third, second, and first paragraphs. Choice *D* is the best answer because it follows a sort of problem/solution structure in writing.

45. D: Choice *D* is the best answer choice because the gerunds are all in parallel structure: "Playing baseball, swimming, and dancing." Choice *A* is incorrect because "Play baseball" does not match the parallel structure of the other two gerunds. Choice *B* is incorrect because the answer uses semicolons instead of commas, which is incorrect. Semicolons are used to separate independent clauses. Choice *C* is incorrect because "to swim" and "to dance" eschew parallel structure.

Science

1. A: To determine density of the rocks, Scientist A must measure both the mass and volume of each rock. These two factors are the only things that will determine the density of the rocks, so Choice *A* is correct. The color, shape, or fragility of the rocks do not affect the density of the rocks, so Choices *B, C, and D* would not help test the hypothesis.

2. B: Each organelle of a cell performs a different job to keep the cell alive and functioning. The nucleus, Choice *B,* is the control center of the cell and ensures that all other organelles are operating properly. The endoplasmic reticulum, Choice *A,* is involved in the synthesis of proteins and lipids. Vacuoles, Choice *C,* help with storage and waste disposal. Ribosomes, Choice *D,* help make proteins.

3. Photosynthesis: Plants combine carbon dioxide and water with light energy to produce glucose. This process is known as photosynthesis. Glycolysis, Choice *A,* is the conversion of glucose into smaller molecules that can be used for cellular respiration. Mitosis, Choice *C,* is the process of single cell reproduction. Krebs cycle, Choice *D,* is a cycle that occurs during aerobic respiration in animal cells to produce energy.

4. D: Mitosis is the process of reproduction that occurs when a single cell replicates itself and then splits into two cells. The cells that are produced are identical to each other. During the process of meiosis, two non-identical parent cells combine with each other, their genetic material is mixed up, and four non-identical daughter cells are produced, Choice *A.*

5. B: The offspring dark grey bean plants would be represented by the genotype yy. There are two crosses that produce this genotype of offspring out of four total possibilities; therefore, ½ of the offspring would be dark grey bean plants.

6. When two phenotypes are co-dominant, they are both expressed equally. The offspring plants with a genotype of Yy would produce both dark grey and light grey beans. They would NOT produce a third phenotype that is a mix of the original two phenotypes, which occurs in incomplete dominance.

7. Antibodies: Vaccines are inactive or dead strains of live viruses. Once they are injected into the blood stream, the body produces antibodies against the virus. These antibodies are stored in the brain so that when the body is infected with an active, live strain of the virus, it can quickly produce antibodies against it and prevent the disease affecting the body.

8. C: Speciation is the development of a new species from an existing species. It occurs when a group of individuals from the species becomes isolated due to an environmental or other factor that prevents it from associating with the rest of the original species, Choice *D.* The isolated individuals begin to develop new traits that are better suited for their new environment, including changes to their reproductive traits,

Choices *A* and *B*. They maintain the ability to reproduce, but they are no longer able to produce viable offspring with individuals from the original species. Therefore, Choice *C* is correct.

9. "Produced" should cover the kidney. "Exits" should cover the urethra. The image depicts the renal system, which filters waste out of the blood and produces urine. The kidneys are where the filtration occurs and urine is produced. The urine then moves through the ureters into the bladder, where it is stored temporarily. When the bladder is ready to be emptied, urine moves out of the body through the urethra.

10. A: The anther and filament make up the stamen. These are the male reproductive parts of the flower. The ovary, stigma, and style, Choices *B, C,* and *D,* make up the pistil, which are the female reproductive parts of the flower. Pollen moves from the anther to the stigma of either the same flower or a different flower to produce more flowers.

11. D: Taxonomy is the classification of organisms. Each level of taxonomy for animals is subclassified into a new category. The order is as follows, from broadest to most specific: Domain, Kingdom, Phylum, Class, Order, Family, Genus, and Species.

12. B: A person with type O blood can only accept type O blood from a donor. Therefore, Kevin is the only family member with a blood type that is compatible with Sarah, Choice *B.* A person with type O blood, however, is considered a universal donor and can donate blood to a person with any blood type.

13. C: When a woman with Rh negative blood is pregnant with a baby that has Rh positive blood, the mother must get a transfusion with blood that is Rh positive so that her antibodies do not attack the baby's red blood cells. The blood that is transfused must be Rh positive. Lucy has type AB blood, so can receive any blood type from a donor. Charlotte has type A blood, so can only receive type A blood from a donor. Therefore, they can both receive type A Rh positive blood, and Choice *C* is correct.

14. D: Homeostasis is the state where the body is in equilibrium. When the body sweats, it is trying to cool the body down and bring it back to its regular temperature. Being at regular temperature is part of being in equilibrium, Choice *D.* Movement of the skeletal muscles involves using energy and sending signals to the brain to make the motions, so that is not a state of equilibrium, Choices *A* and *B.* The heart is always pumping blood to the body, so that does not specifically help the body come back to homeostasis, Choice *C.*

15. A: Protozoa are unicellular, Choice *B,* eukaryotes. They are most notable for their ability to move around independently using their cilia, Choice *C.* They also mostly produce asexually by binary fission, Choice *D.*

16. C: During mitosis, two identical daughter cells are produced from one parent cell. Metaphase is the phase during which the chromosomes that have been replicated align at the equator of the cell, Choice *C.* During prophase, the chromosomes pair up and align together, Choice *A.* During telophase, Choice *B,* the cytoplasm begins to split and the cell pinches in the middle to begin forming two separate cells. Anaphase, Choice *D,* is when the sister chromatids pull apart from each other and line up at the poles of the cell.

17. The four steps of the Scientific Method are Observation, Hypothesis Formation, Experimentation, and Conclusion. Observation is when you use your own senses to discover or measure something around yourself. Hypothesis Formation is when you make a prediction about the observation. The hypothesis should be a conclusive sentence about the observation and should be able to be measured.

Experimentation are the steps that are needed to test the hypothesis. It should include step by step directions from setting up the experiment to taking the measurements and recording the data collected. The Conclusion should state whether or not the hypothesis can be accepted or if it should be rejected.

18. B: Gregor Mendel came up with several laws that describe genetics and heredity. The Law of Independent Assortment is the one that states that alleles on different genes are not linked. As alleles are separated and passed on to offspring, they do not affect the selection of other alleles and are not affected by other alleles either. The Law of Dominance, Choice *A*, states that the dominant phenotype will be visible in offspring that have a hybrid genotype. The Law of Segregation, Choice *C*, states that alleles on the same gene separate when gametes are formed but come together again during fertilization. Therefore, each offspring has two alleles on each gene.

19. B: Looking at the graph, the fastest growing cells are represented by the line that has the highest number of cells at day 12. It is marked by the darkest shaded line. Looking at the legend, the darkest shaded line represents Cell Type B.

20. D: Cartilage, ligaments, and tendons are all part of the skeletal system. Veins are part of the circulatory system, so Choice *B* is incorrect. Tendons connect muscles to bones, Choice *D*. Cartilage, Choice *A*, is a flexible connective tissue that covers and protects the ends of long bones. Ligaments, Choice *C*, is a tough connective tissue that connects two different bones.

21. A radioisotope's half-life is the amount of time that it takes for the radioisotope to decay so that only half the starting amount is left. For example, if a radioisotope's half-life is six hours and you have five milligrams to start with, after six hours, you will have 2.5 milligrams of the radioisotope left. This information can be used to date fossils by measuring the amount of radioactive material left in certain rocks and determining how long the rock has been decaying for.

22.

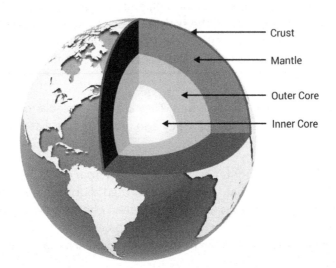

23. C: Seismic waves are energy waves caused by the moving of tectonic plates beneath the Earth's surface that may cause breaks in rocks or explosions. Primary waves, Choice *D*, are the fastest type of seismic wave and reach the surface the fastest. Secondary waves, Choice *A*, travel through the Earth's interior at about half the speed of primary waves. They can only travel through solid materials, such as rock. Surface waves, Choice *B*, only move along the Earth's surface and not through its interior cores. A tsunami is a tidal wave that is often caused by an earthquake but is not a type of seismic wave.

24. A: Renewable energy is a type of energy that can be replenished and is not depleted when it is used. Wind is an example of renewable energy because there is no limit to the quantity of wind that can be found and used to generate energy. Oil, coal, and natural gas, Choices *B*, *C*, and *D*, are all natural energy sources with limited material and are not considered renewable.

25. D: Currents move in certain directions in the ocean due to the temperature of the ocean in the water. Warm water rises and cold water sinks, causing directional movement between warm and cool areas of the earth. When water has to flow into a small passage, it does not affect the overall direction of the oceanic currents, Choice *A*. Rainfall and wind do not affect the direction of the currents because the oceans are such large bodies of water, so Choices *B* and *C* are not correct.

26. Weather in Alaska could be windy, snowing, sunny, or frigid. The climate in Alaska is long, cold winters due to short periods of sunlight during the winter with cool summers with long periods of daylight.

27. B: The Earth's rotation on its axis is what causes people to experience day and night. The area of the Earth facing the sun experiences day while the area of the Earth facing away from the sun experiences night. As the Earth orbits the sun, we experience different seasons, with winter coming when the Earth is farthest from the sun and summer coming when the Earth is closest to the sun. One full orbit equals one year, Choice *A*. The tilt of the Earth does not affect our experience of day or night, Choice *C*. The moon rotates around the Earth but does not affect day or night, Choice *D*.

28. C: The order of the planets in our solar system starting at the sun and moving farther out is Mercury, Venus, Earth, Mars, Jupiter, Saturn, Uranus, and Neptune. Therefore, Saturn is the planet that neighbors Jupiter, Choice *C*.

29. A: Physical changes in matter happen when the physical form of the substance changes but the chemical composition remains the same. When water freezes, it still has the same chemical formula H_2O, but it has changed from liquid to solid form. When a rake rusts, the metal of the rake reacts with oxygen and water in the atmosphere to form rust, Choice *B*. When a yellow liquid and a blue liquid are combined, a chemical reaction is occurring to create the resulting green color, Choice *C*. Burning of fossil fuels combines oxygen with the fuel, Choice *D*, which is a chemical change.

30. Melting point describes the temperature at which a substance moves from solid form to liquid form. If Metal A has a higher melting point, it means that it remains as a solid at higher temperatures than Metal B can. Malleability is the ability of a solid substance to be deformed when pressure is applied. If Metal A has greater malleability, it means that it is more easily deformed. Thermal conductivity is how efficiently heat can be transferred through the substance. If Metal A has lower thermal conductivity than Metal B, that means that the substance is not made up of elements that can easily pass heat.

31. D: Orbitals are where an atom holds its electrons. Protons and neutrons are held in the nucleus of an atom, Choices *A* and *C*. Each orbital holds a different number of electrons. As the orbitals get farther away from the nucleus of an atom, they are larger and able to hold a greater number of electrons.

32. C: The atomic number of an element represents the number of protons in the nucleus of an atom. In the periodic table, each element has its own box and the atomic number is a whole number located in the top left corner of the box. The abbreviation for nickel is Ni and its atomic number is 28.

33. The abbreviation for Potassium is K and the abbreviation for Krypton is Kr. The elements in the periodic table are arranged according to similarities. On the left, the elements are metals and move towards non-metals on the right. Potassium is a metal, whereas Krypton is a non-metal. Krypton is actually a noble gas. They have different atomic masses and different numbers of valence electrons.

34. Exothermic: Combustion reactions occur when oxygen is added to a substance. The result is that the substance burns. In doing so, heat is released from the chemical reaction, which is why combustion reactions feel warm or hot. When heat is released from a reaction, the reaction is called exothermic. Endothermic reactions need heat added to them for them to proceed.

35. B: Acceleration is calculated by dividing the change in velocity by time:

$$a = \frac{(v2 - v1)}{t}$$

Here, v_1 is 20 miles per hour and v_2 is 45 miles per hour. The time is given in minutes and needs to be converted to hours, so 15 minutes is equal to 0.25 hours.

$$a = \frac{45 - 20}{0.25} = 100 \ miles \ per \ hr2$$

36. D: Potential energy is the energy that an object has within itself that could be generated. Kinetic energy is the energy that an object has because of its motion. A person singing, Choice *A*, is generating sound waves that are moving, which has kinetic energy. A marble rolling down a hill and a car that is driving both are in motion and have kinetic energy, Choices *B* and *C*. A charged battery, Choice *D*, has the potential to give an object energy and have electrons flow through it but only has potential energy until it is connected to something.

37. C: Heat can be transferred through conduction, convection, or radiation from one object to another. Velocity is the speed at which something moves, Choice *C*, and is not the method in which heat transfer occurs. Conduction occurs by physical contact of two objects, Choice *A*. Convection occurs when heat is transferred through a fluid, Choice *B*. Radiation occurs when heat is transferred through microwaves, infrared radiation, or visible light, Choice *D*.

Social Studies

1. A: The first line of the excerpt identifies President Abraham Lincoln as the author of the concept. The president is the leader of the executive branch, holding broad constitutional authority related to the enforcement of laws, particularly during wartime. The Emancipation Proclamation was issued as a wartime measure during the Civil War, and it was enforced by the United States military. Thus, Choice *A* is the correct answer. While the judicial branch could review the constitutionality of the Emancipation Proclamation, the law itself isn't related to exercising judicial review. So Choice *B* is incorrect. The legislative branch makes the laws of the United States, and it consists of the House of Representatives and Senate. The Emancipation Proclamation is an executive action with a specific enforcement task, not a part of the typical lawmaking process. Therefore, Choice *C* and Choice *D* are both incorrect.

2. D: In the middle of the excerpt, President Lincoln characterizes the Emancipation Proclamation as a "fit and necessary war measure." He believed freeing the Confederacy's slaves would destabilize its plantation-based economy. In addition, President Lincoln hoped the chance at freedom would incentivize slaves to take up arms against their masters, threatening the Confederacy's security. Lastly, the

Emancipation Proclamation was implemented during the Civil War, and it was carried out by the United States military. Thus, Choice *D* is the correct answer. Rather than a remedy for past injustices, it was a targeted wartime measure since it didn't apply to Union slave owners. Therefore, Choice *A* is incorrect. Emancipation had far more to do with attacking the Confederacy's security and agricultural production than its tax revenue, so Choice *B* is incorrect. While President Lincoln privately viewed slavery as an immoral institution, the Emancipation Proclamation was much more of a pragmatic policy. Therefore, Choice *C* is incorrect.

3. B: Fighting World War II necessitated the largest mobilization of resources and people in American history. Since nearly 16 million served in the United States forces, women were needed to ensure that industrial production continued at a rapid pace. The "Rosie the Riveter" poster was published to encourage women to join the workforce, particularly in traditionally male-dominated industrial sectors. Thus, Choice *B* is the correct answer. The Nineteenth Amendment (1920) granted women the right to vote, and it was ratified more than two decades prior to the United States entering World War II. Therefore, Choice *A* is incorrect. Traditional household roles didn't meaningfully change until the mid-twentieth century, largely due to the greater professional opportunities women enjoyed during World War II. Therefore, Choice *C* is incorrect. Rosie the Riveter was a marketing tactic originally created by the Office for Emergency Management, and it became a feminist symbol and cultural icon after the poster became widely popular. So Choice *D* is incorrect.

4. D: None of the candidates received both the majority of the popular vote and the majority of electoral votes. Samuel J. Tilden received the majority of the popular vote, and Rutherford B. Hayes received the majority of electoral votes. Thus, Choice *D* is the correct answer. Rutherford B. Hayes received the majority of electoral votes, so the Republican candidate won the presidential election. Therefore, Choice *A* is incorrect. Samuel J. Tilden won the popular vote; therefore, Choice *B* is incorrect. Peter Cooper garnered significantly more votes than Green Clay Smith, so Choice *C* is incorrect.

5. A: The Electoral College governs presidential elections, and it requires a candidate to win an absolute majority of electoral votes in order to win a presidential election. If no candidate wins an absolute majority, there is a runoff in the House of Representatives. Rutherford B. Hayes earned an absolute majority of electoral votes, so he won the presidential election of 1876. Thus, Choice *A* is the correct answer. Samuel J. Tilden missed having an absolute majority by a single electoral vote, so Choice *C* is incorrect. Peter Cooper and Green Clay Smith didn't win any electoral votes; therefore, Choice *B* and Choice *D* are both incorrect.

6. B: The statements express a relationship between efficiency, costs, and total production. When combined together, the statements describe how efficiency increases and costs decrease when the total amount of production increases. This is the definition for economies of scale. Thus, Choice *B* is the correct answer. Comparative advantage relates to how some individuals and entities can be far more efficient at specific activities. Since specific activities aren't mentioned in the statement, Choice *A* must be incorrect. The law of demand relates to how the demand for a good decreases as prices increase. The statements discuss the cost of production, not price, so Choice *C* is incorrect. Similarly, Choice *D* is incorrect because the law of supply relates to how the supply of a good increases as its price increases.

7. C: According to these two statements, if a restaurant increases the total production of chicken entrées, then the production costs of chicken entrées would decrease due to the resulting gains in efficiency. This reasoning accurately adheres to both the first and third statements. Thus, Choice *C* is the correct answer. If a widget factory boosts its efficiency, then the production costs per unit would most likely decrease based on the first statement. So Choice *A* is incorrect. Likewise, Choice *B* is incorrect because increased efficiency would most likely result in reduced costs, not stagnating costs. Choice *D* is incorrect because it doesn't state whether experimenting with the new methods will be more efficient. In addition, the statements don't mention the effect of efficiency on total production.

8. D: The map shows the state boundaries for Utah, Arizona, and Nevada. Zion National Park is one of the largest public lands in southern Utah, and it is located directly south of Cedar City. Thus, Choice *D* is the correct answer. Kaibab National Forest is directly north of the Grand Canyon, and it is in northern Arizona. Therefore, Choice *A* is incorrect. Lake Mead National Recreation Area is in southeastern Nevada close to the state border with Arizona, so Choice *B* is incorrect. Choice *C* is incorrect because the Vermillion Cliffs National Monument is located in northern Arizona and forms part of Arizona's border with Utah.

9. C: The Grand Staircase-Escalante National Monument is shown in the northwest section of the map, in Utah. Utah State Route 12 crosses through the Grand Staircase-Escalante National Monument at multiple points. Thus, Choice *C* is the correct answer. U.S. Route 93 is in Nevada and runs through Las Vegas. Therefore, Choice *A* is incorrect. U.S. Route 180 is in southern Arizona and runs through Kaibab National Forest, so Choice *B* is incorrect. Utah State Route 18 is to the west of Grand Staircase-Escalante National Monument by St. George. Therefore, Choice *D* is incorrect.

10. C: The production method depicted in the photograph is an assembly line. The Ford Motor Company introduced assembly lines with moving belts for the purpose of increasing the scale of production. Every worker was responsible for a single task, facilitating the mass manufacturing of automobiles. Since mass manufacturing was highly efficient, it greatly reduced the costs of production. Thus, Choice *C* is the correct answer. One of an assembly line's primary benefits is that it doesn't require skilled labor because workers are repeating the same basic task. Therefore, Choice *A* is incorrect. Mass manufacturing reduced customization due to the use of interchangeable parts; therefore, Choice *B* is incorrect. Ford assembly lines did ultimately lead to improvements in the quality, reliability, and durability of automobiles; however, these improvements aren't a typical benefit of assembly lines. Other innovations heavily contributed to the superior qualities of Ford automobiles, so Choice *D* is incorrect.

11. A: Net imports are defined as the difference between imports and exports, and this figure is represented as a line cutting through the bars on the graph. The approximate net imports in the years provided in the answer choices are as follows: 1979, 1,200 billion cubic feet (Choice *A*); 1995, 2,700 billion cubic feet (Choice *B*); 2003, 3,300 billion cubic feet (Choice *C*); and 2011, 2,000 billion cubic feet (Choice *D*). Of those years, net imports of natural gas were the lowest in 1979. Thus, Choice *A* is the correct answer.

12. D: The map illustrates North American territorial holdings in 1775. The shading is different for the Eastern Seaboard and the Midwest for a variety of reasons. Great Britain originally colonized most of the Eastern Seaboard, while France and Spain seized territory in present-day Canada and the American Midwest. Following its victory in the Seven Years' War (French and Indian War), Great Britain annexed former French Canadian territory and gained control over Spanish Florida. This is why the shading of British territory is expressed differently on the map. Thus, Choice *D* is the correct answer. Spain never controlled all of the American Midwest, so Choice *A* is incorrect. The United States didn't gain its independence from Great Britain until 1783; therefore, Choice *B* is incorrect. The Midwest wasn't a buffer zone because Great Britain claimed much of the area, which directly abutted Spain's territorial claims. So Choice *C* is incorrect.

13. A: Great Britain introduced the Proclamation Line of 1763 in the immediate aftermath of the Seven Years' War (French and Indian War). The Proclamation Line of 1763 prohibited the American colonists from settling west of the Appalachian Mountains, which infuriated the colonists. The American colonists had fought in the French and Indian War under the assumption it would open a new frontier region. Consequently, American colonists viewed the Proclamation Line of 1763 as a betrayal, and colonial leaders organized the patriot movement to advocate for independence. Thus, Choice *A* is the correct answer. Great Britain fulfilled its promises to recognize territorial claims of its American Indian allies, like the Iroquois, through the enactment of the Royal Proclamation of 1763. In effect, the Proclamation Line of 1763 cooled the simmering tensions between the European colonizers and American Indian tribes, which is why the British implemented the Proclamation Line of 1763. So Choice *B* is incorrect. Spain never fully explored, let alone settled, most of the territory it claimed in North America, so Choice *C* is incorrect. France had already ceded its North American territory to Britain by 1775; therefore, Choice *D* is incorrect.

14. C: The Tenth Amendment further defines how the federal government shares power with states and the people. According to the Tenth Amendment, any power not granted to the federal government is reserved for the states and people unless otherwise prohibited in the Constitution. The concept of federalism is defined as the sharing of power between federal and subnational political entities. Thus, Choice *C* is the correct answer. Citizenship refers to the status and rights of people who a country recognizes as its legal members, so Choice *A* is incorrect. Equality typically indicates that people enjoy the same rights and opportunities; therefore, Choice *B* is incorrect. Sovereignty refers to a political entity's supreme authority, particularly in relation to the entity's ability to govern its claimed territory. Although the Tenth Amendment references the sharing of political power, the Amendment is more related to federalism than sovereignty. Therefore, Choice *D* is incorrect.

15. B: The states on the map are shaded based on median household income, and the shading scheme is provided in a legend found on the lower right side of the map. According to the map, Colorado is in the group of states with the highest median household income. Thus, Choice *B* is the correct answer. Arizona and Maine are in the next highest groups of states with a median household income between $50,000 and $59,999. So Choice *A* and Choice *C* are both incorrect. Choice *D* is incorrect because West Virginia is in the group of states with a median household income of less than $45,000, meaning it's one of the poorest states in the country.

16. A: The correct answer has to be a category on which the United Kingdom spent a higher percentage of total expenditures than the United States. According to the table, alcoholic beverages and tobacco products accounted for 4.8% of total expenditures in the United Kingdom and 1.9% of total expenditures in the United States in 2009. Thus, Choice *A* is the correct answer. Automobile purchases accounted for 5.1% of total expenditures in the United Kingdom and 6.1% of total expenditures in the United States in 2009, so Choice *B* is incorrect. Education accounted for 1.8% of total expenditures in the United Kingdom and 2.4% of total expenditures in the United States in 2009, so Choice *C* is incorrect. Transportation accounted for 15.2% of total expenditures in the United Kingdom and 17.5% of total expenditures in the United States in 2009, so Choice *D* is incorrect.

17. D: The correct answer must be a category that increased as a percentage of total expenditures between 2009 and 2014 in the United States, United Kingdom, and Japan. Housing increased as a percentage of total expenditures in the United States (26.3% to 26.6%), the United Kingdom (27.1% to 28.5%), and Japan (22.2% to 23.7%). Thus, Choice *D* is the correct answer. Clothing decreased as a percentage of total expenditures, making it the perfect inverse of the correct answer. So Choice *A* is incorrect. Likewise, the category of culture, entertainment, and recreation also decreased in all three countries; therefore, Choice *B* is incorrect. Healthcare is closer since it increased as a percentage of total expenditures in the United States and the United Kingdom; however, healthcare held steady as a percentage of total expenditures (4.3%) in Japan. Therefore, Choice *C* is incorrect.

18. C: The table lists some of the powers for the three branches of the federal government. Under the Legislative Branch column, it lists the power to regulate interstate commerce, which is derived from Article I, Section 8, of the United States Constitution. If the corporation is distributing products nationwide, then the products must necessarily cross state borders, which is interstate commerce by definition. Thus, Choice *C* is the correct answer. The executive branch's powers mostly involve enforcing laws and military matters, so Choice *A* is incorrect. Likewise, the judicial branch could only review laws and disputes involving interstate commerce, not the commerce itself, so Choice *B* is incorrect. Choice *D* is incorrect because interstate commerce is squarely within the purview of the legislative branch.

19. D: The United States Constitution doesn't grant the federal government any powers related to conducting elections. According to the Tenth Amendment, unmentioned powers are reserved to the states and people. As a result, state and local governments enjoy the power to conduct elections. Thus, Choice *D* is the correct answer. The executive branch is responsible for enforcing federal laws, not state and local laws, so it isn't involved in elections. Therefore, Choice *A* is incorrect. The judicial branch can hear disputes about election laws, but it doesn't pass or enforce laws. Therefore, Choice *B* is incorrect. The legislative branch enjoys the power to pass laws, but it doesn't have the constitutional authority to pass laws related to elections. So Choice *C* is incorrect.

20. B: At the end of World War II, Berlin was divided into zones split between the Soviet Union, the United States, and Western European powers. However, Berlin was located within Soviet-controlled East Germany, so millions of East Germans immigrated to West Berlin or fled the country to escape a stagnant economy and oppressive political regime. In response, the Soviet Union constructed the Berlin Wall to prevent these defections. Additionally, the Soviet Union accused the United States of interfering with Soviet policies in East Berlin. Thus, Choice *B* is the correct answer. The Berlin Wall was constructed decades after the Allied Powers divided Berlin into four zones, so Choice *A* is incorrect. The Berlin Wall symbolized the worsening of diplomatic relations during the Cold War, not the reconstruction of Germany; therefore, Choice *C* is incorrect. While the United States struggled to send supplies to West Berlin during the Berlin Blockade (1948–1949), the Berlin Wall didn't directly interfere with American supply lines. Therefore, Choice *D* is incorrect.

21. A: The Fifth Amendment to the United States Constitution grants defendants legal protection against self-incrimination at trial, and the Supreme Court case of *Miranda v. Arizona* (1966) is about whether people should have the same protection during police questioning. Based on the information contained in the excerpt's second paragraph, self-incrimination occurs when someone provides the police with evidence that exposes them to criminal liability at trial. Thus, Choice *A* is the correct answer. The excerpt doesn't mention an intent requirement and neither does the legal definition of self-incrimination. So Choice *B* is incorrect. Self-incrimination is related to both police questioning and trials, which occur before and after the prosecutor's decision to bring charges, respectively. Therefore, Choice *C* is incorrect. Self-incrimination is when someone doesn't protect themselves from criminal liability; therefore, Choice *D* is incorrect.

22. B: The first sentence of the second paragraph lists four things the police must legally tell people before questioning them; if the police don't say these thing, the evidence gathered cannot be used at trial. First, the person must be warned they have the right to remain silent, so, Choice *D* is incorrect. Second, the person must be warned about the consequences of speaking to police, such as their statements being used against them in a court of law. So, Choice *A* is incorrect. Third, the person must be told that they have a right to an attorney, so Choice *C* is incorrect. Fourth, the person must be told an attorney will be appointed if they cannot afford to hire one. The Supreme Court doesn't mention the consequences of someone waiving their right to counsel. Thus, Choice *B* is the correct answer.

23. A: The graph compares voting rates in presidential election based on race and Hispanic origin. The voting rate for non-Hispanic black voters fell significantly further than any other group, declining from 66.6% in 2012 to 59.6% in 2016. Thus, Choice *A* is the correct answer. The voting rate for Hispanic voters declined from 48% in 2012 to 47.6% in 2016, a difference of 0.4%. So, Choice *B* is incorrect. Similarly the voting rate for non-Hispanic, non-white, and non-black voters only declined by 0.6%. Therefore, Choice *C* is incorrect. The voting rates for non-Hispanic white voters increased from 64.1% in 2012 to 65.3% in 2016; therefore, Choice *D* is incorrect.

24. C: The voting rate for non-Hispanic black voters was 66.6% in 2012. Not only was this the highest voting rate for non-Hispanic black voters on the graph, it was also the only year in which the non-Hispanic black voting rate exceeded the non-Hispanic white voting rate. Thus, Choice *C* is the correct answer. The voting rate for non-Hispanic black voters was 60.3% in 2004, so Choice *A* is incorrect. The voting rate for non-Hispanic black voters was 65.2% in 2008, so Choice *B* is incorrect. The voting rate for non-Hispanic black voters was 59.6% in 2016, so Choice *D* is incorrect.

25. B: The correct answer is Choice *B*. From 2008 to 2009, employment decreased overall, from 113,188,642 to 106,947,104. However, in the healthcare industry, employment increased from 13,228,745 to 13,494,501. Choice *A* is incorrect; employment did not increase overall, and the healthcare industry did not decrease between these years. Choice *C* is incorrect; employment in skilled nursing care facilities increased from 1,616,656 to 1,641,479. Finally, looking at Choice *D*, let's find the percentage employment increased in offices of physicians. Taking the difference between the two years, we have 2,209,350 − 2,235,673 = 26,323. Then, let's divide the increase by the original number: 26,323 ÷ 2,209,350=.0119. Then, multiply the answer by 100: .0119 × 100 = 1.19. There is only a 1.19% increase in employment in this area, so Choice *D* is incorrect.

26. B: All the events in the timeline are related to the Civil War, and they are listed in chronological order. The Battle of Antietam is the fourth event on the timeline, and it occurred on September 17, 1862. The Battle of Fort Sumter is the third event on the timeline, and it occurred on April 12, 1861. So the Battle of Antietam occurred after the Battle of Fort Sumter. Thus, Choice *B* is the correct answer. The Atlanta Campaign is the eighth event on the timeline, and it was launched on May 7, 1864. So Choice *A* is incorrect. The Battle of Gettysburg is the sixth event on the timeline, and the Union forces prevailed on July 3, 1863. Therefore, Choice *C* is incorrect. The Emancipation Proclamation is the fifth event on the timeline, and it was issued five days after the Battle of Antietam. Therefore, Choice *D* is incorrect.

27. C: The correct answer will be the pair of events with the shortest timespan between them. President Lincoln won his second term in office on November 8, 1864, and Robert E. Lee surrendered on April 9, 1865. So slightly more than five months passed between these events, and this is a significantly shorter gap than the time between the events in the other answer choices. Thus, Choice *C* is the correct answer. Nearly twenty-two months passed between the Battle of Antietam and Battle of Gettysburg, so Choice *A* is incorrect. Approximately nineteen months passed between the establishment of the Confederacy and President Lincoln issuing the Emancipation Proclamation. Therefore, Choice *B* is incorrect. There were approximately three and a half years between the presidential election of 1860 and the Atlanta Campaign; therefore, Choice *D* is incorrect.

28. C: The cartoon has two important context clues to help determine which war bears the closest relationship to it. First, the cartoon depicts Congress spraying gasoline on a house fire labeled "Free Speech," and the hose is labeled "Sedition Laws." Second, according to the caption, the cartoon was published in 1920. World War I (1914–1918) featured a heated debate over the limits of free speech because the conflict was very unpopular in the United States. To silence the anti-war movement, Congress passed the Espionage Act of 1917 and the Sedition Act of 1918 to limit free speech. Thus, Choice *C* is the correct answer. The Korean War (1950–1953) occurred three decades after the cartoon's publication, and Congress didn't pass sedition laws before the war. So Choice *A* is incorrect. The Vietnam War (1955–1975) began even later in history than the Korean War, and it also didn't feature sedition laws. Therefore, Choice *B* is incorrect. Similarly, World War II (1939–1945) occurred decades after the cartoon was published, and Congress didn't need to limit free speech to build a consensus to intervene against the Axis Powers. Therefore, Choice *D* is incorrect.

29. D: The excerpt is the Preamble to the United States Constitution. The Framers open the Preamble by listing several goals for the Constitution, including a desire to "form a more perfect Union." The Framers were clearly trying to address the failures of the Articles of Confederation, particularly through the empowerment of a federal government. In this context, a "more perfect Union" is a general reference to the new federalist system of government. Thus, Choice *D* is the correct answer. While the Framers did include a series of complex checks and balances in the Constitution, the Preamble doesn't mention these specific details. Therefore, checks and balances are more accurately characterized as a mere component of the "more perfect Union." So Choice *A* is incorrect. The Constitution doesn't explicitly prohibit secession; therefore, Choice *B* is incorrect. Likewise, the Constitution doesn't require states to cede all claims of sovereignty; instead, it creates a power-sharing arrangement between the central and subnational governments under a federalist system. So, Choice *C* is incorrect.

30. A: The Enlightenment (1715–1789) most influenced the Framers of the Constitution. Amongst other ideas and values, the Enlightenment promoted the importance of liberty, individual civil rights, religious tolerance, and constitutional government. The Preamble of the United States Constitution reflects the Framers' ideological commitments in its references to liberty, justice, and constitutionalism. Thus, Choice *A* is the correct answer. Existentialism addressed the anxiety caused by the modern world, and it became popular in the late nineteenth century and early twentieth century. Therefore, Choice *B* is incorrect. Romanticism was a nineteenth century artistic movement that grew in response to Enlightenment's emphasis on logic and reason by focusing on emotions, so Choice *C* is incorrect. The Renaissance was the dominant cultural movement of the fifteenth and sixteenth centuries. While the Renaissance's humanist philosophers laid the foundation for the Enlightenment, the Framers had a much more direct relationship with the Enlightenment. Therefore, Choice *D* is incorrect.

31. B: The Ganges River can be found in northeast India near India's borders with China, Nepal, and Bangladesh. Kanpur sits directly on the Ganges River, almost halfway between its most northern and southern points. Thus, Choice *B* is the correct answer. Amritsar is located in northern India, so Choice *A* is incorrect. Kolkata is the next closest to the Ganges River, but it's much farther from the Ganges River than Kanpur. Therefore, Choice *C* is incorrect. Srinagar is located to the north of Amritsar; therefore, Choice *D* is incorrect.

32. A: The Himalayan Mountains separate the Indian subcontinent from East Asia, and they function as a natural boundary for the countries of China, India, and Nepal. For example, the Himalayas run the length of the entire China-Nepal border. Thus, Choice *A* is the correct answer. Although Bangladesh shares a border with India, Bangladesh is located south of the Himalayas. So, Choice *B* is incorrect. Nepal and India both have territory that abuts the Himalayas, but the border between these two countries is south of the Himalayas. Therefore, Choice *C* is incorrect. The India-Pakistan border is hundreds of miles away from the Himalayas; therefore, Choice *D* is incorrect.

33. D: Martin Luther King Jr. was one of the most prominent and effective leaders in the Civil Rights Movement, which advocated for racial equality. Along with ending racial segregation and discrimination, the Civil Rights Movement sought to end the disenfranchisement of Black Americans through the repeal of Southern states' Jim Crow laws. As a result, Martin Luther King Jr. pressed the Kennedy and Johnson administrations to enforce Black Americans' voting rights as previously granted under the Fifteenth Amendment. His advocacy eventually succeeded when Congress passed the Voting Rights Act of 1965. Thus, Choice *D* is the correct answer. Martin Luther King Jr. supported the other three pieces of liberal legislation, but those didn't directly involve racial equality, which was his primary aim. The Equal Pay Act of 1963 addressed the gender pay gap and sex discrimination, so Choice *A* is incorrect. The Food Stamp Act of 1964 sought to improve the nutrition of low-income households. Therefore, Choice *B* is incorrect. The Public Works and Economic Development Act of 1965 created the Economic Development Administration to increase employment and stimulate economic growth in low-income communities through government grants. Therefore, Choice *C* is incorrect.

34. B: Martin Luther King Jr. delivered his "I Have a Dream" speech during the March on Washington for Jobs and Freedom, which was held on August 28, 1963. Several hundred thousand protestors descended on Washington to advocate for racial equality, and Martin Luther King Jr. delivered his famous speech in front of the Lincoln Memorial. Thus, Choice *B* is the correct answer. The March for Our Lives was held on March 24, 2018, and it called for federal legislation to end gun violence. Therefore, Choice *A* is incorrect. The Million Man March was modeled after the March on Washington for Jobs and Freedom, and it also took place at the National Mall. However, the Million Man March took place on October 16, 1995. Therefore, Choice *C* is incorrect. Martin Luther King Jr. was an outspoken anti-war activist, and his widow Coretta Scott King helped lead the Moratorium to End the War in Vietnam, which took place on October 16, 1969. However, this protest also occurred after the assassination of Martin Luther King Jr. So Choice *D* is incorrect.

35. D: The map illustrates how the Department of the Interior classifies America's watersheds into twelve distinct regions, and the Rio Grande is located in the Southwestern United States. The Upper Colorado Basin extends southward to the Mexico-United States border, and it contains a portion of the Rio Grande watershed, which can be identified on the map based on its shading and labeling. Thus, Choice *D* is the correct answer. The California-Great Basin is located to the west of the Rio Grande watershed; therefore, Choice *A* is incorrect. The Lower Colorado Basin is the next closest watershed, but the Rio Grande ends near the Lower Colorado Basin's eastern boundary. Therefore, Choice *B* is incorrect. The Missouri Basin is located to the north of the Rio Grande watershed, so Choice *C* is incorrect.

36. C: Canada is located to the north of the United States, so any watershed on the Canada-United States border must be located in the northern United States. The North Atlantic-Appalachian region is located in the Northeastern United States, and it shares both land and maritime borders with Canada. Thus, Choice *C* is the correct answer. The Arkansas-Rio Grande-Texas-Gulf region is in the Southwestern United States, bordering Mexico. Therefore, Choice *A* is incorrect. The Mississippi Basin extends farther north, but its northern boundary borders the Great Lakes region and not Canada. So Choice *B* is incorrect. The South Atlantic Gulf region is located in the Southeastern United States, so Choice *D* is incorrect.

Practice Test #2

Mathematical Reasoning

No Calculator Questions

1. Kevin owns a traveling car wash. He charges a flat rate of $4 per car and then an additional fee of $8 per minivan, $13 per truck, and $5 per sedan. Suppose that a used car lot wants Kevin to come wash all the cars on their lot. The variable M represents the number of minivans, T represents the number of trucks, and S represents the number of sedans. Which equation represents the total cost, C, of the car wash service?

    a. $C = 4(8M + 13T + 5S)$
    b. $C = 12M + 17T + 9S$
    c. $C = 8M + 13T + 15S$
    d. $C = 8(4M + 13T + 5S)$

2. Which point is plotted on the following coordinate plane?

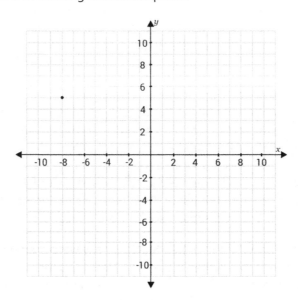

    a. (5, -8)
    b. (8, 5)
    c. (-8, 5)
    d. (-8, -5)

3. Solve the inequality $5(x + 6) - 50 < 4x - 2$.

    a. $x < 18$
    b. $x > 18$
    c. $x < 8$
    d. $x > 28$

4. The following two points are plotted on a number line: -18.2 and 6.3. What is the distance between these two points?

    a. -24.5
    b. 11.9
    c. 24.5
    d. -11.9

5. If $c = -8$ and $d = -2$, what is $|c + d|$?

    a. 6
    b. 8
    c. 10
    d. 12

Calculator Questions

6. If $g(x) = 5x^2 + 2.5x - 8$, what is $g(2.25)$?

    a. 30.9375
    b. 38.9375
    c. 22.9375
    d. 19.6875

7. If the area of a circle is 30.25π, what is the radius of the circle?

    a. 5.5
    b. 5.5π
    c. 15.125
    d. 15.125π

8. In order to get a grade of B in a class, Mark needs to have an 80 average on his 5 tests. He received the following grades on his first 4 tests: 85, 76, 56, and 88. What is the lowest score he would need to receive on the fifth test to get a B in the class?

    a. 90
    b. 95
    c. 85
    d. 80

9. What is the probability of selecting a red marble from a bag that contains 7 red marbles, 6 blue marbles, and 12 green marbles?

    a. 0.7
    b. 0.28
    c. 0.39
    d. 0.07

10. A university has 1275 students and 40% of them are science majors. Therefore, _____ students are science majors.

11. One side of a rectangular room is 7.7 ft long, and the other side is 11.2 ft long. What is the area of the room?
  a. 86.24 square feet
  b. 86.24 feet
  c. 37.8 square feet
  d. 37.8 feet

12. Joseph made $132.75 working 15 hours last week at his part-time job. Therefore, he averaged an hourly wage of $ _____ per hour.

13. The sides of a cube are 4.13 cm long. When you round to the nearest hundredth, the cube has a volume of _____ cubic cm.

14. John drove 9 hours to get to New York City. Over the first 6 hours, he drove 66.5 mph. Then, over the final 3 hours, he drove 55.3 mph. What was his average speed (in mph) over the entire 9-mile trip? Round to the nearest tenth.

15. Penny works as a server at a local restaurant. Last Saturday she worked 7.5 hours and made $65.50 in tips. Her hourly wage is $3.50 per hour. How much did she make in total? Answer in dollars.

16. At a local pet store, 70% of the animals available to adopt are dogs and 30% are cats. Dogs cost $100 to adopt and cats cost $80 to adopt. What is the average cost of all the animals? Answer in dollars.

Calculator Questions

17. What is the length of side x in the following right triangle?

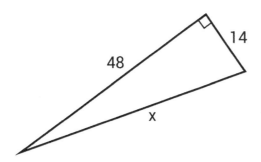

    a. 42
    b. 52
    c. 50
    d. 2500

18. Each time a toy is manufactured, there is a 6% chance that it will be defective. What is the probability that 3 toys in a row will be defective?
    a. 0.000216
    b. 0.0216
    c. 0.06
    d. 0.0036

19. What is the surface area of the following cylinder? Round to the nearest tenth.

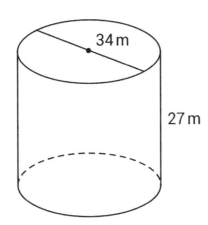

    a. 4699.82 sq. m
    b. 13031.33 m
    c. 4699.82 m
    d. 24513.85 m

20. Katy is raising money to purchase meals for a family in need. Each meal costs $4.75 and there is a flat fee of $15 for delivery. How many meals can she buy if she raises $410? Meals can only be purchased as whole meals.

    a. 82

    b. 83

    c. 84

    d. 86

21. Simplify $(y^8)(y^{13})$

    a. y^{104}

    b. y^{21}

    c. $2y^{21}$

    d. y^5

22. Simplify $\dfrac{z^{93}}{z^{31}}$

    a. z^{62}

    b. z^3

    c. z^{31}

    d. z^{124}

23. At a mall, 45% of the stores are clothing stores. If there are 60 stores in the mall, how many are not clothing stores?

    a. 27

    b. 33

    c. 30

    d. 23

24. Steve rented a delivery truck to deliver his flower arrangements. He returned the truck at 9 a.m. and returned it at 6 p.m. He paid a total of $140.85 for the truck. What was the rental cost per hour?

    a. $140.85

    b. $15.65

    c. $14.09

    d. $17.61

25. Katy works at a shoe store. She makes $15 an hour and works 20 hours each week. Also, she earns a 6% commission on all her sales. If Katy sells $2,400 worth of shoes in a single week, how much are her total earnings for the week?

    a. $300

    b. $144

    c. $444

    d. $2100

26. There are x trees along the running trail in the local park. The number of trees at the zoo is 34 more than 4 times the number of trees along the trail. Which expression represents the number of trees at the zoo?

    a. $34 > 4x$

    b. $34 - 4x$

    c. $34 + 4x$

    d. $4x > 34$

27. What is the value of $6x^3 - 8y^3$ if $x = 1.25$ and $y = 2.1$?
    a. 62.36925
    b. -1988.65725
    c. -4139.757
    d. -62.36925

28. Keisha is planning a company picnic for the end of quarter celebration. She has a budget of $1,275 for renting a pavilion and purchasing catered food. There are 45 people in the office that will attend the picnic, and the cost of renting a pavilion is $310. Which of the following inequalities shows how to find the amount of money, x, that Annie can spend on a catered lunch for each employee?
    a. $45x + 310 \geq 1275$
    b. $45x - 310 \leq 1275$
    c. $45x - 310 \geq 1275$
    d. $45x + 310 \leq 1275$

29. Solve the equation $5(x - 4) + 17 = 5x - 16$.
    a. No solution
    b. $x = 16$
    c. $x = -13$
    d. All real numbers

30. Simplify $(x^5)^{10}$
    a. x^5
    b. x^{15}
    c. x^{50}
    d. x^2

31. What is the probability of selecting a quarter out of a bag that contains 5 quarters and 10 dimes?
    a. $\frac{1}{2}$
    b. $\frac{1}{4}$
    c. $\frac{1}{3}$
    d. $\frac{1}{5}$

32. Which of the following is equivalent to $(5a + 6b)(7a + 2b)$?
    a. $35a + 52ab + 12b$
    b. $12a^2 + 20ab + 8b^2$
    c. $35a^2 + 12b^2$
    d. $35a^2 + 52ab + 12b^2$

33. In the summer, Kyle earns $12.50 an hour mowing lawns and works 20 hours each week. He was offered a job washing dishes in the winter. In this job, he would earn $10.00 an hour. How many hours would Kyle need to work each week in the winter washing dishes to earn the same amount as he does each week in the summer mowing lawns?
    a. 20
    b. 22
    c. 25
    d. 30

34. If a cube has a volume of 64 cubic inches, what is the length of one of its sides?
    a. 4 inches
    b. 8 inches
    c. 12 inches
    d. 16 inches

35. Convert $\frac{11}{19}$ to a decimal. Round your answer to the nearest hundredth.
    a. 0.57
    b. 0.58
    c. 0.11
    d. 1.72

36. Which of the following is equal to the expression $5(a + b) - 3a + 6b$?
    a. $8a + 11b$
    b. $2a + 7b$
    c. $2a + 6b$
    d. $2a + 11b$

37. Kristin baked some brownies for the bake sale. The total cost of baking the brownies was $11.75. She is going to sell each brownie for $2.75 each. If she sells b brownies, which of the following expressions represents her total profit?
    a. $11.75 - 2.75b$
    b. $2.75b - 11.75$
    c. $9b$
    d. 14.5

38. If a sphere has a radius of 2 inches, what is the surface area of the sphere?
    a. 4π cubic inches
    b. 8π cubic inches
    c. 16π cubic inches
    d. 8π square inches

39. Solve the following equation: $x^2 + 6x + 8 = 0$.
    a. $x = 2, 4$
    b. $x = 8, 1$
    c. $x = -8, -1$
    d. $x = -2, -4$

40. Find and simplify the product of $3x + 1$ and $2x - 4$.

    a. $6x^2 - 10x - 4$

    b. $6x^2 - 10x + 4$

    c. $6x^2 + 10x - 14$

    d. $5x^2 - 10x - 4$

41. Simplify $(5x^3 - 2x^2 + 19x - 11) + (3x^3 - 7x^2 + 20x - 12)$.

    a. $8x^3 + 9x^2 + 39x + 23$

    b. $8x^3 - 9x^2 + x - 23$

    c. $8x^3 - 9x^2 + 39x - 23$

    d. $2x^3 - 9x^2 + 3x - 23$

42. Find the equation of a line that passes through the points $(-2, 2)$ and $(4, 8)$.

    a. $y = x + 4$

    b. $y = x + 8$

    c. $y = 2x + 2$

    d. $y = 2x + 4$

43. Keith received a shipment of 4 boxes; each box contains 38 pounds of organic ground beef. Keith paid a total of $1,482 for the ground beef. What was the price per pound of ground beef?

    a. $39

    b. $11.75

    c. $9.75

    d. $370.50

44. A taxi driver charges a $5 flat fee upon pick-up, then $2.25 per mile or fraction of a mile. If you only have $23, what is the longest possible ride you can take?

    a. 8 miles

    b. 10 miles

    c. 10.2 miles

    d. 12.2 miles

45. Which of the following values would make $\frac{x^2+1}{x-4}$ undefined?

    a. $x = -4$

    b. $x = 1$

    c. $x = -1$

    d. $x = 4$

46. What is the probability of rolling a number larger than 4 on a six-sided die?

    a. $\frac{1}{6}$

    b. $\frac{1}{3}$

    c. $\frac{1}{2}$

    d. $\frac{2}{3}$

Reasoning Through Language Arts

Questions 1–7 are based on the following passage:

But more basically, I am in Birmingham because injustice is here. Just as the prophets of the eighth century B.C. left their villages and carried their "thus saith the Lord" far beyond the boundaries of their home towns, and just as the Apostle Paul left his village of Tarsus and carried the gospel of Jesus Christ to the far corners of the Greco-Roman world, so am I compelled to carry the gospel of freedom beyond my own home town. Like Paul, I must constantly respond to the Macedonian call for aid.

Moreover, I am cognizant of the interrelatedness of all communities and states. I cannot sit idly by in Atlanta and not be concerned about what happens in Birmingham. Injustice anywhere is a threat to justice everywhere. We are caught in an inescapable network of mutuality, tied in a single garment of destiny. Whatever affects one directly, affects all indirectly. Never again can we afford to live with the narrow, provincial "outside agitator" idea. Anyone who lives inside the United States can never be considered an outsider anywhere within its bounds ....

In any nonviolent campaign there are four basic steps: 1) collection of the facts to determine whether injustices are alive; 2) negotiation; 3) self-purification; and 4) direct action. We have gone through all of these steps in Birmingham ... Birmingham is probably the most thoroughly segregated city in the United States. Its ugly record of police brutality is known in every section of the country. Its unjust treatment of Negroes in the courts is a notorious reality. There have been more unsolved bombings of Negro homes and churches in Birmingham than in any city in this nation. These are the hard, brutal, and unbelievable facts. On the basis of these conditions Negro leaders sought to negotiate with the city fathers. But the political leaders consistently refused to engage in good faith negotiation.

Then came the opportunity last September to talk with some of the leaders of the economic community. In these negotiating sessions certain promises were made by the merchants—such as the promise to remove the humiliating racial signs from the stores. On the basis of these promises Reverend Shuttlesworth and the leaders of the Alabama Christian Movement for Human Rights agreed to call a moratorium on any type of demonstrations. As the weeks and months unfolded we realized that we were the victims of a broken promise. The signs remained. As in so many experiences in the past, we were confronted with blasted hopes, and the dark shadow of a deep disappointment settled upon us. So we had no alternative except that of preparing for direct action, whereby we would present our very bodies as a means of laying our case before the conscience of the local and national community. We were not unmindful of the difficulties involved. So we decided to go through the process of self-purification. We started having workshops on nonviolence and repeatedly asked ourselves the questions, "Are you able to accept the blows without retaliating?" "Are you able to endure the ordeals of jail?"

Excerpt from Martin Luther King Jr.'s "Letter from a Birmingham Jail," 1963

1. What is the main idea of this passage?
   a. Birmingham was one of the most segregated cities in the United States.
   b. Injustice anywhere is a threat to justice everywhere.
   c. Direct action is necessary to get results.
   d. One cannot trust civic leaders to negotiate.

2. Which of these is NOT one of the four basic steps to a nonviolent campaign:
    a. Collection of the facts to determine whether injustices are alive
    b. Negotiation
    c. Direct action
    d. Ideological purification

3. In the passage, what does the word "moratorium" mean:
    a. Temporary stoppage
    b. Temporary increase
    c. Continuation
    d. Resumption

4. Which one of these is NOT a reason for King to travel to Birmingham:
    a. It had more unsolved bombings
    b. Its record of police brutality
    c. It was his hometown
    d. Its treatment of Black people in the courts

5. In the sentence: "We are caught in an inescapable network of mutuality, tied in a single garment of destiny," *garment* is a:
    a. Simile
    b. Allusion
    c. Metaphor
    d. Symbolism

6. Based on the passage, what left King and other Black leaders no choice except to act directly?
    a. Bombings continued
    b. Humiliating racial signs remained in windows
    c. Merchants refused to meet with them
    d. There was no time to gather facts and information

7. That King repeatedly brings up religion is MOST likely related to which of the following?
    a. He knew that religion would be a strong ethical appeal to individuals reading the text.
    b. He was a reverend and would therefore use scripture.
    c. He worked with the Alabama Christian Movement and knew they would appreciate his inclusion of scripture.
    d. He was writing to religious leaders who would respond best to the inclusion of religious references.

8. Maritza wanted to go to the park to swim with her friends, but when she got home, she realized that nobody was there to take her.

Follow the numbered instructions to transform the sentence above into a new sentence.
    1. Replace the phrase "wanted to go" with "went."
    2. Replace the word "but" with the word "and."
    3. Replace the word "home" with "there."
    4. Replace the phrase "nobody was there" with "they had thrown."
    5. Take out the phrase "to take."
    6. Add the words "a surprise party!" at the end of the sentence.

a. Maritza went to the park to swim with her friends, and when she got home, she realized they had thrown her a surprise party!

b. Maritza wanted to go to the park to swim with her friends, but when she got home, she realized they had thrown her a surprise party!

c. Maritza went to the park to swim with her friends, and when she got there, she realized that they had thrown her a surprise party!

d. Maritza went to the park to swim with her friends, and when she got there, she realized that nobody had thrown her a surprise party!

9. According to the pie chart below, which browser is most used on Wikimedia in October 2011?

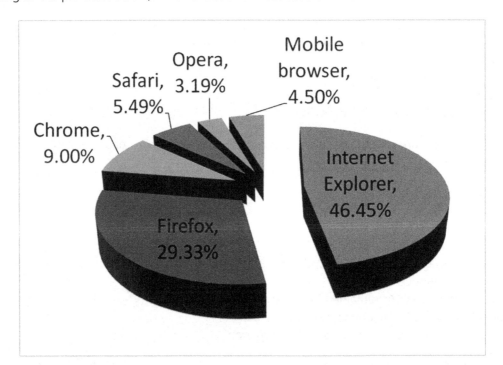

Use the following chart to answer questions 10 and 11

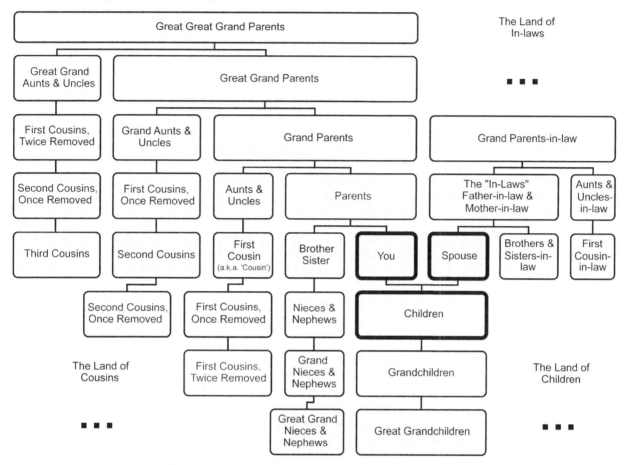

By User Apartment02, Based on a chart by User Hotcactuspepper, 2 January 2008, Wikimedia Commons, https://commons.wikimedia.org/wiki/File:Relatives_Chart.svg

10. What does the chart mean by "Grand nieces and nephews" in relation to "You"?
    a. They are the grandchildren of your brother or sister.
    b. They are the children of your brother and sister.
    c. They are the first cousins of your brother and sister.
    d. They are the second cousins of your brother and sister.

11. What does the chart mean by "First cousin-in-law" in relation to "You"?
    a. They are the parents of your spouse.
    b. They are the aunts and uncles of your spouse.
    c. They are the cousins of your spouse.
    d. They are the children of your spouse.

Questions 12–17 are based on the following passage:

The lover of country life who looks upon rural objects in the true spirit, and, for the first time surveys the cultivated portions of the United States, will be struck with the incongruous appearance and style of our farm houses and their contiguous buildings; and, although, on examination, he will find many, that in their interior accommodation, and perhaps relative arrangement to each other, are tolerably suited to the business and convenience of the husbandman, still, the feeling will prevail that there is an absence of method, congruity, and correct taste in the architectural structure of his buildings generally, by the American farmer.

We may, in truth, be said to have no architecture at all, as exhibited in our agricultural districts, so far as any correct system, or plan is concerned, as the better taste in building, which a few years past has introduced among us, has been chiefly confined to our cities and towns of rapid growth. Even in the comparatively few buildings in the modern style to be seen in our farming districts, from the various requirements of those buildings being partially unknown to the architect and builder, who had their planning—and upon whom, owing to their own inexperience in such matters, their employers have relied—a majority of such dwellings have turned out, if not absolute failures, certainly not what the necessities of the farmer has demanded. Consequently, save in the mere item of outward appearance—and that, not always—the farmer and cottager have gained nothing, owing to the absurdity in style or arrangement, and want of fitness to circumstances adopted for the occasion.

We have stated that our prevailing rural architecture is discordant in appearance; it may be added, that it is also uncouth, out of keeping with correct rules, and, ofttimes offensive to the eye of any lover of rural harmony. Why it is so, no matter, beyond the apology already given—that of an absence of cultivation, and thought upon the subject. It may be asked, of what consequence is it that the farmer or small property-holder should conform to given rules, or mode, in the style and arrangement of his dwelling, or out-buildings, so that they be reasonably convenient, and answer his purposes? For the same reason that he requires symmetry, excellence of form or style, in his horses, his cattle, or other farm stock, household furniture, or personal dress. It is an arrangement of artificial objects, in harmony with natural objects; a cultivation of the sympathies which every rational being should have, more or less, with true taste; that costs little or nothing in the attainment, and, when attained, is a source of gratification through life. Every human being is bound, under ordinary circumstances, to leave the world somewhat better, so far as his own acts or exertions are concerned, than he found it, in the exercise of such faculties as have been given him. Such duty, among thinking men, is conceded, so far as the moral world is concerned; and why not in the artificial? So far as the influence for good goes, in all practical use, from the building of a temple, to the knocking together of a pig-stye—a labor of years, or the work of a day—the exercise of a correct taste is important, in a degree.

Excerpt from <u>Rural Architecture</u> by Lewis Falley Allen

12. What is the main idea of this passage?
   a. Farmers need protection from the government in keeping their farmhouses, and they need more grant money to rebuild better farms.
   b. Urban architecture is better than rural architecture because more funds are being poured into the building of cities.
   c. Rural architecture in America is inconsistent in style and construct and is thus offensive to the farmer's need of symmetry and order.
   d. When you compare the style of European farms to American farms, one can see that American farms are superior in their construction.

13. What is the author's tone in the passage?
   a. Passionate
   b. Critical
   c. Neutral
   d. Joyous

14. What is the meaning of the word *incongruous* at the beginning of the third paragraph?
    a. Out of place
    b. Perfectly symmetrical
    c. Having abundance
    d. Having the characteristic of beauty

15. What does the author say about when human beings leave the world?
    a. They should leave it better than when they found it.
    b. They should look back on the architecture and be proud of it.
    c. They are responsible for the debts they leave behind.
    d. Their family will look to the work they have left behind.

16. What does the author imply about America's rural architecture?
    a. It is brilliant and cultivated
    b. It has great potential
    c. It is humble and practical
    d. It is poor and uncivilized

17. What is the meaning of the word *discordant* at the beginning of the third paragraph?
    a. Pleasantness
    b. Harmonizing
    c. Conflicting
    d. Bold

18. **Chapter** is to **book** as
    a. Book is to story.
    b. Fable is to myth.
    c. Paragraph is to essay.
    d. Dialogue is to play.

19. **Dress** is to **garment** as
    a. Diesel is to fuel.
    b. Month is to year.
    c. Suit is to tie.
    d. Clothing is to wardrobe.

Questions 20–24 are based on the following passage.

The world war represents not the triumph, but the birth of democracy. The true ideal of democracy—the rule of a people by the *demos*, or group soul—is a thing unrealized. How then is it possible to consider or discuss an architecture of democracy—the shadow of a shade? It is not possible to do so with any degree of finality, but by an intention of consciousness upon this juxtaposition of ideas—architecture and democracy—signs of the times may yield new meanings, relations may emerge between things apparently unrelated, and the future, always existent in every present moment, may be evoked by that strange magic which resides in the human mind.

Architecture, at its worst as at its best, reflects always a true image of the thing that produced it; a building is revealing even though it is false, just as the face of a liar tells the thing his words endeavor to conceal. This being so, let us make such architecture as is ours declare to us our true estate.

The architecture of the United States, from the period of the Civil War, up to the beginning of the present crisis, everywhere reflects a struggle to be free of a vicious and depraved form of feudalism, grown strong under the very ægis of democracy. The qualities that made feudalism endeared and enduring; qualities written in beauty on the cathedral cities of mediaeval Europe—faith, worship, loyalty, magnanimity—were either vanished or banished from this pseudo-democratic, aridly scientific feudalism, leaving an inheritance of strife and tyranny—a strife grown mean, a tyranny grown prudent, but full of sinister power the weight of which we have by no means ceased to feel.

Power, strangely mingled with timidity; ingenuity, frequently misdirected; ugliness, the result of a false ideal of beauty—these in general characterize the architecture of our immediate past; an architecture "without ancestry or hope of posterity," an architecture devoid of coherence or conviction; willing to lie, willing to steal. What impression such a city as Chicago or Pittsburgh might have made upon some denizen of those cathedral-crowned feudal cities of the past we do not know. He would certainly have been amazed at its giant energy, and probably revolted at its grimy dreariness. We are wont to pity the mediaeval man for the dirt he lived in, even while smoke greys our sky and dirt permeates the very air we breathe: we think of castles as grim and cathedrals as dim, but they were beautiful and gay with color compared with the grim, dim canyons of our city streets.

Excerpt from <u>Architecture and Democracy</u> by Claude Bragdon

20. Which of the following does the author NOT consider to be a characteristic of modern architecture?
    a. Power, strangely mingled with timidity
    b. Ugliness as a result of the false ideal of beauty
    c. Giant energy with grimy and dreariness
    d. Cathedral-crowned grim and dim castles

21. By stating that "Architecture, at its worst as at its best, reflects always a true image of the thing that produced it," the author most likely intends to suggest that:
    a. People always create buildings to look like themselves.
    b. Architecture gets more grim, drab, and depressing as the years go by.
    c. Architecture reflects—in shape, color, and form—the attitude of the society which built it.
    d. Modern architecture is a lot like democracy because it is uniform yet made up of more pieces than your traditional architecture.

22. The author refers to "mediaeval man" in the fourth paragraph in order to:
    a. Make the audience look at feudalism with a sense of nostalgia and desire.
    b. Make the audience feel gratitude for modern comforts such as architecture.
    c. Make the audience realize the irony produced from pitying him.
    d. Make the audience look back at feudalism as a time which was dark and dreary.

23. Based on the discussion in paragraph 3, which of the following would be considered architecture from medieval Europe?
    a. Canyon
    b. Castle
    c. Factory
    d. Skyscraper

24. The author's attitude toward modern architecture can best be characterized as:
   a. Narcissistic
   b. Aggrieved
   c. Virtuous
   d. Sarcastic

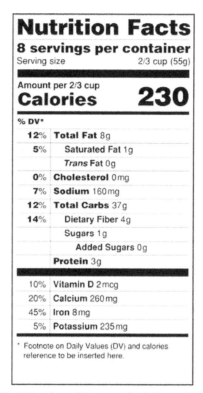

U.S. Food and Drug Administration,
https://www.fda.gov/Food/GuidanceRegulation/GuidanceDocumentsRegulatoryInformation/LabelingNutrition/ucm385663.htm

25. A customer who eats two servings of the above food would consume how many carbohydrates?
   a. 74mg
   b. 74g
   c. 460g
   d. 8g

Questions 26–32 are based on the following passage:

To the Greeks and Romans rhetoric meant the theory of oratory. As a pedagogical mechanism it endeavored to teach students to persuade an audience. The content of rhetoric included all that the ancients had learned to be of value in persuasive public speech. It taught how to work up a case by drawing valid inferences from sound evidence, how to organize this material in the most persuasive order, how to compose in clear and harmonious sentences. Thus to the Greeks and Romans rhetoric was defined by its function of discovering means to persuasion and was taught in the schools as something that every free-born man could and should learn.

In both these respects the ancients felt that poetics, the theory of poetry, was different from rhetoric. As the critical theorists believed that the poets were inspired, they endeavored less to teach men to be poets than to point out the excellences which the poets had attained. Although

275

these critics generally, with the exceptions of Aristotle and Eratosthenes, believed the greatest value of poetry to be in the teaching of morality, no one of them endeavored to define poetry, as they did rhetoric, by its purpose. To Aristotle, and centuries later to Plutarch, the distinguishing mark of poetry was imitation. Not until the renaissance did critics define poetry as an art of imitation endeavoring to inculcate morality . . .

The same essential difference between classical rhetoric and poetics appears in the content of classical poetics. Whereas classical rhetoric deals with speeches which might be delivered to convict or acquit a defendant in the law court, or to secure a certain action by the deliberative assembly, or to adorn an occasion, classical poetic deals with lyric, epic, and drama. It is a commonplace that classical literary critics paid little attention to the lyric. It is less frequently realized that they devoted almost as little space to discussion of metrics. By far the greater bulk of classical treatises on poetics is devoted to characterization and to the technique of plot construction, involving as it does narrative and dramatic unity and movement as distinct from logical unity and movement.

Excerpt from <u>Rhetoric and Poetry in the Renaissance: A Study of Rhetorical Terms in English Renaissance Literary Criticism</u> by DL Clark

26. What does the author say about one way in which the purpose of poetry changed for later philosophers?
   a. The author says that at first, poetry was not defined by its purpose but was valued for its ability to be used to teach morality. Later, some philosophers would define poetry by its ability to instill morality. Finally, during the Renaissance, poetry was believed to be an imitative art, but was not necessarily believed to instill morality in its readers.
   b. The author says that the classical understanding of poetry dealt with its ability to be used to teach morality. Later, philosophers would define poetry by its ability to imitate life. Finally, during the Renaissance, poetry was believed to be an imitative art that instilled morality in its readers.
   c. The author says that at first, poetry was thought to be an imitation of reality, then later philosophers valued poetry more for its ability to instill morality.
   d. The author says that the classical understanding of poetry was that it dealt with the search for truth through its content; later, the purpose of poetry was through its entertainment.

27. What does the author of the passage say about classical literary critics in relation to poetics?
   a. That rhetoric was more valued than poetry because rhetoric had a definitive purpose to persuade an audience, and poetry's wavering purpose made it harder for critics to teach.
   b. That although most poetry was written as lyric, epic, or drama, the critics were most focused on the techniques of lyric and epic and their performance of musicality and structure.
   c. That although most poetry was written as lyric, epic, or drama, the critics were most focused on the techniques of the epic and drama and their performance of structure and character.
   d. That the study of poetics was more pleasurable than the study of rhetoric due to its ability to assuage its audience, and the critics, therefore, focused on what poets did to create that effect.

28. What is the primary purpose of this passage?

a. To contemplate the differences between classical rhetoric and poetry and to consider their purposes in a particular culture.

b. To inform the readers of the changes in poetic critical theory throughout the years and to contrast those changes to the solidity of rhetoric.

c. To educate the audience on rhetoric by explaining the historical implications of using rhetoric in the education system.

d. To convince the audience that poetics is a subset of rhetoric as viewed by the Greek and Roman culture.

29. The word *inculcate* in the second paragraph can be best interpreted as referring to which of the following?

a. Imbibe

b. Instill

c. Implode

d. Inquire

30. Which of the following most closely resembles the way in which the passage is structured?

a. The first paragraph presents us with an issue. The second paragraph offers a solution to the problem. The third paragraph summarizes the first two paragraphs.

b. The first paragraph presents us with definitions and examples of a particular subject. The second paragraph presents a second subject in the same way. The third paragraph offers a contrast of the two subjects.

c. The first paragraph presents an inquiry. The second paragraph explains the details of that inquiry. The last paragraph offers a solution.

d. The first paragraph presents two subjects alongside definitions and examples. The second paragraph presents a comparison of the two subjects. The third paragraph presents a contrast of the two subjects.

31. Given the author's description of the content of rhetoric in the first paragraph, which one of the following is most analogous to what it taught? (The sentence is shown below.)

It taught how to work up a case by drawing valid inferences from sound evidence, how to organize this material in the most persuasive order, how to compose in clear and harmonious sentences.

a. As a musician, they taught me that the end product of the music is everything—what I did to get there was irrelevant, whether it was my ability to read music or the reliance on my intuition to compose.

b. As a detective, they taught me that time meant everything when dealing with a new case, that the simplest explanation is usually the right one, and that documentation is extremely important to credibility.

c. As a writer, they taught me the most important thing about writing was consistently showing up to the page every single day, no matter where my muse was.

d. As a football player, they taught me how to understand the logistics of the game, how my placement on the field affected the rest of the team, and how to run and throw with a mixture of finesse and strength.

32. Which of the following words, if substituted for the word *treatises* in paragraph two, would LEAST change the meaning of the sentence?
    a. Commentary
    b. Encyclopedias
    c. Sermons
    d. Anthems

The following two questions are based on the following excerpt:

> The other of the minor deities at Nemi was Virbius. Legend had it that Virbius was the young Greek hero Hippolytus, chaste and fair, who learned the art of venery from the centaur Chiron, and spent all his days in the greenwood chasing wild beasts with the virgin huntress Artemis (the Greek counterpart of Diana) for his only comrade.

Excerpt from <u>The Golden Bough</u> by Sir James George Frazer

33. Based on a prior knowledge of literature, the reader can infer this passage is taken from which of the following?
    a. A eulogy
    b. A myth
    c. A historical document
    d. A technical document

34. In the blank provided, write down the answer to: What is the meaning of the word "comrade" as the last word in the passage?

The next two questions are based on the following excerpt:

> When I wrote the following passages, or rather the bulk of them, I lived alone, in the woods, a mile from any neighbor, in a house which I had built myself on the shore of Walden Pond, in Concord, Massachusetts, and earned my living by the labor of my hands only. I lived there two years and two months. At present I am a sojourner in civilized life again.

from <u>Walden</u> by Henry David Thoreau

35. The text is most likely to be found in a(n) _____.
    a. Introduction
    b. Appendix
    c. Dedication
    d. Glossary

36. What does the word *sojourner* most likely mean at the end of the passage?
    a. Illegal alien
    b. Temporary resident
    c. Lifetime partner
    d. Farm crop

The next two questions are based on the following passage.

I do not believe there are as many as five examples of deviation from the literalness of the text. Once only, I believe, have I transposed two lines for convenience of translation; the other deviations are (*if* they are such) a substitution of an *and* for a comma in order to make now and then the reading of a line musical. With these exceptions, I have sacrificed *everything* to faithfulness of rendering. My object was to make Pushkin himself, without a prompter, speak to English readers. To make him thus speak in a foreign tongue was indeed to place him at a disadvantage; and music and rhythm and harmony are indeed fine things, but truth is finer still. I wished to present not what Pushkin would have said, or should have said, if he had written in English, but what he does say in Russian. That, stripped from all ornament of his wonderful melody and grace of form, as he is in a translation, he still, even in the hard English tongue, soothes and stirs, is in itself a sign that through the individual soul of Pushkin sings that universal soul whose strains appeal forever to man, in whatever clime, under whatever sky.

Preface for <u>Poems by Alexander Pushkin</u> by Ivan Panin

37. From clues in this passage, what type of work is the author doing?
    a. Translation work
    b. Criticism
    c. Historical validity
    d. Writing a biography

38. According to the author, what is the most important aim of translation work?
    a. To retain the beauty of the work.
    b. To retain the truth of the work.
    c. To retain the melody of the work.
    d. To retain the form of the work.

Questions 39–41 are based on the following passage:

Peach (Amygdalus persica).—In the last chapter I gave two cases of a peach-almond and a double-flowered almond which suddenly produced fruit closely resembling true peaches. I have also given many cases of peach-trees producing buds, which, when developed into branches, have yielded nectarines. We have seen that no less than six named and several unnamed varieties of the peach have thus produced several varieties of nectarine. I have shown that it is highly improbable that all these peach-trees, some of which are old varieties, and have been propagated by the million, are hybrids from the peach and nectarine, and that it is opposed to all analogy to attribute the occasional production of nectarines on peach-trees to the direct action of pollen from some neighbouring nectarine-tree. Several of the cases are highly remarkable, because, firstly, the fruit thus produced has sometimes been in part a nectarine and in part a peach; secondly, because nectarines thus suddenly produced have reproduced themselves by seed; and thirdly, because nectarines are produced from peach-trees from seed as well as from buds. The seed of the nectarine, on the other hand, occasionally produces peaches; and we have seen in one instance that a nectarine-tree yielded peaches by bud-variation. As the peach is certainly the oldest or primary variety, the production of peaches from nectarines, either by seeds or buds, may perhaps be considered as a case of reversion. Certain trees have also been described as indifferently bearing peaches or nectarines, and this may be considered as bud-variation carried to an extreme degree.

39. Which statement is NOT a detail from the passage?
    a. At least six named varieties of the peach have produced several varieties of nectarine.
    b. It is not probable that all of the peach-trees mentioned are hybrids from the peach and nectarine.
    c. An unremarkable case is the fact that nectarines are produced from peach-trees from seed as well as from buds.
    d. The production of peaches from nectarines might be considered a case of reversion.

40. Which of the following most closely reveals the author's tone in this passage?
    a. Enthusiastic
    b. Objective
    c. Critical
    d. Desperate

41. Which of the following is an accurate paraphrasing of the following phrase?

Certain trees have also been described as indifferently bearing peaches or nectarines, and this may be considered as bud-variation carried to an extreme degree.

    a. Some trees are described as bearing peaches and some trees have been described as bearing nectarines, but individually the buds are extreme examples of variation.
    b. One way in which bud-variation is said to be carried to an extreme degree is when specific trees have been shown to casually produce peaches or nectarines.
    c. Certain trees are indifferent to bud-variation, as recently shown in the trees that produce both peaches and nectarines in the same season.
    d. Nectarines and peaches are known to have cross-variation in their buds, which indifferently bears other sorts of fruit to an extreme degree.

The following passage is based on the following four questions:

Quilts, as known and used to-day, may be divided into two general classes, washable and non-washable, depending upon the materials of which they are made. The methods for constructing each class are the same, and are so very simple that it seems hardly necessary to explain them.

The name quilt implies two or more fabrics held together with many stitches. Webster defines a quilt as "Anything that is quilted, especially as a quilted bedcover or a skirt worn by women; any cover or garment made by putting wool, cotton, etc., between two cloths and stitching them together." The verb, to quilt, he defines as "To stitch or to sew together at frequent intervals in order to confine in place the several layers of cloth and wadding of which a garment, comforter, etc., may be made. To stitch or sew in lines or patterns."

The "Encyclopædia Britannica" is a little more explicit and also gives the derivation of the name, quilt, as follows: "Probably a coverlet for a bed consisting of a mass of feathers, down, wool, or other soft substances, surrounded by an outer covering of linen, cloth, or other material." In its earlier days the "quilt" was often made thick and sewed as a form of mattress. The term was also given to a stitched, wadded lining for body armour. "The word came into English from old French *cuilte*. This is derived from Latin *culcitra*, a stuffed mattress or cushion. From the form *culcitra* came old French *cotra*, or *coutre* whence *coutre pointe*, this was corrupted into counterpoint, which in turn was changed to counterpane. The word 'pane' is also from the

Latin *pannus*, a piece of cloth. Thus 'counterpane,' a coverlet for a bed, and 'quilt' are by origin the same word."

Broadly speaking, from these definitions, any article made up with an interlining may be called a quilt. However, usage has restricted the meaning of the word until now it is applied to a single form of bed covering. In the United States the distinction has been carried even farther and a quilt is understood to be a light weight, closely stitched bedcover. When made thicker, and consequently warmer, it is called a "comfort."

Excerpt from <u>Quilts: Their Story and How to Make Them</u> by Marie D. Webster

42. What is the main purpose of this passage?
   a. To persuade the audience to buy a quilt.
   b. To explain the definition and etymology of quilts.
   c. To praise quilters for their hard work and skill.
   d. To tell a story about the history of quilts.

43. What kind of evidence does the author use in the passage?
   a. Illustrations
   b. Statistics
   c. Paraphrase
   d. Direct quotes

44. What does the author say when comparing the two dictionary definitions of the word *quilt*?
   a. The author says that Webster goes off track with its definition and is incorrect.
   b. The author says that Webster gives a more specific definition than the Encyclopedia Britannica.
   c. The author says that the Encyclopedia Britannica gives a more specific definition than Webster.
   d. The author says that the Encyclopedia Britannica goes off track with its definition and is incorrect.

45. What does the author say about comforters, or the "comfort"?
   a. Comforts refer to the cotton between the two blankets being stitched together.
   b. Comforts are when family comes to you when you're in a time of need.
   c. Comforts are blankets that are stitched together that are thicker than quilts.
   d. Comforts are a type of eggnog that people make during the holidays.

Extended Response

Analyze the argument in the two articles. Develop an argument where you explain which argument is better supported and why. Be sure to present specific evidence from both passages to support the argument. Note that you shouldn't talk about the argument you most agree with; rather, you should talk about which argument has better support and why.

Topic: Should we invest in space travel?

Passage 1

Since the Space Race of the 1960s, the United States has spent billions on space exploration. While some argue this money would be better spent elsewhere, our exploration of space and the technologies we've developed through this exploration make our daily lives better.

Additionally, the scientific knowledge we gain about our world and universe is invaluable. First, without space exploration, we'd miss out on life-changing technologies like cellular phones and GPS. Not only does the research change our daily lives, but when we look at concrete, tangible impacts, we also see surgical and agricultural advancements. Similarly, the growth of robotics is directly tied to our space exploration. In fact, robots are used to mine valuable resources that are dwindling on Earth. Finally, the data we gather regarding climate change and our ability to monitor changes informs our decisions to mitigate long-term impact.

In regard to nontangibles, space exploration stokes national interest in science, teaches us about the human body and the aging process, and allows us to foster relationships with the international community.

Despite recent calls to stem NASA funding, the benefits of space exploration may hold both the knowledge and resources we need to not just survive, but to flourish.

Passage 2

NASA's estimated budget for 2020 is $22.6 billion dollars. While this is a fraction of our trillion-dollar budget, it is money we should spend on national needs. From our crumbling infrastructure and educational deficits to issues with clean water and brownfield sites, all could be improved with capital investments.

While opponents boast of the benefits space exploration has given science and technology, and they are not wrong, one of the best aspects is the way that knowledge is shared among our international partners, in part because of our role in developing the International Space Station (ISS). As a result, we could easily diminish our spending on space exploration while still benefiting from the research and technologies developed during space travel and time at the ISS. Finally, because there is so much value in space exploration, we see private industry and investors ready to step in and continue the work our government started. In fact, Elon Musk's Space X has already begun completing missions for NASA like deliveries to the ISS at a fraction of the cost to U.S. taxpayers.

Our national divestment in space exploration doesn't mean the end of the benefits, it simply means attention can be paid to matters closer to home.

Science

1. Which part of the cardiac system is responsible for transporting non-oxygenated blood from the heart to the respiratory system?
   a. Capillaries
   b. Pulmonary vein
   c. Aorta
   d. Pulmonary artery

2. In which organ do nutrients get absorbed into the bloodstream?
   a. Gallbladder
   b. Small intestine
   c. Kidney
   d. Stomach

3. What type of rock is most likely to be seen near a volcano?
    a. Igneous
    b. Clastic
    c. Metamorphic
    d. Sedimentary

4. In his standard day, Tom drives his car to work. He keeps the lights on in all the rooms of his house for 4 hours every day. He looks at social media on his phone for entertainment. He keeps the temperature of his house at 72 °C. Which is NOT a way in which he could reduce his personal energy usage?
    a. Ride a bicycle to work
    b. Turn the lights off in rooms he is not occupying
    c. Increase the temperature of his house by 2 degrees in the winter
    d. Read a book for entertainment

5. **Fill in the blank.**
Label Earth's atmospheric layers in the diagram below.

- Exosphere
- Stratophere
- Thermosphere
- Troposphere
- Mesosphere

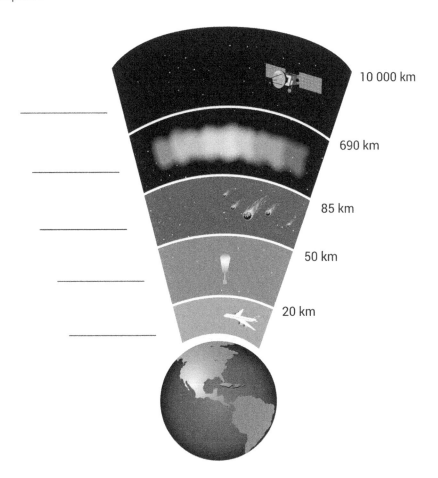

6. Eva observes big puffy clouds in the sky. She listens to the weather forecast and hears a rainstorm is predicted for the next day where she lives. What type of clouds is she seeing in the sky?

    a. Cumulus

    b. Rain

    c. Cirrus

    d. Stratus

7. What happens during a lunar eclipse?

    a. The moon casts a shadow on the sun

    b. The Earth casts a shadow on the moon

    c. The moon, sun, and Earth create a 90-degree angle

    d. The moon is on the opposite side of the sun as the Earth

8. Which artifact of our solar system is a small asteroid that lands on the Earth's surface?

    a. Meteoroid

    b. Comet

    c. Meteor

    d. Meteorite

9. **Fill in the blank.**

Izzy is holding two balls of the same size. One is filled with sand that has a mass of 100 grams. The other is filled with plastic beads that have a mass of 50 grams. The ball filled with _____ is less dense.

- Sand
- Plastic beads

10. Which description best fits matter in the form of liquid?

    a. Particles vibrate back and forth and do not take the shape of their container

    b. Particles are close together and cannot change positions

    c. Particles are close together and can change positions

    d. Particles are moving fast, far apart from each other

11. Looking at the noble gases of the periodic table, what is the atomic mass of helium?

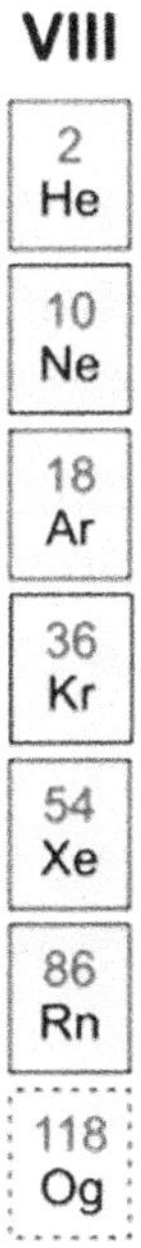

VIII

| 2 |
| He |

| 10 |
| Ne |

| 18 |
| Ar |

| 36 |
| Kr |

| 54 |
| Xe |

| 86 |
| Rn |

| 118 |
| Og |

   a. 39.95
   b. 4
   c. 2
   d. 54

12. **Fill in the blank.**
In a redox reaction such as the combination of magnesium and oxygen to become magnesium oxide, oxygen is the _____ and accepts the electrons.

- Oxidizing agent
- Reducing agent

13. Consider the following reaction for questions 13 and 14: $2Al(OH)_3 + 3H_2SO_4 \rightarrow Al_2(SO_4)_3 + 6H_2O$.

How many oxygen atoms are on the left side of the equation?
   a. 18
   b. 5
   c. 12
   d. 6

14. Yussef has 20 moles of $Al(OH)_3$ and 12 moles of H_2SO. How many moles of $Al_2(SO_4)_3$ can he produce?
   a. 12
   b. 3
   c. 4
   d. 20

15. Mark is adding Substance A to water. He has added 50 grams to 100 mL of hot water. The solution is still clear but is giving off an odor. Which scenario best explains what is happening?
  a. Substance A is precipitating.
  b. The solution is rotting.
  c. The solution has not reached its boiling point.
  d. The solution has not reached its saturation point.

16. **Short answer.**
Explain how momentum is transferred when a golfer hits a golf ball with his club.

17. **Short Answer.**
Using Newton's Law of Force, calculate the force (in Newtons) acting on a 70-kg pound person riding a 5-kg bicycle that is accelerating at a rate of 13 m/s^2. Show your work.

18. Which of the following sound waves has the lowest frequency?
  a. Ultrasound at 106 Hz
  b. Doorbell at 400 Hz
  c. Flute playing a note at 2000 Hz
  d. Viola playing at 130 Hz

Questions 19 and 20 refer to the image below.

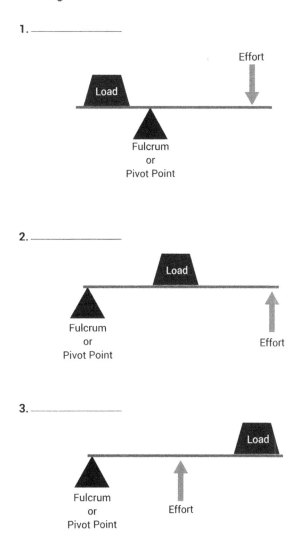

19. **Fill in the blank.**
 Label the three classes of levers in the image above.

- First-class
- Second-class
- Third-class

20. In the top lever image, how could you make the load easier to lift with less effort?
    a. Increase the weight of the load
    b. Move the fulcrum closer to the load
    c. Shorten the length of the bar on top of the fulcrum
    d. Move the fulcrum closer to where the effort is being applied

21. You have a box that weighs 15 pounds. You can choose to move it a longer distance across a lower incline or a shorter distance across a higher incline. Which one would be more work?

    a. It would not take any work to move the box.

    b. Moving up the lower incline

    c. Moving up the higher incline

    d. They would be the same amount of work.

22. **Short answer.**

Describe two of the most limiting factors of the carrying capacity of a population.

Questions 23 and 24 refer to the image below.

Desert Biome Food Web

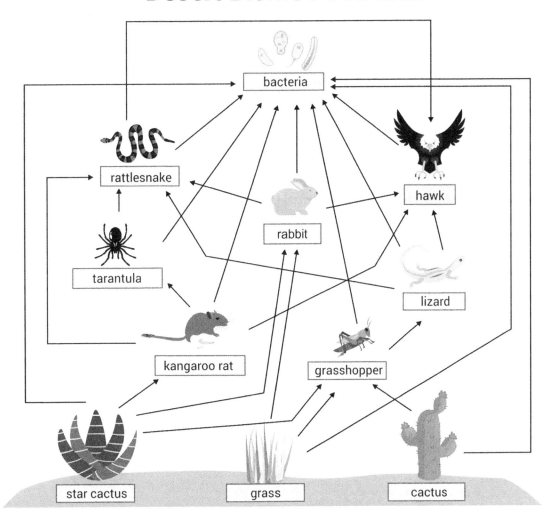

23. Which organism preys on all of the other organisms in the food web?
    a. Bacteria
    b. Rabbit
    c. Cactus
    d. Grasshopper

24. Which organism would suffer if there were no longer cacti in the dessert?
    a. Grass
    b. Kangaroo rat
    c. Grasshopper
    d. Tarantula

25. Which is an example of a species that has a sense that humans do not have?
    a. Crickets chirping to talk to one another
    b. Birds detecting the magnetic field of the Earth to find their way
    c. A dog whining after stepping on something sharp
    d. Deer stopping when they see the headlights of a car

26. **Fill in the blank.**

The _____ of a plant consists of hollow cells that can carry water from the soil into the roots to provide water to the above ground portion of the plant.

- Xylem
- Phloem
- Seed
- Angiosperm

27. Which is a fungus that is eaten as a commercial food product?
    a. Bread
    b. Blue cheese
    c. Tomato
    d. Potato

28. **Fill in the blanks.**

Humans are also known as Homo Sapiens. Which classifications of modern taxonomy are used in this designation?

- Kingdom
- Phylum
- Class
- Order
- Family
- Genus
- Species

Homo Sapiens

_____ _____

29. Which type of parasite lives inside the host's body?
    a. Spider
    b. Ectoparasite
    c. Brood parasite
    d. Endoparasite

30. Which is a true statement about viruses?
    a. They have the ability to perform basic cellular tasks independently.
    b. They can reproduce on their own.
    c. They are host-specific and only attaching to one type of cell.
    d. Antibiotics can kill them.

31. **Short answer.**
Tom is a male and is color-blind. His wife carries a recessive gene for color-blindness. Will 100% of their offspring be color-blind? Explain your answer.

32. Which type of mutation would alter the proteins that were encoded by the gene?
    a. Base substitution
    b. Insertion
    c. Deletion
    d. All of the above

33. Which substance is NOT required for plants to complete photosynthesis?
    a. Glucose
    b. Sunlight
    c. Carbon Dioxide
    d. Water

34. At the end of meiosis, how many chromosomes does each human daughter cell have?
    a. 22
    b. 23
    c. 46
    d. 12

35. Which substrate would this enzyme activate?

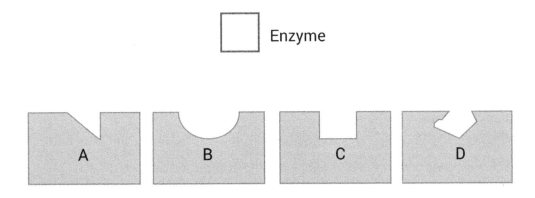

a. A
b. B
c. C
d. D

36. **Fill in the blank.**

The chemical reaction that Frank is running for his chemistry lab is not moving fast enough to detect any changes. He decided to add _____ to the reaction to detect the change faster.

- A catalyst
- More product solutions
- Glucose molecule

37. Which feature is found in plant cells but not animal cells?
   a. Mitochondrion
   b. Nucleus
   c. DNA
   d. Cell wall

Social Studies

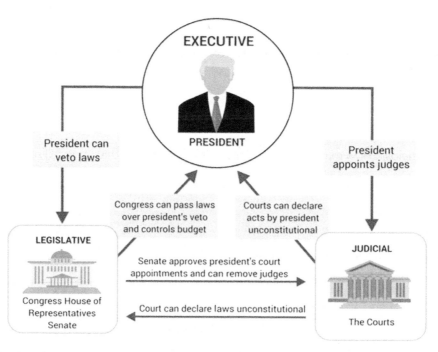

1. Which constitutional concept does the diagram depict?
   a. Checks and balances
   b. Concurrent powers
   c. Constitutional review
   d. Separation of powers

2. Which additional power does the executive branch hold over the judicial branch?
   a. Appropriation
   b. Discretionary power
   c. Judicial review
   d. Pardon power

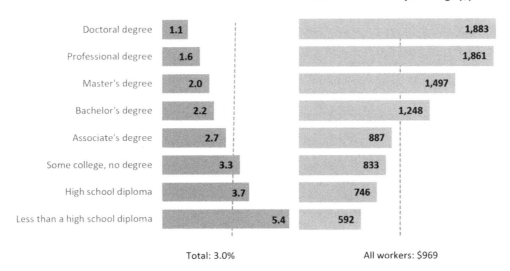

Unemployment rates and earnings by educational attainment, 2019

Unemployment Rates and Earnings Based on Highest Level of Education (2019), U.S. Bureau of Labor

3. Which degree does NOT result in higher median weekly earnings compared to all other workers?

    a. Associate's degree
    b. Bachelor's degree
    c. Doctoral degree
    d. Master's degree

4. What is the unemployment rate for people who attended some college but never earned a college degree?

    a. 2.7%
    b. 3.3%
    c. 3.7%
    d. 5.4%

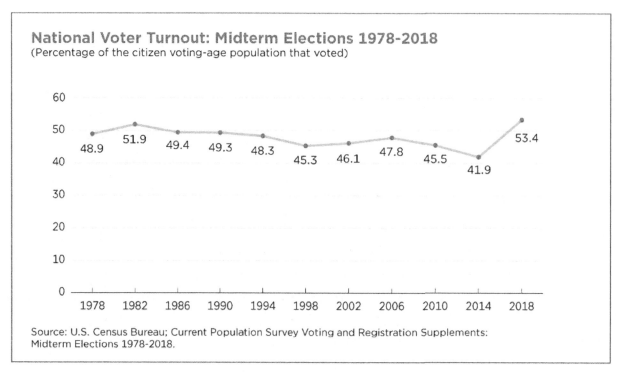

National Voter Turnout: Midterm Elections 1978-2018
(Percentage of the citizen voting-age population that voted)

Source: U.S. Census Bureau; Current Population Survey Voting and Registration Supplements: Midterm Elections 1978-2018.

5. Voter turnout for the midterm elections declined compared to the previous election in which year?
   a. 1982
   b. 1994
   c. 2002
   d. 2006

6. Which statement provides the correct definition of a midterm election?
   a. Midterm elections solely refer to congressional elections that occur every two years.
   b. Midterm elections are heavily influenced by the sitting president's approval rating, and they occur every six years.
   c. Midterm elections occur in the middle of presidential terms and include both congressional and subnational races.
   d. Midterm elections occur when every seat in the House of Representatives is up for election.

The following is the full text of the First Amendment to the United States Constitution:

> Congress shall make no law respecting an establishment of religion, or prohibiting the free exercise thereof; or abridging the freedom of speech, or of the press; or the right of the people peaceably to assemble, and to petition the government for a redress of grievances.

7. Which right is NOT guaranteed under the First Amendment?
   a. Free exercise of religion
   b. Freedom of assembly
   c. Freedom of speech
   d. Right to an attorney

Next!, Joseph Keppler (1904),

8. Which statement best describes how Standard Oil is depicted in the political cartoon?
   a. Standard Oil is depicted as an entity that is undermining democracy and the free market.
   b. Standard Oil is depicted as an entity that primarily poses a direct and immediate threat to the working classes.
   c. Standard Oil is depicted as a multinational corporation that leverages its financial power to gain global influence.
   d. Standard Oil is depicted as a power-hungry monopoly that already controls the White House.

9. How does the political cartoon portray Standard Oil's relationship to representative government in the United States?
   a. Standard Oil is portrayed as the consequence of corrupt government in the United States.
   b. Standard Oil is portrayed as more of a threat to economic elites than to representative government.
   c. Standard Oil is portrayed as a dire threat to free and independent government in the United States.
   d. Standard Oil is portrayed as a corrupt political institution that has unlawfully seized power.

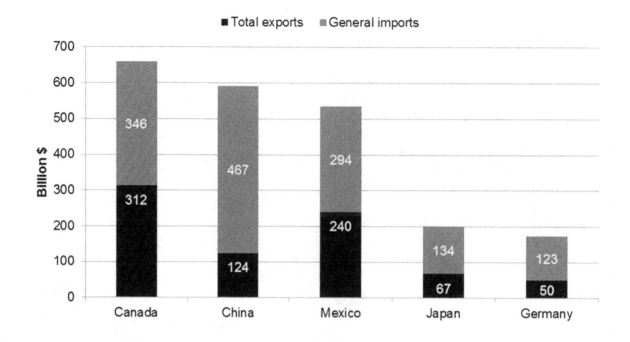

American Economic Relationships with its Five Largest Trading Partners in 2014, Cathy Jabara
United States International Trade Commission (2015),

10. According to the graph, with which country does the United States have the largest total trade deficit?
    a. Canada
    b. China
    c. Japan
    d. Mexico

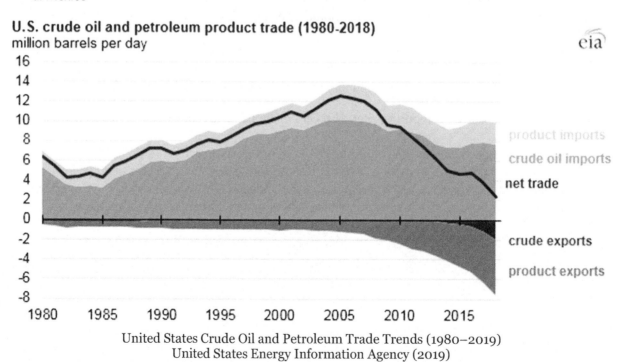

U.S. crude oil and petroleum product trade (1980-2018)
million barrels per day

United States Crude Oil and Petroleum Trade Trends (1980–2019)
United States Energy Information Agency (2019)

11. Which statement best explains why petroleum imports to the United States declined as petroleum exports from the U.S. increased?

    a. The United States suffered from a steep decline in global petroleum production, which devalued its domestic refineries.

    b. The United States exported more petroleum products to stabilize the petrodollar in the global market.

    c. The United States dramatically increased its domestic petroleum production during this time period.

    d. The United States stopped providing subsidies to domestic refineries and imported more petroleum products.

Review the following timeline about the Civil Rights Movement:

- 1948: President Truman issues an executive order to integrate the U.S. military.

- 1954: The Supreme Court finds segregation in public schools to be unconstitutional in *Brown v. Board of Education.*

- 1955: Rosa Parks refuses to relinquish her seat on a bus, triggering the Montgomery bus boycott.

- 1957: President Eisenhower orders federal troops to keep order when an Arkansas public school is integrated.

- 1961: Activists take Freedom Rides across the South to protest segregation.

- 1963: Martin Luther King delivers "I Have a Dream" speech during the March on Washington.

- 1964: President Johnson signs the Civil Rights Act of 1964.

- 1965: President Johnson signs the Civil Rights Act of 1965.

12. Which event occurred before President Eisenhower intervened to integrate an Arkansas public school?
    a. Civil Rights Act of 1965
    b. Freedom rides
    c. "I Have a Dream" speech
    d. Integration of U.S. military

13. When did Supreme Court issue its decision in *Brown v. Board of Education?*
    a. 1957
    b. 1955
    c. 1954
    d. 1948

Buzz Aldrin Walks on Moon, NASA (1969),

14. Which factors most contributed to the event depicted in the photograph?
    a. The United States and the Soviet Union developing ballistic missiles during the nuclear arms race
    b. The United States and the Soviet Union entering a relatively peaceful period of relations known as détente
    c. The United States and the Soviet Union agreeing to join the United Nations Committee on the Peaceful Uses of Outer Space
    d. The United States and the Soviet Union ending discussions to launch a joint mission to the Moon

Across the Continent: "Westward the Course of Empire Takes its Way"
Frances Flora Bond Palmer (1868),

15. Which type of settlement is the town depicted in this painting?
   a. Boomburb
   b. Boomtown
   c. Exurb
   d. Suburb

16. Which statement most accurately describes the relationship between railroads and Western expansion in American history?
   a. Railroads facilitated migration and transportation to the Western frontier, but they undermined the economic development of Western cities.
   b. Railroads connected large cities, which laid the foundation for the development of suburbs.
   c. Railroads efficiently transported goods and people over land, which weakened maritime trading networks.
   d. Railroads functioned as a consistent source of economic growth and displaced American Indians.

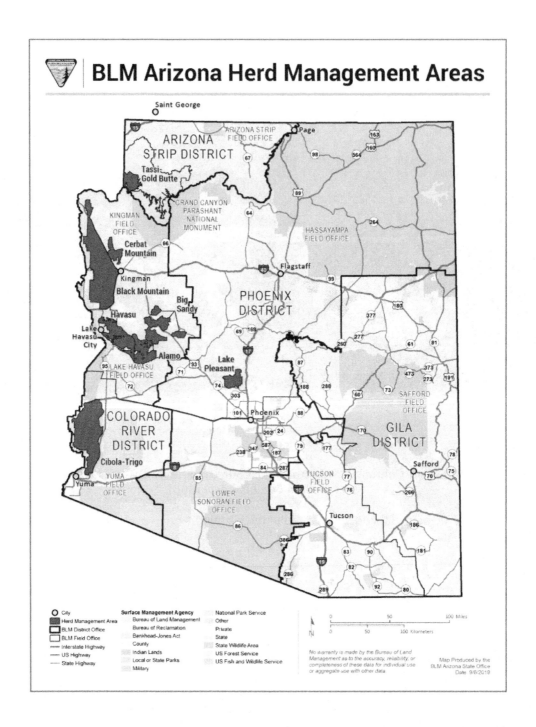

Regulatory Scheme for Arizona's Herd Management Areas, BLM Arizona State Office (2019)

17. Which field office is located the closest to a herd management area?
    a. Hassayampa Field Office
    b. Safford Field Office
    c. Tucson Field Office
    d. Yuma Field Office

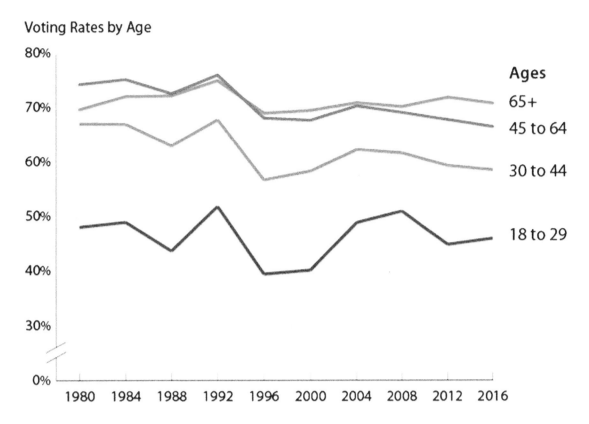

Voting Rates by Age

Ages
65+
45 to 64
30 to 44
18 to 29

Comparison of Voting Rates by Age Group (1980–2016), U.S. Census Bureau (2017)

18. Which age group had increased voter turnout in 2016?
　　a. 65+
　　b. 45 to 64
　　c. 30 to 44
　　d. 18 to 29

19. Which election had two age groups with nearly identical voter turnout?
　　a. 2000
　　b. 2004
　　c. 2008
　　d. 2012

Labor force participation rate of women by age
1948-2016 annual averages

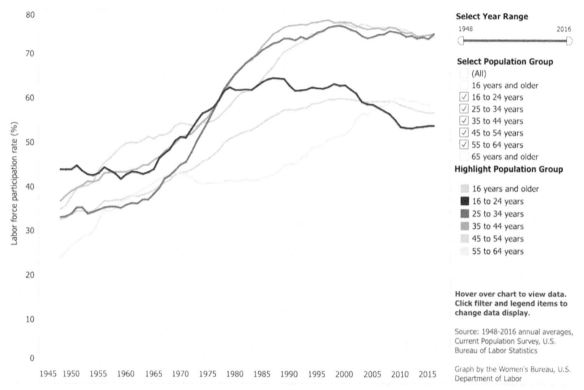

Labor Participation of American Women (1948–2016), U.S. Department of Labor (2018),

20. Which age group experienced the most rapid decline in labor participation between 2000 and 2010?
   a. 16 to 24
   b. 25 to 34
   c. 35 to 44
   d. 45 to 54

21. Which statement best explains why nearly all the age groups experienced a sharp increase in labor participation after 1960?
   a. Women received the right to vote, resulting in greater independence for women.
   b. The economy veered toward collapse, forcing households to obtain a second source of income.
   c. The women's liberation movement normalized contraception and popularized feminism.
   d. Women gained a variety of legal rights, particularly in relation to marital property and divorces.

After the War a Medal and Maybe a Job, John Sloan, Library of Congress (1914)

22. Which public opinion, popular before the United States entered World War I, is depicted in the cartoon?

    a. The United States should transition from capitalism to socialism in order to better protect workers.

    b. The United States should remain neutral because the working class would pay the price of an intervention.

    c. The United States should enter World War I to protect its European allies and defend freedom.

    d. The United States should respect the sacrifices made by the veterans of World War I.

The following is an excerpt from *Executive Order 13653* *(2013):*

> The impacts of climate change—including an increase in prolonged periods of excessively high temperatures, more heavy downpours, an increase in wildfires, more severe droughts, permafrost thawing, ocean acidification, and sea-level rise—are already affecting communities, natural resources, ecosystems, economies, and public health across the Nation. These impacts are often most significant for communities that already face economic or health-related challenges, and for species and habitats that are already facing other pressures....

> The Federal Government must build on recent progress and pursue new strategies to improve the Nation's preparedness and resilience. In doing so, agencies should promote: (1) engaged and strong partnerships and information sharing at all levels of government; (2) risk-informed decision making and the tools to facilitate it; (3) adaptive learning, in which experiences serve as opportunities to inform and adjust future actions; and (4) preparedness planning.

23. Which statement best characterizes how this policy seeks to alter the interactions between government institutions?

    a. Government institutions should establish interagency partnerships to share information and coordinate their response to the crisis.

    b. Government institutions should create intra-agency channels of communications to optimize their preparedness plans.

    c. Government institutions should engage in adaptive learning only in scenarios where they have specialized expertise.

    d. Government institutions should prioritize assistance to communities that are already facing economic and health-related issues.

24. Who held the constitutional authority to implement the policy change described in Executive Order 13653 (2013)?

    a. Chief Justice John Roberts

    b. President Barack Obama

    c. Senate Majority Leader Mitch McConnell

    d. Speaker of the House John Boehner

Terrorist Attack on the Pentagon, photo by Jim Watson, United States Navy, 2001

25. Which non-state actor launched the attack depicted in this photograph?

    a. Al-Qaeda

    b. Boko Haram

    c. ISIS

    d. The Taliban

Table 1. Change in Reported Voting Totals, by Age, Race and Hispanic-Origin: 2012-2016

(In thousands)

Race and Hispanic-Origin	Change in reported voters	Change in citizen voting-age population
All ages		
Total Individuals	4,589	8,978
White non-Hispanic	2,808	1,588
Black non-Hispanic	-765	1,773
Other race non-Hispanic	1,051	2,284
Hispanic	1,494	3,333
18-29 year olds		
Total Individuals	1,081	1,267
White non-Hispanic	695	*-301
Black non-Hispanic	*-211	422
Other race non-Hispanic	196	384
Hispanic	402	761
30-44 year olds		
Total Individuals	*177	982
White non-Hispanic	*142	*-628
Black non-Hispanic	-413	*292
Other race non-Hispanic	*199	533
Hispanic	*251	783
45-64 year olds		
Total Individuals	*-345	*905
White non-Hispanic	-857	-1,314
Black non-Hispanic	-398	*374
Other race non-Hispanic	366	724
Hispanic	544	1,119
65 years and older		
Total Individuals	3,673	5,824
White non-Hispanic	2,828	3,829
Black non-Hispanic	*258	686
Other race non-Hispanic	290	641
Hispanic	298	669

Source: Current Population Survey Voting and Registration Supplements: 2012 and 2016

* Indicates that the 2012 and 2016 estimates were not statistically different from each other.

Comparison of Changes in Reported Voting to Changes in Voting-Age Population from 2012 to 2016

26. Which demographic group experienced the steepest decline in reported voting?
   a. Black non-Hispanic 30-to-44-year-olds
   b. Hispanic 45-to-64-year-olds
   c. Other race non-Hispanic 18-to-29-year-olds
   d. White non-Hispanic 65 years and older

27. Which group of citizens whose age qualifies them to vote experienced the greatest increase in population?

    a. 18-to-29-year-olds

    b. 30-to-44-year-olds

    c. 45-to-64-year-olds

    d. 65 years and older

Farms with Female Principal Operator, by County, 2012

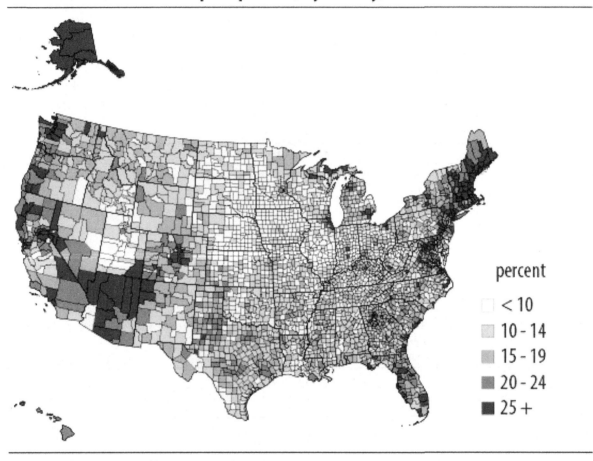

Source: USDA NASS, 2012 Census of Agriculture.

Percent of Farms with a Female Principal Operator (2012), USDA National Agricultural Statistics Service

28. Which region had the highest percentage of farms with a female principal operator?

    a. Arizona

    b. Iowa

    c. Missouri

    d. Utah

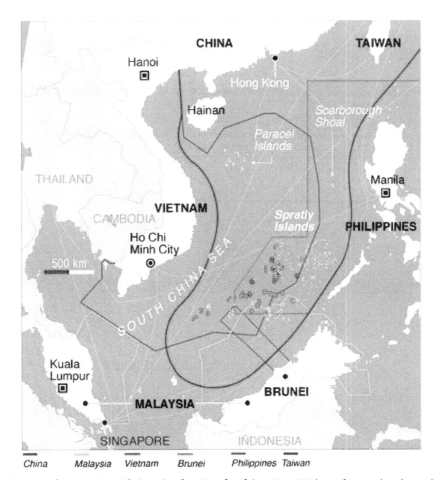

Disputed Maritime Claims in the South China Sea, Voice of America (2012)

29. What is the complete list of all countries that claim the Paracel Islands?
    a. China
    b. China and Taiwan
    c. China, Taiwan, and Vietnam
    d. China, Taiwan, Vietnam, and Philippines

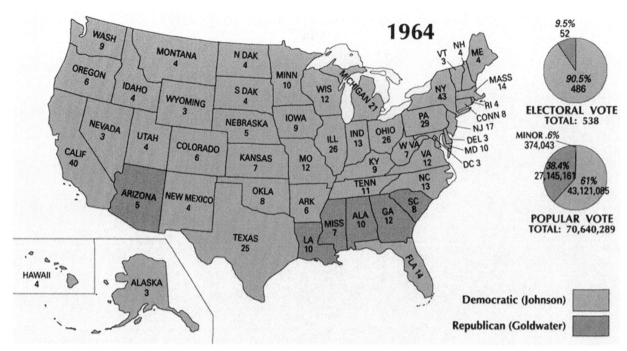

Results for the 1964 Presidential Election

30. Which political institution does the map illustrate?
   a. Election board
   b. Electoral College
   c. Executive branch
   d. Ranked voting system

31. What do the numbers attached to the states represent?
   a. The number scheme expresses the margin of victory in each state for the preceding presidential election.
   b. The number scheme represents how many electors each state sends to the presidential nominating convention.
   c. The number scheme expresses the total population of voting-age citizens living in each state.
   d. The number scheme is based on the total number of congressional seats for each state.

War Poster, Frederic Cooper, U.S. Food Administration, 1917

32. Which statement best explains the government's motivation for publishing this poster?
    a. The government sought to improve citizens' diets as a public health measure.
    b. The government sought to stimulate growth in critical agricultural industries.
    c. The government sought to redirect agricultural resources toward the war effort.
    d. The government sought to increase citizens' savings so they would buy more war bonds.

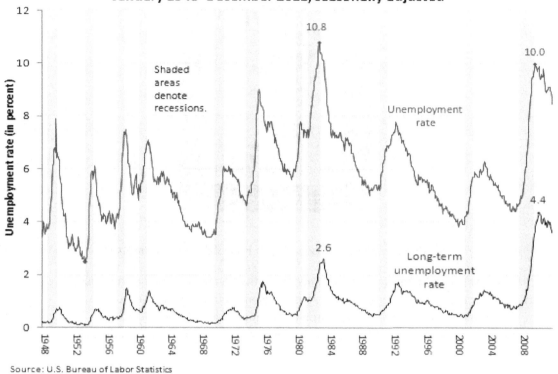

Unemployment rate and long-term unemployment rate, January 1948–December 2011, seasonally adjusted

Shaded areas denote recessions.

Source: U.S. Bureau of Labor Statistics

Comparison of Unemployment Rates During Recessions (1948–2011)
U.S. Bureau of Labor Statistics (2012)

33. The unemployment rate increased the most during which recession?
    a. 1960–1961
    b. 1969–1970
    c. 1981–1982
    d. 2001–2002

34. The long-term unemployment rate peaked in which year?
    a. 1982
    b. 1993
    c. 2001
    d. 2009

The following is an excerpt from <u>Obergefell v. Hodges</u> (2015):

> A first premise of the Court's relevant precedents is that the right to personal choice regarding marriage is inherent in the concept of individual autonomy. This abiding connection between marriage and liberty is why *Loving* invalidated interracial marriage bans under the Due Process Clause....
>
> Like choices concerning contraception, family relationships, procreation, and childrearing, all of which are protected by the Constitution, decisions concerning marriage are among the most intimate that an individual can make. See *Lawrence*, supra, at 574. Indeed, the Court has noted it would be contradictory "to recognize a right of privacy with respect to other matters of family life and not with respect to the decision to enter the relationship that is the foundation of the family in our society."

35. Based on the excerpt, what did the Supreme Court most likely conclude in *Obergefell v. Hodges* (2015)?
 a. The Supreme Court most likely found marriage to be a fundamental constitutional right.
 b. The Supreme Court most likely overturned the *Loving* decision.
 c. The Supreme Court most likely relied on *Lawrence* to halt the expansion of individual rights.
 d. The Supreme Court most likely limited privacy rights under the Due Process Clause.

36. Which statement provides the most likely meaning of "relevant precedents" as used in the excerpt?
 a. Relevant precedents refer to a specific legal framework used to analyze due process claims.
 b. Relevant precedents consist of legal principles and/or rules from earlier cases.
 c. Relevant precedents connect marriage and liberty to limit individual freedom.
 d. Relevant precedents differ from legal precedents because they always contain a single premise.

Answer Explanations #2

Mathematical Reasoning

1. B: There is a flat fee of $4 per car and a specific price per car. These specific prices are $12 per minivan, $17 per truck, and $9 per sedan. Each price is multiplied times the total number of each type of car, respectively, and added together to obtain the total cost.

2. C: The point is 8 units to the left of the y-axis, so the x-coordinate is -8. The point is 5 units above the x-axis, so the y-coordinate is 5. Therefore, the ordered pair is (-8, 5).

3. A: To solve $5(x + 6) - 50 < 4x - 2$, first distribute the 5, resulting in:

$$5x + 30 - 50 < 4x - 2$$

Then combine like terms to obtain:

$$5x - 20 < 4x - 2$$

Adding 20 to both sides results in $5x < 4x + 18$. Finally, subtracting $4x$ from both sides gives the solution $x < 18$.

4. C: The distance between two points on a number line is equal to the absolute value of the difference between the two points. Therefore, the distance is:

$$|-18.2 - 6.3| = |-24.5| = 24.5|$$

5. C: $|c + d| = |-8 + (-2)| = |-10| = 10$.

6. C: $g(2.25) = 5(2.25)^2 + (2.5)(2.25) - 8$. Following the order of operations, exponents are completed first and then multiplication second. Finally, addition and subtraction are completed from left to right. Therefore:

$$g(2.25) = 25.3125 + 5.625 - 8 = 22.9375$$

7. A: The area of a circle is πr^2, where r is the radius of the circle. In this case, the area is 30.25π, which means that $r^2 = 30.25$. Taking the square root of 30.25 results in the radius $r = 5.5$.

8. B: Let x be the score on the fifth test. Therefore, to get a B in the class, $\frac{85+76+56+ 88+x}{5} > 80$. Therefore, $\frac{305+x}{5} > 80$. Multiplying times 5 on both sides results in $305 + x > 400$. Subtracting 305 from each side results in $x > 95$. Therefore, Mark must obtain at least a 95 on the last test to receive a B in the class.

9. B: The probability is equal to the number of red marbles divided by the total number of marbles. The equation for the total number of marbles is $7 + 6 + 12 = 25$. Therefore, the probability of selecting a red marble is $\frac{7}{25} = 0.28$.

10. 510: To find 40% of 1,275, first find the decimal form of 40%. Move the decimal point two places to the left, resulting in 0.4. Then, multiply 0.4 times 1,275, resulting in 510. The correct answer is 510.

11. A: To find the area of a rectangle, multiply the two sides together. The units are square units. Therefore, the area of this room is:

$$7.7(11.2) = 86.24 \text{ square feet}$$

12. 8.85: To find his hourly wage, divide the total amount by the number of hours. Therefore, he made $\frac{\$132.75}{15} = \$8.85$ per hour. The correct answer is 8.85.

13. 70.44: The volume of a cube is equal to s^3, where s is the length of one side. For this cube, the volume is:

$$V = 4.13^3 = 70.444997 \ldots \text{ cubic cm}$$

Rounded to the nearest hundredth gives the answer of 70.44.

14. Over the first 6 hours, John drove 66.5 mph. Therefore, he drove $6(66.5) = 399$ miles in the first 6 hours. Then, over the final 3 hours, he drove 55.3 mph. Therefore, he drove $3(55.3) = 165.9$ miles in the final 3 hours. He drove a total of $399 + 165.9 = 564.9$ total miles. His average speed was $\frac{564.9}{9} = 62.8$ mph, rounded to the nearest tenth. The correct answer is **62.8.**

15. From her hourly wage, Penny made:

$$\$3.50(7.5) = \$26.25$$

Adding $65.50 in tips, she made:

$$\$26.25 + 65.50 = \$91.75 \text{ total}$$

The correct answer is **91.75.**

16. This is a weighted average problem. The average cost of a dog carries a weight of 70%, and the average cost of a cat carries a weight of 30%. The overall average cost is equal to the sum of each weight times the cost of each animal. The dog's contribution to the overall average is:

$$70\% \times 100 = 0.7 \times 100 = \$70$$

The cat's contribution to the overall average is:

$$30\% \times 80 = 0.3 \times 80 = \$24$$

Add these two amounts to obtain the overall average:

$$\$70 + \$24 = \$94$$

The correct answer is **94.**

17. C: Because it is a right triangle, the Pythagorean theorem can be used. The missing side length is the hypotenuse. Therefore, $14^2 + 48^2 = x^2$, or $2500 = x^2$. Take the square root of 2,500 to obtain $x = 50$.

18. A: The probability of the occurrence of consecutive independent events is equal to the product of their probabilities. The probability of each toy being defective is $6\% = 0.06$. Therefore, the probability of 3 toys in a row being defective is:

$$(0.06)(0.06)(0.06) = 0.06^3 = 0.000216$$

19. A: For a radius r and height h, the surface area of a cylinder is $SA = 2\pi rh + 2\pi r^2$. In this cylinder, the diameter is given as 34 m. Therefore, the radius is half of the diameter, which is 17 m. Its height is 27 m. Its surface area is $2\pi rh + 2\pi r^2 = 2\pi(17)(34) + 2\pi(17)^2 = 4699.82$ square meters, rounded to the nearest tenth.

20. B: If Katy raises $410, after the $15 delivery charge, she will have $410 - \$15 = \$395$ left. If each meal is $4.75, dividing $395 by $4.75 results in $\frac{395}{4.75} \approx 83.16$. Because they are sold in full meals only, round 83.16 down to 83. Katy only has enough money for 83 entire meals.

21. B: To combine exponential expressions through multiplication with the same base, add the exponents and keep the same base. Therefore, the answer is:

$$y^{13+8} = y^{21}$$

22. A: To divide exponential expressions with the same base, subtract the exponents and keep the same base. Therefore, the answer is:

$$z^{93-31} = z^{62}$$

23. B: Of the 60 stores, 45 are clothing stores. Therefore, $0.45 \times 60 = 27,$ so 27 stores are clothing stores. The rest are not clothing stores: $60 - 27 = 33$. Therefore, 33 stores are clothing stores.

24. B: To find the hourly rate of the van, divide the total cost by the number of hours Steve rented it. The total number of hours was 9 (9 a.m. to 6 p.m.). Therefore, the hourly rate of the van was:

$$\frac{\$140.85}{9} = \$15.65$$

25. C: From Katy's hourly wage, she earned $15(20) = \$300$. From commission, she earned 6% of $2,400, or $0.06(2400) = \$144$. Therefore, she earned $300 + \$144 = \$444$ total that week.

26. C: The phrase "more than" represents addition in mathematics. Four times the number of trees along the trail is equal to $4x$. Therefore, the expression representing the number of trees at the zoo is $34 + 4x$.

27. D: To evaluate the expression, substitute 1.25 in for x and 2.1 in for y. Therefore, $6(1.25)^3 - 8(2.1)^3$ needs to be evaluated. Following order of operations, the exponents need to be evaluated first. Therefore, the expression is equal to:

$$6(1.953125) - 8(9.261)$$

Next, evaluate the multiplication, obtaining:

$$11.71875 - 74.088 = -62.36925$$

28. D: If the cost of the lunch is x, then, because there will be 45 employees attending the lunch, the cost of catering will be $45x$. The sum of this amount and $310, which is the cost of the pavilion, has to be less than or equal to the budgeted amount of $1,275. Therefore, the correct inequality is:

$$45x + 310 < 1275$$

29. A: To solve the equation, first distribute the 5 to remove the parentheses:

$$5x - 20 + 17 = 5x - 16$$

Collecting like terms results in:

$$5x - 3 = 5x - 16$$

Subtracting $5x$ from both sides results in an untrue equation: $-3 = -16$. Therefore, this equation is never true. Hence, there is no solution.

30. C: An exponential expression raised to another power is simplified by multiplying the exponents. Therefore:

$$(x^5)^{10} = x^{50}$$

31. C: The probability is equal to the number of quarters divided by the total number of coins in the bag. The total number of coins in the bag is represented by the equation $5 + 10 = 15$. When one coin is selected, the probability of selecting a quarter is $\frac{5}{15} = \frac{1}{3}$.

32. D: To find the equivalent expression, multiply out the product through the use of the FOIL technique.

$$(5a + 6b)(7a + 2b) = (5a)(7a) + (5a)(2b) + (6b)(7a) + (6b)(2b)$$

$$35a^2 + 10ab + 42ab + 12b^2$$

$$35a^2 + 52ab + 12b^2$$

33. C: In the summer, Kyle makes $12.50(20) = $250 each week mowing lawns. Therefore, he would need to work $250 \div 10 = 25$ each week in the winter washing dishes to make $250 each week.

34. A: The volume of a cube with side s is equal to $V = s^3$. In this scenario, $64 = s^3$. A number multiplied times itself three times that equals 64 is 4. The cube root of 64 is 4. Therefore, each side in this cube is 4 inches long.

35. B: To convert a fraction to a decimal, divide the numerator by the denominator. In this case, $\frac{11}{19} \approx$ 0.57894737 ... Rounding to the nearest tenth means rounding to two place values to the right of the decimal. Therefore, the fraction is equal to 0.58.

36. D: First, clear the fraction by using the distributive law. Therefore:

$$5(a + b) - 3a + 6b = 5a + 5b - 3a + 6b$$

Combining like terms results in $2a + 11b$.

37. B: Kristin's total profit is equal to her revenue minus her cost. Her revenue is $2.75b$, which is the price per brownie times the total number of brownies sold. Her cost is 11.75. Therefore, her revenue is $2.75b - 11.75$.

38. C: The formula for the surface area of a sphere with radius r is $SA = 4\pi r^2$. In this example, $r = 2$. Therefore:

$$SA = 4\pi(2^2) = 16\pi$$

The units are cubic inches.

39. D: In order to solve a quadratic equation, factor the quadratic expression and set each factor equal to zero. Then, solve each individual equation.

$$x^2 + 6x + 8 = (x + 4)(x + 2) = 0$$

Therefore, $x + 4 = 0$ and $x + 2 = 0$. The solutions are $x = -4, -2$.

40. A: The product can be written as:

$$(3x + 1)(2x - 4)$$

In order to simplify this expression, the FOIL method is used.

$$(3x + 1)(2x - 4) = (3x)(2x) - (3x)(4) + (1)(2x) - (1)(4)$$

$$6x^2 - 12x + 2x - 4$$

$$6x^2 - 10x - 4$$

41. C: In order to add two polynomials, drop the parentheses and combine like terms. Like terms have the same variable raised to the same exponent.

$$(5x^3 - 2x^2 + 19x - 11) + (3x^3 - 7x^2 + 20x - 12)$$

$$(5 + 3)x^3 + (-2 - 7)x^2 + (19 + 20)x + (-11 - 12)$$

$$8x^3 - 9x^2 + 39x - 23$$

42. A: To find the equation of a line through two points, first find the slope. The slope is:

$$m = \frac{y_2 - y_1}{x_2 - x_1} = \frac{8 - 2}{4 - (-2)} = \frac{6}{6} = 1$$

Then, find the y-intercept. Using the slope intercept form of a line, $y = mx + b$, for this line $y = x + b$. Plugging in one of the points results in $8 = 4 + b$. Therefore, $b = 4$. The equation of this line is:

$$y = x + 4$$

43. C: To find the price per pound, you need to know how many pounds of ground beef Keith purchased. There were 4 boxes, and each box contained 38 pounds of ground beef. Therefore, he purchased $4(38) = 152$ pounds of ground beef. Dividing the total cost by the number of pounds results in:

$$\$1482 \div 152 = \$9.75 \text{ per pound}$$

44. A: If you have $23, after the flat fee of $5, you have $18 left. Dividing this amount by the cost per mile of $2.25, you can ride $18 ÷ $2.25 = 8 miles in the taxi.

45. D: The expression would be undefined if the denominator were equal to zero. Setting $x - 4 = 0$ results in a solution of $x = 4$.

46. B: There are two dice rolls that are larger than 4 on a six-sided die: 5 and 6. There are 6 possible outcomes on a six-sided die. Therefore, the probability of rolling a 5 or a 6 is $\frac{2}{6} = \frac{1}{3}$.

Reasoning Through Language Arts

1. B: King comes right out and says, "injustice anywhere is a threat to justice everywhere," thereby clarifying his message and explaining why he has traveled to Birmingham. While Choices *A*, *C*, and *D* are discussed in the passage, they are not the main focus. Choice *A* is an example of why King feels the need to travel to Birmingham. Choice *C* is what he argues we must do to respond. And, finally Choice *D* is a problem he runs into specifically when attempting to negotiate rather than act.

2. D: King mentions Choices *A*, *B*, and *C* in the text, but the final step is self-purification, not Choice *D*.

3. A: A moratorium is a temporary stoppage, in this case of the protests, so business leaders could address concerns.

4. C: All but Choice *C* are listed in the passage as specific reasons for King to visit Birmingham and draw attention to the issues there.

5. C: A metaphor is a comparison of two things that are not alike and is not literally possible. In other words, people are not literally in a garment, but their futures are likely woven together like clothing.

6. B: While other problems might have continued in the Birmingham community, the only one that King mentions specifically in the passage is "The signs remained." Therefore, Choice *B* is the correct answer.

7. A: While Choices *B*, *C*, and *D* are all true, King's audience was much wider than those in the religious community and his role was more than that of a religious leader. Instead, he chooses this tactic to appeal to Southerners for whom religion was an important aspect of their lives. He can, in this way, appeal to their values and make connections between those values and his goals.

8. C: Maritza went to the park to swim with her friends, and when she got there, she realized that they had thrown her a surprise party!" Following the directions carefully will result in this sentence. All the other sentences are close, but Choices *A, B,* and *D* leave at least one of the steps out.

9. According to the pie chart, Internet Explorer (I.E.) is the most used browser in October 2011. Following that is Firefox with 23.6% usage, Chrome with 20.6% usage, and Safari with 11.2% usage.

10. A: They are the grandchildren of your brother and sister. Nieces and nephews refer to the children of your brother and sister, and then your grand nieces and nephews are the grandchildren of your brother and sister, and the children of your nieces and nephews.

11. C: They are the cousins of your spouse. The parents of your spouse are called your Mother-in-law or Father-in-law. The aunts and uncles of your spouse are called Aunts and Uncles-in-law. The children of your spouse would be your children or your step-children.

12. C: The main idea of the passage is Choice *C:* Rural architecture in America is inconsistent in style and construct and is thus offensive to the farmer's need of symmetry and order. The first paragraph explains the inconsistency of rural architecture. The second paragraph expands on this point and acknowledges that better architecture exists in the cities. The third paragraph concludes how the farmer's need for symmetry and order is ignored in rural architecture.

13. B: The author's tone in the passage is *critical,* Choice *B.* We see the author criticizing and offering their opinion on the current state of rural architecture. There is a mix of passion and neutrality in the author's tone, but neither one wins out, so Choices *A* and *C* are not the best answers. *Joyous* is also incorrect, as there is more of a critical tone than a tone of joy.

14. A: The meaning of the word *incongruous* means absurd, out of place, or unequal. The author mentions the style of the farmhouses being "incongruent" and remarks that there is an "absence of method" to their construction.

15. A: The author says that human beings should leave the world better than when they found it. This is found in the second paragraph toward the bottom: "Every human being is bound, under ordinary circumstances, to leave the world somewhat better, so far as his own acts or exertions are concerned, than he found it."

16. D: The author implies that America's rural architecture is poor and uncivilized. The author says in the third paragraph that rural architecture is "discordant in appearance" and also "uncouth," which means awkward or uncultivated.

17. C: The word *discordant* means conflicting, or *not* in harmony. The author uses this word to describe the rural architecture, meaning that it is inconsistent in style and clashes with any kind of logical construction.

18. C: This is a part/whole analogy. A chapter a section, or portion, of a book. All of the choices relate to literary topics, but the best option is paragraph is to essay. A paragraph is a section, or building block of an essay in much the same way that a chapter is in a book. The word pairs in Choices *A, B,* and *E* are best described as near synonyms, but not necessarily parts of one another. Choice *D,* dialogue is to play, does include more of a part to the whole relationship, but dialogue is the way the story is conveyed in a play (like sentences in a book). A better matching analogy to chapter is to book would be scene or act is to play, since plays are divided into scenes or larger acts.

19. A: This is a type of category analogy. A dress is a type of garment. Garment is the broad category, and dress is the specific example used. Diesel is a type of fuel, so it holds the same relationship. Month is not a type of year; it's part of the year. A suit is not a type of tie, but it might be worn with a tie. Clothing is not a type of wardrobe; it is stored in a wardrobe. Lastly, a coat isn't a type of winter; it is a garment worn in the winter.

20. D: The author does not consider modern architecture to be "cathedral-crowned grim and dim castles." The author is speaking of feudal architecture when they say this phrase. Choices *A, B,* and *C* are all mentioned in the text as characteristics of modern architecture, especially in the cities of Chicago and Pittsburgh.

21. C: Architecture reflects—in shape, color, and form—the attitude of the society which built it. Choice *A* is too specific and is taken too literal. It is not that the architecture represents the builders; it represents the builder's culture. Choice *B* is also incorrect; we do see the words "at its worst as at its best," but the meaning of this is skewed in Choice *B*. Choice *D* is incorrect; the statement does not suggest this analysis.

22. C: The author refers to the "mediaeval man" in order to make the audience realize the irony produced from pitying him. The author says that we pity him for the dirt he lived in; however, we inhale smoke and dirt from our own skies. Choices *A, B,* and *D* are incorrect, as they just miss the mark of the statement.

23. B: The text talks about castles and cathedrals as examples of architecture from medieval Europe. Factories and skyscrapers are considered pre-modern or modern architecture. Canyons are not buildings, so Choice *A* is incorrect.

24. B: The author's tone is best described as "aggrieved" because it is annoyed, offended, and disgruntled by the "power, strangely mingled with timidity," the "ingenuity, frequently misdirected" aspects of modern architecture. The author calls out its "giant energy" and "grimy and dreariness," along with its "grim, dim canyons of our city streets."

25. B: 74g. Choice *A* has the correct number, but the unit of measurement is in "mg" instead of "g." Choices *C* and *D* are incorrect.

26. B: The author says that the classical understanding of poetry dealt with its ability to be used to teach morality. Later, philosophers would define poetry by its ability to imitate life. Finally, during the Renaissance, poetry was believed to be an imitative art that instilled morality in its readers. The rest of the answer choices are mixed together from this explanation in the passage. Poetry was never mentioned for use in entertainment, which makes Choice *D* incorrect. Choices *A* and *C* are incorrect for mixing up the chronological order.

27. C: This is the best answer choice as portrayed by the third paragraph. Although most poetry was written as lyric, epic, or drama, the critics were most focused on the techniques of the epic and drama and their performance of structure and character. Choice *A* is incorrect because nowhere in the passage does it say rhetoric was more valued than poetry, although it did seem to have a more definitive purpose than poetry. Choice *B* is incorrect; this almost mirrors Choice *A*, but the critics were *not* focused on the lyric, as the passage indicates. Choice *D* is incorrect because the passage does not mention that the study of poetics was more pleasurable than the study of rhetoric.

28. A: The main idea of the passage is to contemplate the differences between classical rhetoric and poetry and to consider their purposes in a particular culture. Choice *B* is incorrect; although changes in poetics throughout the years is mentioned, this is not the main idea of the passage. Choice *C* is incorrect; although this is partly true—that rhetoric within the education system is mentioned—the subject of poetics is left out of this answer choice. Choice *D* is incorrect; the passage makes no mention of poetics being a subset of rhetoric.

29. B: The correct answer choice is Choice *B, instill*. Choice *A, imbibe,* means to drink heavily, so this choice is incorrect. Choice *C, implode,* means to collapse inward, which does not make sense in this context. Choice *D, inquire,* means to investigate. This option is better than the other options, but it is not as accurate as *instill*.

30. B: The first paragraph presents us with definitions and examples of a particular subject. The second paragraph presents a second subject in the same way. The third paragraph offers a contrast of the two subjects. In the passage, we see the first paragraph defining rhetoric and offering examples of how the Greeks and Romans taught this subject. In the second paragraph, we see poetics being defined along with examples of its dynamic definition. In the third paragraph, we see the contrast between rhetoric and poetry characterized through how each of these were studied in a classical context.

31. D: The best answer is Choice *D:* As a football player, they taught me how to understand the logistics of the game, how my placement on the field affected the rest of the team, and how to run and throw with a mixture of finesse and strength. The content of rhetoric in the passage . . . "taught how to work up a case by drawing valid inferences from sound evidence, how to organize this material in the most persuasive order, and how to compose in clear and harmonious sentences. What we have here is three general principles: 1) it taught me how to understand logic and reason (drawing inferences parallels to understanding the logistics of the game), 2) it taught me how to understand structure and organization (organization of material parallels to organization on the field) and 3) it taught me how to make the end product beautiful (how to compose in harmonious sentences parallels to how to run with finesse and strength). Each part parallels by logic, organization, then style.

32. A: Treatises is most closely related to the word *commentary.* Choice B does not make sense because thesauruses and encyclopedias are not written about one single subject. Choice *C* is incorrect; sermons are usually given by religious leaders as advice or teachings. Choice *D* is incorrect; anthems are songs and do not fit within the context of this sentence.

33. B: A myth. Look for the key words that give away the type of passage this is, such as "deities," "Greek hero," "centaur," and the names of demigods like Artemis. A eulogy is typically a speech given at a funeral, making Choice *A* incorrect. Choices *C* and *D* are incorrect, as "virgin huntresses" and "centaurs" are typically not found in historical or professional documents.

34. Friend. Based on the context of the passage, we can see that Hippolytus was a friend to Artemis because he "spent all his days in the greenwood chasing wild beasts" with her.

35. A: Introduction. The passage tells us how the following passages of the book are written, which acts as an introduction to the work. An appendix comes at the end of a book, giving extra details, making Choice *B* incorrect. Choice *C* is incorrect; a dedication is at the beginning of a book, but usually acknowledges those the author is grateful towards. Choice *D* is incorrect because a glossary is a list of terms with definitions at the end of a chapter or book.

36. B: Temporary resident. Although we don't have much context to go off of, we know that one is probably not a "lifetime partner" or "farm crop" of civilized life. These two do not make sense, so Choices *C* and *D* are incorrect. Choice *A* is also a bit strange. To be an "illegal alien" of civilized life is not a used phrase, making Choice *A* incorrect.

37. A: The author is doing translation work. We see this very clearly in the way the author talks about staying truthful to the original language of the text. The text also mentions "translation" towards the end. Criticism is taking an original work and analyzing it, making Choice *B* incorrect. The work is not being tested for historical validity, but being translated into the English language, making Choice *C* incorrect. The author is not writing a biography, as there is nothing in here about Pushkin himself, only his work, making Choice *D* incorrect.

38. B: To retain the truth of the work. The author says that "music and rhythm and harmony are indeed fine things, but truth is finer still," which means that the author stuck to a literal translation instead of changing up any words that might make the English language translation sound better.

39. C: This question requires close attention to the passage. Choice *A* can be found where the passage says "no less than six named and several unnamed varieties of the peach have thus produced several varieties of nectarine, so this choice is incorrect. Choice *B* can be found where the passage says "it is highly improbable that all these peach-trees . . . are hybrids from the peach and nectarine." Choice *D* is incorrect because we see in the passage that "the production of peaches from nectarines, either by seeds or buds, may perhaps be considered as a case of reversion." Choice *C* is the correct answer because the word "unremarkable" should be changed to "remarkable" in order for it to be consistent with the details of the passage.

40. B: The author's tone in this passage can be considered objective. An objective tone means that the author is open-minded and detached about the subject. Most scientific articles are objective. Choices *A, C,* and *D* are incorrect. The author is not very enthusiastic on the paper; the author is not critical, but rather interested in the topic. The author is not desperate in any way here.

41. B: Choice *B* is the correct answer because the meaning holds true even if the words have been switched out or rearranged some. Choice *A* is incorrect because it has trees either bearing peaches or nectarines, and the trees in the original phrase bear both. Choice *C* is incorrect because the statement does not say these trees are "indifferent to bud-variation," but that they have "indifferently [bore] peaches or nectarines." Choice *D* is incorrect; the statement may use some of the same words, but the meaning is skewed in this sentence.

42. B: The main purpose of this passage is to explain the definition and etymology of quilts. The second and third paragraph describes the etymology and definition of *quilt* from two different sources.

43. D: The author uses direct quotes in this passage to provide evidence. Specifically, they use direct quotes from two different encyclopedia sources. Direct quotes are a great way to provide evidence to a text because you can incorporate them into your own sentence, and they add credibility if they are from a trusted source.

44. C: The author says that the Encyclopedia Britannica gives a more specific definition than Webster. This is found in the third paragraph where it says, "The 'Encyclopædia Britannica' is a little more explicit," which means that it's more specific than Webster.

45. C: Comforts are blankets that are stitched together that are thicker than quilts. This is found in the very last paragraph when the author says, "In the United States the distinction has been carried even farther and a quilt is understood to be a light weight, closely stitched bedcover. When made thicker, and consequently warmer, it is called a "comfort." This describes a separate blanket than the quit called a "comfort," today known as "comforters."

Science

1. D: The pulmonary artery transports non-oxygenated blood from the heart to the lungs. Once there, the blood can get oxygenated and then transported back to the heart through the pulmonary vein, Choice *B,* and then circulated to the rest of the body. Capillaries, Choice *A,* are the smallest blood vessels that carry blood to tissues and allow for transfer of oxygen and nutrients. The aorta, Choice *D,* is the body's largest artery that carries oxygenated blood away from the heart.

2. B: The body gets nutrients through the consumption of food. The stomach, Choice *D*, is responsible for mashing up the food. Then the slurry travels to the small intestine, Choice *B*, where nutrients pass through the small intestine wall into the bloodstream. The gallbladder, Choice *A*, stores bile that is produced by the liver. The kidney, Choice *C*, filters out waste from the bloodstream.

3. A: There are three types of rocks: igneous, metamorphic, and sedimentary. Igneous rocks, Choice *A*, are formed from the cooling of lava or magma. Therefore, these would most likely be found near a volcano where lava is found. Metamorphic rocks, Choice *C*, are formed from the transformation of existing rocks—they are transformed by large amounts of pressure or very high temperatures, causing physical and chemical changes in the rock. Sedimentary rocks, Choice *D*, are formed from the accumulation of lots of small particles that become fused together. Clastic rocks, Choice *B*, are a type of sedimentary rock.

4. C: Tom can reduce his personal energy usage by cutting down on the number of things he does that use up non-renewable energy sources, such as electricity and gas. Riding a bicycle to work, Choice *A*, would reduce the amount of gas he uses in his car. Turning the lights off, Choice *B*, reduces the amount of electricity he uses. Reading a book, Choice *D*, instead of using his phone for entertainment would also reduce his electricity usage. Increasing the temperature of his house in the winter, Choice *C*, would cause the furnace to use more natural gas or oil and increase this energy usage.

5.

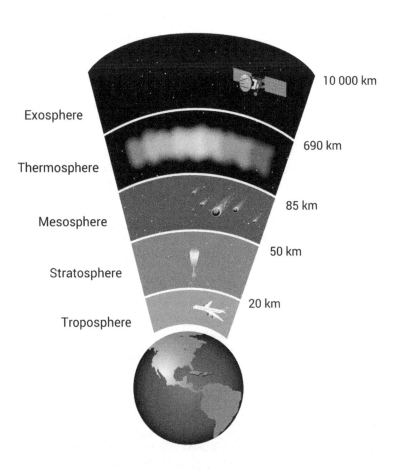

6. A: Cumulus clouds, Choice *A*, are the big puffy clouds with flat bottoms that are low in the sky. They often lead to clouds that make precipitation that reaches the ground. Cirrus clouds, Choice *C*, are thin, wispy clouds that are high in the sky. Although they do create precipitation since they are made of ice crystals, the precipitation vaporizes in the air before it reaches the ground. Stratus clouds, Choice *D*, are clouds that are low in sky and level on top of each other horizontally. Rain clouds are not a specific, scientific type of cloud, Choice *B*.

7. B: During a lunar eclipse, the Earth is in between the moon and the sun. The sun usually illuminates the moon, but during a lunar eclipse, the sun causes the Earth to cast a shadow on the moon so only part of the moon is visible. The moon is not a source of light and cannot create a shadow, so Choice *A* is incorrect. Choices *C* and *D* are incorrect because the positions of the moon, sun, and Earth would not allow a shadow to be cast on the moon.

8. D: A meteoroid, Choice *A*, is a small rocky body that is moving in the solar system. When it crosses the Earth's atmosphere, it becomes a meteor, Choice *C*. Once the meteor hits the Earth's surface, it becomes a meteorite, Choice *D*. A comet, Choice *B*, is like a small snowball traveling through the solar system. When it passes by the sun and warms up, it begins to release gases and looks like it has a tail.

9. Plastic beads: Density is the amount of mass per volume of an object. Items that have a larger mass in a smaller volume are denser than things that have a smaller mass in a larger volume. When the volume is the same for two objects, the object that has a larger mass will be denser. Since the ball filled with sand has a larger mass of 100 grams versus the plastic beads that are 50 grams, the ball with the sand will be denser.

10. C: The molecules of a liquid are close together and take the shape of the container they are in. They slip over each other during pouring and when the liquid is moving around so they can change positions. The molecules of solids are close together, cannot change position, only slightly vibrate within their structure, and cannot take the shape of their container. So, Choices *A* and *B* describe solids. Gases have molecules that move very fast and are far apart from each other, which is described by Choice *D*.

11. B: The atomic mass of an element on the periodic table is listed below the element's abbreviation. It represents the total mass of protons and neutrons of the atom. Looking at the box for Helium, the atomic mass is 4.

12. Oxidizing agent: In a redox reaction, the oxidizing agent is the molecule that accepts electrons and the reducing agent is the molecule that donates the electrons. When magnesium and oxygen are combined, the oxygen molecule accepts the electrons, so it is the oxidizing agent.

13. A: The left side of the equation contains $2Al(OH)_3$ and $3H_2SO_4$. Each aluminum hydroxide has 3 oxygen atoms and there are 2 molecules, so that makes 6 oxygen atoms. Each sulfuric acid has 4 oxygens and there are 3 molecules, so that makes 12 oxygen atoms. Therefore, in total, there are 18 oxygen atoms on the left side of the equation.

14. C: The limiting reagent of a reaction is the one that will be completely used up first. In the equation $2Al(OH)_3 + 3H_2SO_4 \rightarrow Al_2(SO_4)_3 + 6H_2O$, for every 2 moles of aluminum hydroxide, you need 3 moles of sulfuric acid to produce one mole of aluminum sulfate and 6 moles of water. If Yussef has 20 moles of aluminum hydroxide, he would need 30 moles of sulfuric acid to completely use up the aluminum hydroxide. However, he only has 12 moles of sulfuric acid. With 12 moles of sulfuric acid, he only needs 8 moles of aluminum hydroxide. The 12 moles of sulfuric acid will run the reaction 4 times, producing 4 moles of aluminum sulfate and 24 moles of water.

15. D: If the solution is still clear, it has not yet reached its saturation point for the Substance A. There is no visible precipitation of Substance A, so Choice *A* is incorrect. The odor does not indicate that the solution is rotting, so Choice *B* is incorrect. Adding a substance to hot water does not make it reach its boiling point, Choice *C*.

16. Momentum is conserved, or has a constant value, before and after the collision. When the golfer swings his club and the ball is still, the club has all of the momentum, which is equal to the sum of all of the masses of the objects multiplied by their velocity. After the ball is hit, some of the club's momentum is transferred to the ball. The club moves slower and the ball moves faster because it has less mass.

17. Force is equal to mass multiplied by acceleration, $F = m \times a$. The mass is equal to the bicycle plus the rider, $70 + 5 = 75\,kg$.

$$F = 75\,kg \times 13\,m/s^2 = 975\,kg - m/s^2 = 975\,N$$

18. A: The frequency of a sound wave is how many times it completes a cycle in a given amount of time. A hertz is equal to the frequency per second, so something that has 1 Hz cycles one time in 1 second. Higher frequency sound waves cycle more times per second than lower frequency sound waves. Given the answer options, an ultrasound at 106 Hz has the lower frequency, Choice *A*.

19.

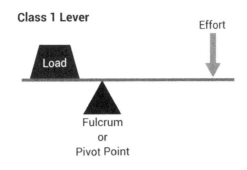

Class 1 Lever

Load

Effort

Fulcrum
or
Pivot Point

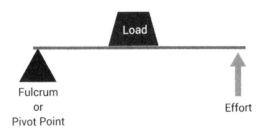

Class 2 Lever

Load

Fulcrum
or
Pivot Point

Effort

Class 3 Lever

Load

Fulcrum
or
Pivot Point

Effort

20. B: The farther away the fulcrum is from the load, the more effort it will take to lift the load. Moving the fulcrum closer to the load makes it easier to life the load, Choice *B*. Increasing the weight of the load would use more effort to lift the load, Choice *A*. Shortening the length of the bar would not significantly impact the amount of effort needed to lift the load, Choice *C*. Moving the fulcrum closer to where the effort is applied would increase the amount of effort needed to lift the load, Choice *D*.

21. D: Although it takes less force to move up a lower incline, the lower incline is longer, so the force would need to be applied for a longer time. More force is required to move up a higher incline, but since it is shorter, the force would not be applied for as long. Work is equal to the force times the distance. Therefore, Choice *D* is correct and the same amount of work would be done in both scenarios.

22. The carrying capacity of a population is the maximum population that an environment can sustain. It is limited by all of the factors that the environment contains. Most importantly, the food resources and physical space limit the carrying capacity. If a population grows and there isn't enough food for everyone to eat, the population size will begin to decrease. If there isn't enough physical space for everyone in a population, they will not be able to grow and thrive and the population size will also decrease.

23. A: In a food web, the arrows point from the prey to the predator. Every animal and plant on the chart has an arrow pointing towards bacteria because bacteria can infect and prey on all of the animals and plants. The rabbit preys on two plants, Choice *B*. The Cactus does not prey on anything, Choice *C*. The grasshopper preys on the plants, Choice *D*.

24. C: Predator-prey relationships are delineated through webs because predators often prey on more than one species. The grasshopper preys on star cactus, grass, and cactus, so if there were no longer cacti in the dessert, they would lose one of their food sources and suffer. Grass, kangaroo rats, and tarantulas do not prey on cacti, so they would not suffer from the loss of cacti in the dessert, Choices *A*, *B*, and *D*.

25. B: Humans cannot inherently detect the magnetic field of the Earth, so Choice *B* is correct because birds do have that sense. Humans have five senses: sight, hearing, taste, touch, and smell. Choice *A* describes the sense of hearing with crickets communicating to each other by chirping. Choice *C* describes the sense of touch because the dog got hurt by touching something sharp. Choice *D* describes the sense of sight because the deer are seeing the bright headlights.

26. Xylem: The xylem of the plant carries water upwards, against gravity, through the plant. Phloem transports soluble organic compounds throughout the plant. Seeds are produced after fertilization of the plant. Angiosperms are plants that have eggs that are fertilized. The eggs turn into seeds after fertilization.

27. B: Fungi are organisms that produce spores and feed on organic matter. They include molds, yeasts, and mushrooms. Bread, Choice *A*, comes from grains. Tomatoes and potatoes, Choices *C* and *D*, are the products of plants. Blue cheese is a commercial food that is actually a mold, Choice *B*.

28. Homo is the genus and Sapiens is the species.

29. D: Endoparasites are the type of parasite that live inside their host's body. An example is a tapeworm. Ectoparasites, Choice *B*, live on the outside of their host and feed off of it from there. An example is a flea. A brood parasite, Choice *C*, manipulates its host so that the host starts taking care of the parasitic offspring instead of its own. An example is a cowbird. Cowbirds lay eggs in catbird nests so that the catbirds will take care of the offspring when they hatch.

30. C: Viruses are host-specific and are programmed to only infect one type of cell. For example, common cold viruses infect the upper respiratory tract cells whereas a meningitis virus infects the tissue around the brain and spinal cord. Viruses do not have all of the working parts of a cell, so they rely on their hosts to help them reproduce and perform basic cellular tasks, Choices *A* and *B*. Antibiotics only work against infections from bacteria, not from viruses, Choice *D*.

31. No. Color-blindness is a recessive sex-linked gene on the X chromosome. Although the gene is recessive, if a male has the gene, there is no other dominant allele on another X chromosome to block its expression. Looking at the Punnett Square for their cross-breeding, only 50% of their female offspring and 50% of their male offspring would be color-blind.

	X	x_c
x_c	Xx_c	xx_c
Y	XY	x_cY

32. D: Mutations occur when there is a change in the base pair sequence of the gene. All three choices would alter the sequence of the base pairs and therefore, change the proteins that were encoded after the mutation. Choice *A*, a base substitution, would swap out one base for another base. An insertion mutation, Choice *B*, would add an extra base to the gene. A deletion mutation, Choice *C*, would take a base out of the gene.

33. A: Glucose is a product of photosynthesis. It is not required to start the process of photosynthesis or used anywhere in the rest of the process. Sunlight, carbon dioxide, and water, Choices, *B*, *C*, and *D*, are all needed for photosynthesis to occur.

34. B: Meiosis is the process of reproduction in human cells. Each human gamete has 23 chromosomes. The chromosomes from each parent cell are combined and then duplicated, giving a total of 92 chromosomes. These chromosomes are then divided twice to give four daughter cells, each with 23 chromosomes.

35. C: Enzymes only work with specific substrates. The activation site and shape of the enzyme must match exactly with the substrate site. The shape of the enzyme in the picture would only fit with substrate C. Choice *A* would need a triangle shaped enzyme. Choice *B* would need a circular enzyme. Choice *D* would need an irregular-shaped enzyme.

36. Catalyst: Catalysts are responsible for making reactions move faster. They lower the activation energy of the reaction and increase the rate of the reaction but are not consumed by the reaction while doing so. More product solutions would not increase the rate of the reaction. Glucose molecules may change the original chemical reaction that Frank was doing.

37. D: Plant cells have cell walls around their cell membranes, which add structure and strength to each individual cell. Animal cells only have a cell membrane. Both animal and plant cells have mitochondria, nuclei, and DNA in their cells, Choices *A*, *B*, and *C*.

Social Studies

1. A: The diagram illustrates relationships between the three branches of the federal government. Arrows point from one branch to another, and the powers they're labeled with all restrict the targeted branch in some way. Therefore, the diagram depicts checks and balances; this constitutional concept involves giving a political entity powers for the purpose of counterbalancing the powers of another political entity. Thus, the correct answer is Choice *A*. The states and the federal government share concurrent powers, so Choice

B is incorrect. Constitutional review is another term for judicial review, which is the judicial branch's primary power. Therefore, Choice *C* is incorrect. Although checks and balances are part of the constitutional separation of powers, checks and balances are more directly related to how the powers depicted in the diagram interfere with how another branch operates. Therefore, Choice *D* is incorrect.

2. D: The question asks for a check that the executive branch holds over the judicial branch, aside from the power to appoint judges. The executive branch also enjoys the ability to pardon people accused and/or convicted of a federal crime. In other words, the executive branch can invalidate the decisions of federal courts. Thus, the correct answer is Choice *D*. The legislative branch exclusively controls appropriation, which relates to budgeting, spending, and taxing. Therefore, Choice *A* is incorrect. The executive branch holds discretionary powers in how it chooses to enforce the law, but this is a general constitutional concept and not a specific check on the judicial branch. Therefore, Choice *B* is incorrect. Judicial review refers to the judicial branch's check on the executive branch and legislative branch, so Choice *C* is incorrect.

3. A: The median weekly earnings for all workers are $969 per week, and it is represented by the dotted line on the chart. People holding only an associate's degree earn $887 per week, which amounts to less than the median weekly earnings for all workers. Thus, the correct answer is Choice *A*. People holding only a bachelor's degree earn $1,248 per week; therefore, Choice *B* is incorrect. People holding only a doctoral degree earn $1,883 per week; therefore, Choice *C* is incorrect. People holding only a master's degree earn $1,497 per week; therefore, Choice *D* is incorrect.

4. B: The left side of the bar graph depicts unemployment rates. People who have attended some college without earning a degree make up the third category from the bottom, and this group's unemployment rate is 3.3%. Thus, the correct answer is Choice *B*. Choice *A* is incorrect because 2.7% is the unemployment rate for people holding only an associate's degree. Choice *C* is incorrect because 3.7% is the unemployment rate for people holding only a high school diploma. Choice *D* is incorrect because 5.4% is the unemployment rate for people holding less than a high school diploma.

5. B: The graph illustrates voting rates for midterm elections between 1978 and 2018, and the question asks for the year in which the voting rate was lower than in the previous election. The voting rate was 48.3% in 1994, which was 1% less than the voting rate in the 1990 election. Thus, the correct answer is Choice *B*. The voting rate increased from 48.9% in 1978 to 51.9% in 1982, so Choice *A* is incorrect. Similarly, the voting rate increased from 45.3% in 1998 to 46.1% in 2002, so Choice *C* is incorrect. Lastly, the voting rate increased from 46.1% in 2002 to 47.8% in 2006, so Choice *D* is incorrect.

6. C: Midterm elections occur every four years, and they are held in the middle of a presidential term. Every seat in the House of Representatives is up for grabs in midterm elections, and approximately one third of Senate seats are being voted on at the same time. In addition, many state and local elections are held during midterm elections. Thus, the correct answer is Choice *C*. Midterm elections aren't solely congressional elections, and they don't occur every two years. Therefore, Choice *A* is incorrect. The sitting president's approval rating is often the most important factor for candidates competing in midterm elections, but midterm elections aren't held every six years. So, Choice *B* is incorrect. Choice *D* is incorrect because it is not the best or most complete definition.

7. D: The First Amendment is the source of five individual rights and freedoms in the United States Constitution. First, the First Amendment's prohibition on Congress' ability to regulate religious practice and creation of the right to freely exercise religion combine to establish the freedom of religion (Choice *A*). Second, the First Amendment creates the freedom of assembly (Choice *B*), which guarantees people's right to hold meetings and form associations. Third, the First Amendment prohibits the government from interfering with people's freedom of speech (Choice C). Fourth, the First Amendment guarantees the freedom of the press. Fifth, the First Amendment guarantees the right to petition the government, which allows people to protest and request assistance without being punished. Rather than the First Amendment, the Sixth Amendment creates the right to an attorney in criminal cases. Thus, the correct answer is Choice D.

8. A: John D. Rockefeller established Standard Oil, and it grew to become one of the most valuable corporations in world history due to its monopolistic control of nearly the entire American oil industry. The political cartoon uses an octopus' tentacles to symbolize the far-reaching influence of Standard Oil, including political corruption and disruption of the free market. Thus, the correct answer is Choice A. While the cartoonist likely believes Standard Oil harmed the working classes, this cartoon shows Standard Oil crushing political institutions and businessmen. Therefore, Choice B is incorrect. Standard Oil was one of the world's first multinational corporations, but the cartoon doesn't address its effect on global institutions and markets. So, Choice C is incorrect. Compared to how the octopus is strangling the United States Capitol on the left side of the cartoon, the cartoon portrays it as merely reaching toward the White House, which is why the cartoon was entitled "Next!" Therefore, Choice D is incorrect.

9. C: The political cartoon heavily criticizes Standard Oil's relationship to representative government in the United States. The Octopus has one tentacle wrapped around the United States Capitol on the left side and another tentacle wrapped around a generic state capitol building on the right side. In addition, the octopus is preparing to attack the White House next. These tentacles symbolize how Standard Oil is strangling free and independent government. Thus, the correct answer is Choice C. Standard Oil is portrayed as the source of corruption rather than a consequence, so Choice A is incorrect. Choice B is incorrect because Standard Oil is shown attacking multiple government institutions in the cartoon. The cartoon definitely portrays Standard Oil as corrupt, lawless, and greedy. However, Standard Oil isn't a political institution, so Choice D is incorrect.

10. B: The graph expresses the total exports and general imports for United States in regard to five countries, and the question asks which country the United States has the largest total trade deficit with. A trade deficit is used in economics to calculate by how much a country's imports exceed its exports. According to the graph, American imports from China totaled $467 billion, and American exports to China totaled $124 billion. As a result, the United States' trade with China resulted in a $343 billion trade deficit. In comparison, the United States had a $34 billion trade deficit with Canada (Choice A), a $54 billion trade deficit with Mexico (Choice D), a $67 billion trade deficit with Japan (Choice C), and a $73 billion trade deficit with Germany. Therefore, the United States' trade deficit with China dwarfs its trade deficits with the other countries. Thus, the correct answer is Choice B.

11. C: The graph illustrates American trade patterns for petroleum products and crude oil from 1948 to 2018. Imports are foreign products and oil transported from a foreign country to the United States, and exports are foreign products and oil transported from the United States to a foreign country. Between 2005 and 2018, American imports of petroleum products and crude oil both declined, while American exports of petroleum products and crude oil increased considerably. These trends demonstrate a reduced reliance on foreign trade, so the most likely explanation is that American domestic production increased dramatically. Thus, the correct answer is Choice C. A steep decline in global petroleum production could explain the diminished imports, but domestic refineries increased in value, as evidenced in the corresponding increase of American exports. Therefore, Choice A is incorrect. The term *petrodollar* refers to the fact that most international oil deals are conducted in dollars. Although the United States sometimes institutes foreign trade policies to stabilize the value of the dollar, this conclusion isn't supported by the graph. So, Choice B is incorrect. The graph doesn't contain any information about domestic subsidies, and imports of petroleum products declined during the time period in question. Therefore, Choice *D* is incorrect.

12. D: The timeline lists eight events related to the Civil Rights Movement, and the question asks for an event that occurred before President Eisenhower intervened to integrate an Arkansas public school in 1957. President Truman integrated the United States military in 1948. Thus, the correct answer is Choice *D*. The name of the Civil Rights Act of 1965 indicates that it became law later than 1957, so Choice *A* is incorrect. Activists organized Freedom Rides in 1961, so Choice *B* is incorrect. Martin Luther King delivered his "I Have a Dream" speech in 1963, so Choice *C* is incorrect.

13. C: The question asks when the Supreme Court issued its decision in *Brown v. Board of Education*, which reversed its ruling in *Plessy v. Ferguson* (1896) and found racial segregation of public schools to be unconstitutional. According to the timeline, the Supreme Court issued its decision in 1954. Thus, the correct answer is Choice *C*. Choice *A* is incorrect because 1957 is the year President Eisenhower intervened to integrate an Arkansas public school. Choice *B* is incorrect because 1955 is the year Rosa Parks protested segregation on a Montgomery bus. Choice *D* is incorrect because 1948 is the year President Truman integrated the United States military.

14. A: The photograph depicts the American astronaut Buzz Aldrin walking on the Moon as part of NASA's Apollo 11 mission (1969). Before this unprecedented lunar landing, the United States competed with the Soviet Union in the space race during the Cold War. Technological innovations in ballistic missiles triggered the space race; the missiles allowed the global superpowers to launch satellites, probes, and people into space. Thus, the correct answer is Choice *A*. The United States and the Soviet Union somewhat deescalated their ideological conflict during a period known as détente, but this cooling of tensions occurred after the Apollo 11, so Choice *B* is incorrect. The United States and the Soviet Union joined the United Nations Committee on the Peaceful Uses of Outer Space in 1958; however, the United States didn't depend on international cooperation for its space program or Apollo 11. Therefore, Choice *C* is incorrect. President Kennedy proposed a joint American-Soviet lunar landing, but his assassination put an end to this plan. The proposed joint mission didn't contribute to the success of Apollo 11; therefore, Choice *D* is incorrect.

15. B: The painting depicts a frontier settlement with a railroad station, and a small caravan is traveling through its center. Workers in the bottom left corner of the painting are chopping down trees, likely to extract resources and clear space for the settlement to expand. Directly to the north of the settlement there's a road leading to more housing. These are characteristics of a boomtown, which were frontier settlements that rapidly grew based on their connection to a railroad and/or natural resources. Thus, the correct answer is Choice *B*. Boomburb settlements are suburban areas whose population and commercial development increase rapidly while they still retain suburban characteristics. Boomburbs developed in the late twentieth century; therefore, Choice *A* is incorrect. Exurbs are a modern type of high-income and low-density settlement that developed beyond the more densely packed suburbs. So, Choice *C* is incorrect. Suburbs rapidly developed after World War II as middle-class and high-income families sought to escape the dense and polluted cities. Therefore, Choice *D* is incorrect.

16. D: The painting depicts an expanding settlement, and people are waiting for a train that's heading west. Additionally, American Indians can be seen on the right side of the painting. Smoke is blowing off the train into the American Indians, and they are positioned on the opposite side of the tracks as the settlement. These details illustrate and symbolize the relationship between railroads and Western expansion. Railroads transported supplies and migrants across the frontier, provided significant employment opportunities, and supported the development of boomtowns. Furthermore, the construction of railroads played a critical role in driving American Indians off their lands. Thus, the correct answer is Choice *D*. Railroads facilitated migration and transportation, but rather than undermining economic development, they stimulated the growth of Western cities. Therefore, Choice *A* is incorrect. The mass production of automobiles led to the development of suburbs, so Choice *B* is incorrect. Railroads connected maritime networks to inland areas, which greatly benefited maritime commerce. Therefore, Choice *C* is incorrect.

17. D: The map provides information about how the government coordinates the management of herd management areas in Arizona. According to the legend beneath the map, the herd management areas are highlighted in dark blue, and nearly all these areas are located in western Arizona. The Yuma Field Office is located in southwestern Arizona, close to the Cibola-Trigo herd management area. Thus, the correct answer is Choice *D*. The Hassayampa Field Office is to the northeast of the Phoenix District, and it's not close to any herd management areas. Therefore, Choice *A* is incorrect. The Safford Field Office is located in the Gila District, and the closest herd management area is more than 100 miles away. So Choice *B* is incorrect. The Tucson Field Office is located in southeastern Arizona, and it is also more than 100 miles away from the closest herd management area. Therefore, Choice *C* is incorrect.

18. D: The graph depicts the voting rates for four different age groups, and the question asks which group had increased voter turnout in 2016. The voting rate for 18-to-29-year-olds increased from approximately 45% in 2012 to approximately 47% in 2016. Thus, the correct answer is Choice *D*. The voting rate for people over the age of 65 decreased from approximately 72% in 2012 to approximately 70.5% in 2016. So Choice *A* is incorrect. The voting rate for 45-to-64-year-olds declined from approximately 68% in 2012 to approximately 66% in 2016. Therefore, Choice *B* is incorrect. The voting rate for 30-to-44-year-olds slightly decreased from approximately 60% in 2012 to approximately 59.5% in 2016. Therefore, Choice *C* is incorrect.

19. B: The question asks for an election in which two groups had nearly identical voter turnout. Voting rates for people over the age of 65 and for 45-to-64-year-olds were both approximately 70% in 2004. Thus, the correct answer is Choice *B*. The voting rates for people over the age of 65 and 45 to 64 years old were similar in 2000, but people over the age of 65 voted slightly more, as evidenced by the small gap between the lines. Therefore, Choice *A* is incorrect. Similarly, the voting rates for people over the age of 65 and for 45-to-64-year-olds were again similar in 2008, but the rates weren't as close as in 2004. Therefore, Choice *C* is incorrect. There wasn't a pair of age groups with extremely similar voting rates in 2012, so Choice *D* is incorrect.

20. A: The graph compares the labor participation rates of women based on age, and the question asks which age group's labor participation declined the most between 2000 and 2010. The labor participation of women between the ages of 16 and 24 (Choice *A*) fell from 63% to 53.6%, so the decline was 6.4%. The labor participation of women between the ages of 25 and 34 (Choice *B*) decreased from 76.1% to 74.7% for a difference of 2.4%. The labor participation of women between the ages of 35 and 44 (Choice *C*) decreased from 77.2% to 75.2%, amounting to a decline of 2%. The labor participation of women between the ages of 45 and 54 (Choice *D*) decreased from 76.8% to 75.7%, so the decline was 1.1%. Therefore, the decline in labor participation for women between the ages of 16 and 24 was more than three times greater than that of the next-highest group. Thus, the correct answer is Choice *A*.

21. C: The labor participation rates for women skyrocketed after 1960 as traditional household roles declined. Along with several other key factors, increased access to birth control and feminism played a major role in women entering the workforce in record numbers. Birth control greatly expanded women's ability to engage in family planning, and the rising popularity of feminism during the 1960s encouraged more women to seek financial and personal independence. Thus, the correct answer is Choice *C*. The Nineteenth Amendment enacted women's suffrage four decades before the labor participation of women significantly increased. So, Choice *A* is incorrect. While households did increasingly desire a second source of income during this period, the American economy wasn't teetering toward a collapse. Therefore, Choice *B* is incorrect. Women did gain a variety of legal rights during the 1950s and 1960s, including the right to divorce and expanded property rights, but these changes were less directly related to labor participation than birth control and family planning. Therefore, Choice *D* is incorrect.

22. B: The cartoon depicts a prosperous businessman and a mutilated soldier; the businessman offers the soldier a medal instead of a job. On the whole, the cartoon delivers a strong anti-war message in its illustration of how an American intervention would grievously harm the working class. This opinion was prevalent across the United States before the intervention, which is why Congress passed the Espionage Act of 1917 and the Sedition Act of 1918 to suppress dissent. Thus, the correct answer is Choice *B*. Although the cartoon is mocking the businessman, it isn't necessarily critiquing capitalism or advocating for socialism. Therefore, Choice *A* is incorrect. The cartoon is opposed to the United States entering World War I, so Choice *C* is incorrect. In addition, pro-war justifications for the intervention weren't particularly popular opinions. Choice *D* is incorrect because the cartoon is satirizing a businessman paying the soldier with respect instead of money.

23. A: The second paragraph of the excerpt describes the federal government's new policies to prepare the United States for the cataclysmic effects of climate change. The first of the four points in the policy program states that there needs to be a stronger partnership between all levels of the federal government. Thus, the correct answer is Choice *A*. Intra-agency channels of communication exist only within a single government institution. While optimizing intra-agency channels of communication would help the federal government, the policy is explicitly directed at improving interagency communication, which is communication between multiple institutions and layers of government. Therefore, Choice *B* is incorrect. The policy is calling for more adaptive learning, but this directive is broader than as described in Choice *C*. Under this policy, adaptive learning would occur in the context of greater interagency communication, so it would apply to scenarios where one institution is learning from another institution's specialized expertise. So, Choice *C* is incorrect. The first paragraph states that the government should prioritize communities already facing economic and health-related issues; however, this doesn't answer the question about how the policy would alter the interactions between government institutions. Therefore, Choice *D* is incorrect.

24. B: The excerpt is from Executive Order 13653, and only the executive branch holds the power to issue executive orders. The president is the head of the executive branch, and Barack Obama was President of the United States in 2013. Thus, the correct answer is Choice *B*. All the other answer choices were leaders of either the judicial branch or the legislative branch when President Obama issued Executive Order 13653. Choice *A* is incorrect because Chief Justice John Roberts was a member of the United States Supreme Court, which is part of the judicial branch. Choice *C* is incorrect because Senate Majority Leader Mitch McConnell was the leader of the United States Senate, which is part of the legislative branch. Choice *D* is incorrect because Speaker of the House John Boehner was the leader of the House of Representatives, which is part of the legislative branch.

25. A: The photograph illustrates the damage sustained by the Pentagon in the September 11 terrorist attack. Osama bin Laden was the leader of Al-Qaeda, a network of Sunni jihadists, and he masterminded Al-Qaeda's plan to hijack four commercial airliners and fly them into buildings with symbolic meanings, including the Pentagon and the World Trade Center. Thus, Choice *A* is the correct answer. Boko Haram is a Sunni jihadist terrorist group with a base of operations in Nigeria, and it was founded after the

September 11 attacks. Therefore, Choice *B* is incorrect. ISIS began as an offshoot of Al-Qaeda, and it rose to prominence in the chaotic aftermath of the Iraq War, conquering substantial amounts of territory in Iraq and Syria. Therefore, Choice *C* is incorrect. The Taliban is an Islamic organization that controlled a proto-government in Afghanistan. Following the September 11 attacks, the Taliban came under intense scrutiny for providing safe harbor to Al-Qaeda terrorists, including Osama bin Laden, which is why the United States immediately invaded Afghanistan. However, the Taliban weren't closely involved in planning or launching the attack; therefore, Choice *D* is incorrect.

26. A: The question asks which demographic group experienced the largest decline in reported voting, and this data can be found in the table's middle column. Voting among non-Hispanic Black voters between the ages of 30 and 44 years of age (Choice *A*) declined by 413,000. Voting among Hispanic voters between the ages of 45 and 64 (Choice *B*) increased by 544,000. Voting among non-Hispanic, non-Black, and non-white voters between the ages of 18 and 29 (Choice *C*) increased by 196,000. Voting among non-Hispanic white voters over the age of 65 (Choice *D*) declined by 2,828,000. Of these demographic groups, only voting among non-Hispanic Black voters between the ages of 30 and 44 experienced any decline. Thus, the correct answer is Choice *A*.

27. D: The question asks which demographic group of voting-age citizens experienced the largest increase in population. Another way to ask the question is which age group has the maximum possible number of potential voters. Data pertaining to citizen voting-age populations can be found in the table's third column. The citizen voting-age population of 18-to-29-year-olds (Choice *A*) increased by 1,267,000; 30-to-44-year-olds (Choice *B*) increased by 982,000; 45-to-64-year-olds (Choice *C*) increased by 905,000; and people 65 years and older (Choice *D*) increased by 5,824,000. Thus, the correct answer is Choice *D*.

28. A: The map depicts the percentage of farms with female principal operators in 2012, and the legend explains how the areas with the highest percentage will have the darkest shading. According to the map, Arizona (AZ) has the highest percentage of female principal operators of farms. Most of the state exceeds 25% of female principal operators, and there are no areas with less than 10% female principal operators. Thus, the correct answer is Choice *A*. The vast majority of Iowa (IA) has areas with less than 10% female principal operators, so Choice *B* is incorrect. Missouri (MO) is mostly a mixture of areas with less than 10% and 10 to 14% female principal operators. Therefore, Choice *C* is incorrect. The majority of Utah (UT) is composed of areas with less than 10% female principal operators, with some pockets having higher rates. Therefore, Choice *D* is incorrect.

29. C: The map illustrates maritime claims in the South China Sea. The lines on the map represent countries' maritime claims, and these lines begin and end on the countries' coasts. So disputes can be identified when a maritime area falls within multiple territorial lines. The Paracel Islands are located in the northern portion of the South China Sea, and they are claimed by China, Taiwan, and Vietnam. Chinese and Taiwanese claims are represented by the same line on the map because both countries claim to be the rightful ruler of the Chinese mainland and they have identical maritime claims. Thus, the correct answer is Choice *C*. Choice *A* is incorrect because Vietnam and Taiwan both claim the Paracel Islands. Choice *B* is incorrect because it omits the Vietnamese claim. Choice *D* is incorrect because the Philippines' maritime claims don't extend far enough west to reach the Paracel Islands.

30. B: The map illustrates the results of the Electoral College for the 1964 presidential election. As established by the United States Constitution, the Electoral College is responsible for electing the president and vice president. Each state is allotted a certain number of electoral votes, and presidential candidates need to win an absolute majority of electoral votes to win. Thus, the correct answer is Choice *B*. Election boards regulate and conduct oversight over the electoral process, so Choice *A* is incorrect. The executive branch is one of the three branches of the federal government, and it's primarily responsible for enforcing laws. Therefore, Choice *C* is incorrect. Ranked voting is an electoral system in which voters list their preferred candidates in a sequential order rather than selecting a single candidate. The Electoral College allows voters to select only one candidate; therefore, Choice *D* is incorrect.

31. D: The map illustrates the results of the Electoral College for the 1964 presidential election, and each state is labeled with the number of electoral votes it has. For example, California has 40 electoral votes. States' electoral votes are calculated based on the total number of its congressional seats in the United States Senate and House of Representatives. Thus, the correct answer is Choice *D*. The map doesn't provide candidates' margin of victory for each state; therefore, Choice *A* is incorrect. Presidential elections occur after presidential nominating conventions, and, in any event, the map depicts the result of the Electoral College, which isn't directly related to the conventions. Therefore, Choice *B* is incorrect. The total population of voting-age citizens is made up of all the people who are legally eligible to cast votes in an election, so Choice *C* is incorrect.

32. C: The poster implores Americans to conserve wheat, meat, fats, and sugar. Based on the date provided in the caption, the U.S. Food Administration published this poster during World War I (1914–1918). The United States government justified its intervention as a defense of freedom. Since the poster says the goal is to serve the cause of freedom, the rationing is clearly intended to redirect agricultural resources toward the war effort. Thus, the correct answer is Choice *C*. The poster isn't related to public health because the proposed substitutes have comparable nutritional value. Therefore, Choice *A* is incorrect. The poster is concerned with the consumption, not production, of agricultural products; therefore, Choice *B* is incorrect. The federal government issued war bonds during World War I, but that's not the poster's focus. So, Choice *D* is incorrect.

33. C: The graph depicts unemployment and long-term unemployment rates. Recessions are represented as shaded columns, and the unemployment rate is represented by the blue line at the top of the graph. The unemployment rate peaked at 6.5% during the recession of 1960–1961 (Choice *A*); 6% during the recession of 1969–1970 (Choice *B*); 10.8% during the recession of 1981–1982 (Choice *C*); and 5.9% during the recession of 2001–2002 (Choice *D*). Therefore, the recession of 1981–1982 had the highest unemployment rate. Thus, Choice *C* is the correct answer.

34. D: The long-term unemployment rate differs from the unemployment rate; the long-term rate measures workers who are unemployed for 27 weeks or more. The relevant long-term unemployment rates were approximately: 2.4% in 1982 (Choice *A*); 1.6% in 1993 (Choice *B*); 0.4% in 2001 (Choice *C*); and 3.8% in 2009 (Choice *D*). So, 2009 had the highest long-term unemployment rate. Thus, the correct answer is Choice *D*.

35. A: In *Obergefell v. Hodges* (2015), the Supreme Court held that marriage was a fundamental constitutional right, effectively legalizing same-sex marriage. The excerpt contains several clues that this would be the final decision. For example, the Supreme Court connects marriage to liberty, family life, and privacy rights. Thus, the correct answer is Choice *A*. The excerpt states that the *Loving* decision overturned prohibitions on interracial marriage, and the Court cites *Loving* as a precedent to overturn similar prohibitions on same-sex marriage. Therefore, Choice *B* is incorrect. The Supreme Court expanded individual rights in *Obergefell v. Hodges* (2015), and the Court relied on *Lawrence* to show the broad class of existing constitutional rights related to family life. So, Choice *C* is incorrect. Choice *D* is incorrect because the last sentence of the excerpt discusses how privacy rights should be expanded to include marriage.

36. B: The Supreme Court mentions "relevant precedents" in the first sentence. Given how the Supreme Court cites the earlier cases of *Loving* and *Lawrence*, it can be inferred that relevant precedents are legal principles or rules from earlier cases that are applied to current cases. Thus, the correct answer is Choice *B*. Relevant precedents can provide specific legal frameworks, but the legal concept of precedents applies to much more than just due process claims. In fact, precedents serve as the foundation for the entire American legal system. So, Choice *A* is incorrect. Similarly, the concept of relevant precedents is broader than issues related to marriage and liberty. In addition, the Court didn't limit individual freedom in *Obergefell v. Hodges* (2015). Therefore, Choice *C* is incorrect. Relevant precedents are legal precedents that are relevant to a current case, and they often contain more than one premise. Therefore, Choice *D* is incorrect.

Practice Test #3

Mathematical Reasoning

No Calculator Questions

1. What is $\frac{12}{60}$ converted to a percentage?

    a. 0.20
    b. 20%
    c. 25%
    d. 12%

2. Which of the following represents the correct sum of $\frac{14}{15}$ and $\frac{2}{5}$?

    a. $\frac{20}{15}$

    b. $\frac{4}{3}$

    c. $\frac{16}{20}$

    d. $\frac{4}{5}$

3. What is the product of $\frac{5}{14}$ and $\frac{7}{20}$?

    a. $\frac{1}{8}$

    b. $\frac{35}{280}$

    c. $\frac{12}{34}$

    d. $\frac{1}{2}$

4. What is the result of dividing 24 by $\frac{8}{5}$?

    a. $\frac{5}{3}$

    b. $\frac{3}{5}$

    c. $\frac{120}{8}$

    d. 15

5. Subtract $\frac{5}{14}$ from $\frac{5}{24}$. Which of the following is the correct result?

   a. $\frac{25}{168}$

   b. 0

   c. $-\frac{25}{168}$

   d. $\frac{1}{10}$

6. Which of the following is a correct mathematical statement?

   a. $\frac{1}{3} < -\frac{4}{3}$

   b. $-\frac{1}{3} > \frac{4}{3}$

   c. $\frac{1}{3} > -\frac{4}{3}$

   d. $-\frac{1}{3} \geq \frac{4}{3}$

7. Which of the following is INCORRECT?

   a. $-\frac{1}{5} < \frac{4}{5}$

   b. $\frac{4}{5} > -\frac{1}{5}$

   c. $-\frac{1}{5} > \frac{4}{5}$

   d. $\frac{1}{5} > -\frac{4}{5}$

8. How many cases of cola can Lexi purchase if each case is $3.50 and she has $40?

   a. 10
   b. 12
   c. 11.4
   d. 11

9. A car manufacturer usually makes 15,412 SUVs, 25,815 station wagons, 50,412 sedans, 8,123 trucks, and 18,312 hybrids a month. About how many cars are manufactured each month?

   a. 120,000
   b. 200,000
   c. 300,000
   d. 12,000

10. Each year, a family goes to the grocery store every week and spends $105. About how much does the family spend annually on groceries?

   a. $10,000
   b. $50,000
   c. $500
   d. $5,000

11. A grocery store sold 48 bags of apples in one day, and 9 of the bags contained Granny Smith apples. The rest contained Red Delicious apples. What is the ratio of bags of Granny Smith to bags of Red Delicious apples that were sold?

    a. 48:9

    b. 39:9

    c. 9:48

    d. 9:39

12. If Oscar's bank account totaled $4,000 in March and $4,900 in June, what was the rate of change in his bank account over those three months?

    a. $900 a month

    b. $300 a month

    c. $4,900 a month

    d. $100 a month

13. Erin and Katie work at the same ice cream shop. Together, they always work less than 21 hours a week. In a week, if Katie worked two times as many hours as Erin, how many hours did Erin work?

    a. Less than 7 hours

    b. Less than or equal to 7 hours

    c. More than 7 hours

    d. Less than 8 hours

14. Which of the following is the correct decimal form of the fraction $\frac{14}{33}$ rounded to the nearest hundredth place?

```

```

15. Gina took an algebra test last Friday. There were 35 questions, and she answered 60% of them correctly. How many correct answers did she have?

```

```

16. Paul took a written driving test, and he got 12 of the questions correct. If he answered 75% of the questions correctly, how many problems were there in the test?

```

```

Calculator Questions

17. What is the solution to the equation $3(x + 2) = 14x - 5$?

    a. $x = 1$

    b. $x = 0$

    c. All real numbers

    d. There is no solution

18. What is the solution to the equation $10 - 5x + 2 = 7x + 12 - 12x$?
    a. $x = 1$
    b. $x = 0$
    c. All real numbers
    d. There is no solution

19. Which of the following is the result when solving the equation $4(x + 5) + 6 = 2(2x + 3)$?
    a. $x = 26$
    b. $x = 6$
    c. All real numbers
    d. There is no solution

20. Two consecutive integers exist such that the sum of three times the first and two less than the second is equal to 411. What are those integers?
    a. 103 and 104
    b. 104 and 105
    c. 102 and 103
    d. 100 and 101

21. In a neighborhood, 15 out of 80 of the households have children under the age of 18. What percentage of the households have children?
    a. 0.1875%
    b. 18.75%
    c. 1.875%
    d. 15%

22. If a car is purchased for $15,395 with a 7.25% sales tax, what is the total price?
    a. $15,395.07
    b. $16,511.14
    c. $16,411.13
    d. $15,402

23. From the chart below, which two are preferred by more men than women?

Preferred Movie Genres

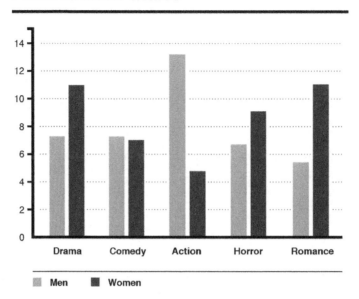

a. Comedy and Action
b. Drama and Comedy
c. Action and Horror
d. Action and Romance

24. Which type of graph best represents a continuous change over a period of time?
a. Bar graph
b. Line graph
c. Pie graph
d. Histogram

25. Using the graph below, what is the mean number of visitors for the first 4 hours?

Museum Visitors

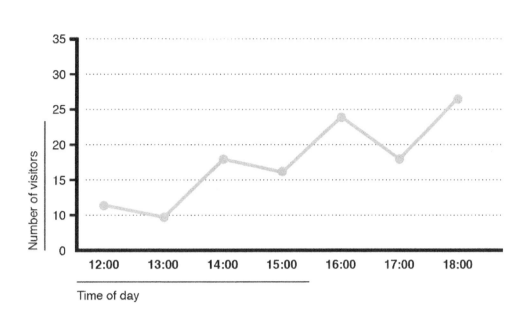

a. 12
b. 13
c. 14
d. 15

26. What is the mode for the grades shown in the chart below?

Science Grades	
Jerry	65
Bill	95
Anna	80
Beth	95
Sara	85
Ben	72
Jordan	98

a. 65
b. 33
c. 95
d. 90

27. What type of relationship is there between age and attention span as represented in the graph below?

Attention Span

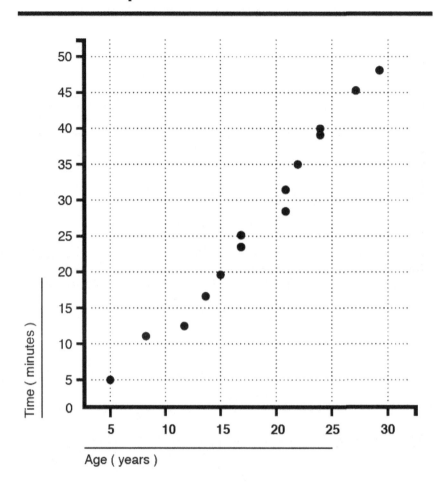

a. No correlation
b. Positive correlation
c. Negative correlation
d. Weak correlation

28. Which of the following relations is a function?
   a. {(1, 4), (1, 3), (2, 4), (5, 6)}
   b. {(-1, -1), (-2, -2), (-3, -3), (-4, -4)}
   c. {(0, 0), (1, 0), (2, 0), (1, 1)}
   d. {(1, 0), (1, 2), (1, 3), (1, 4)}

29. Find the indicated function value:

$$f(5) \text{ for } f(x) = x^2 - 2x + 1$$

    a. 16
    b. 1
    c. 5
    d. Does not exist

30. For which two values of x is $g(x) = 4x + 4$ equal to $g(x) = x^2 + 3x + 2$?
    a. 1, 0
    b. 2, -1
    c. -1, 2
    d. 1, 2

31. The population of coyotes in the local national forest has been declining since 2000. The population can be modeled by the function $y = -(x - 2)^2 + 1600$, where y represents number of coyotes and x represents the number of years past 2000. When will there be no more coyotes?
    a. 2020
    b. 2040
    c. 2012
    d. 2042

32. A ball is thrown up from a building that is 800 feet high. Its position s in feet above the ground is given by the function $s = -32t^2 + 90t + 800$, where t is the number of seconds since the ball was thrown. How long will it take for the ball to come back to its starting point? Round your answer to the nearest tenth of a second.
    a. 0 seconds
    b. 2.8 seconds
    c. 3 seconds
    d. 8 seconds

33. A study of adult drivers finds that it is likely that an adult driver wears his seatbelt. Which of the following could be the probability that an adult driver wears his seat belt?
    a. 0.90
    b. 0.05
    c. 0.25
    d. 0

34. What is the solution to the following system of linear equations?
$$2x + y = 14$$
$$4x + 2y = -28$$

    a. (0, 0)
    b. (14, -28)
    c. All real numbers
    d. There is no solution

35. Which of the following is perpendicular to the line $4x + 7y = 23$?

    a. $y = -\frac{4}{7}x + 23$

    b. $y = \frac{7}{4}x - 12$

    c. $4x + 7y = 14$

    d. $y = -\frac{7}{4}x + 11$

36. What is the solution to the following system of equations?
$$2x - y = 6$$
$$y = 8x$$

    a. (1, 8)

    b. (-1, 8)

    c. (-1, -8)

    d. There is no solution.

37. The percentage of smokers above the age of 18 in 2000 was 23.2%. The percentage of smokers over the age of 18 in 2015 was 15.1%. Find the average rate of change in the percentage of smokers over the age of 18 from 2000 to 2015.

    a. -.54%

    b. -54%

    c. -5.4%

    d. -15%

38. In order to estimate deer population in a forest, biologists obtained a sample of deer in that forest and tagged each one of them. The sample had 300 deer in total. They returned a week later and harmlessly captured 400 deer, and 5 were tagged. Using this information, which of the following is the best estimate of the total number of deer in the forest?

    a. 24,000 deer

    b. 30,000 deer

    c. 40,000 deer

    d. 100,000 deer

39. What is the correct factorization of the following binomial?
$$2y^3 - 128$$

    a. $2(y + 8)(y - 8)$

    b. $2(y - 4)(y^2 + 4y + 16)$

    c. $2(y - 4)(y + 4)^2$

    d. $2(y - 4)^3$

40. What is the simplified form of $(4y^3)^4(3y^7)^2$?

    a. $12y^{26}$

    b. $2304y^{16}$

    c. $12y^{14}$

    d. $2304y^{26}$

41. The number of members of the House of Representatives varies directly with the total population in a state. If the state of New York has 19,800,000 residents and has 27 total representatives, how many should Ohio have with a population of 11,800,000?

    a. 10

    b. 9

    c. 11

    d. 5

42. The following set represents the test scores from a university class: {35, 79, 80, 87, 87, 90, 92, 95, 95, 98, 99}. If the outlier is removed from this set, which of the following is TRUE?

    a. The mean and the median will decrease.

    b. The mean and the median will increase.

    c. The mean and the mode will increase.

    d. The mean and the mode will decrease.

43. Eva Jane is practicing for an upcoming 5K run. She has recorded the following times (in minutes):

       25, 18, 23, 28, 30, 22.5, 23, 33, 20

Use the above information to answer the next three questions to the closest minute. What is Eva Jane's mean time?

    a. 26 minutes

    b. 19 minutes

    c. 25 minutes

    d. 23 minutes

44. What is the mode of Eva Jane's time?

    a. 16 minutes

    b. 20 minutes

    c. 23 minutes

    d. 33 minutes

45. What is Eva Jane's median score?

    a. 23 minutes

    b. 17 minutes

    c. 28 minutes

    d. 19 minutes

46. Use the graph below entitled "Projected Temperatures for Tomorrow's Winter Storm" to answer the question.

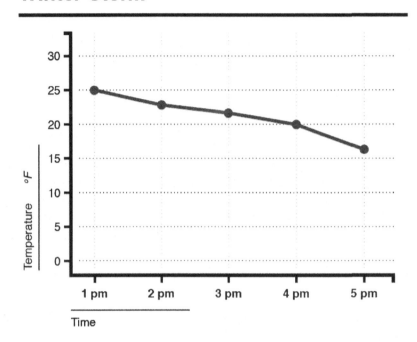

Projected Temperatures for Tomorrow's Winter Storm

What is the expected temperature at 3:00 p.m.?
a. 25 degrees
b. 22 degrees
c. 20 degrees
d. 16 degrees

Reasoning Through Language Arts

Questions 1–9 are based on the following passage:

Happiness lies not in the mere possession of money; it lies in the joy of achievement, in the thrill of creative effort. The joy, the moral stimulation of work no longer must be forgotten in the mad chase of evanescent profits. These dark days, my friends, will be worth all they cost us if they teach us that our true destiny is not to be ministered unto but to minister to ourselves, to our fellow men.

Recognition of that falsity of material wealth as the standard of success goes hand in hand with the abandonment of the false belief that public office and high political position are to be valued only by the standards of pride of place and personal profit; and there must be an end to a conduct in banking and in business which too often has given to a sacred trust the likeness of callous and selfish wrongdoing. Small wonder that confidence languishes, for it thrives only on honesty, on honor, on the sacredness of obligations, on faithful protection, and on unselfish performance; without them it cannot live.

Restoration calls, however, not for changes in ethics alone. This Nation is asking for action, and action now.

Our greatest primary task is to put people to work. This is no unsolvable problem if we face it wisely and courageously. It can be accomplished in part by direct recruiting by the Government itself, treating the task as we would treat the emergency of a war, but at the same time, through this employment, accomplishing great — greatly needed projects to stimulate and reorganize the use of our great natural resources.

Hand in hand with that we must frankly recognize the overbalance of population in our industrial centers and, by engaging on a national scale in a redistribution, endeavor to provide a better use of the land for those best fitted for the land.

Yes, the task can be helped by definite efforts to raise the values of agricultural products, and with this the power to purchase the output of our cities. It can be helped by preventing realistically the tragedy of the growing loss through foreclosure of our small homes and our farms. It can be helped by insistence that the Federal, the State, and the local governments act forthwith on the demand that their cost be drastically reduced. It can be helped by the unifying of relief activities which today are often scattered, uneconomical, unequal. It can be helped by national planning for and supervision of all forms of transportation and of communications and other utilities that have a definitely public character. There are many ways in which it can be helped, but it can never be helped by merely talking about it.

We must act. We must act quickly.

And finally, in our progress towards a resumption of work, we require two safeguards against a return of the evils of the old order. There must be a strict supervision of all banking and credits and investments. There must be an end to speculation with other people's money. And there must be provision for an adequate but sound currency ....

Through this program of action we address ourselves to putting our own national house in order and making income balance outgo. Our international trade relations, though vastly important, are in point of time, and necessity, secondary to the establishment of a sound national economy. I favor, as a practical policy, the putting of first things first. I shall spare no effort to restore world trade by international economic readjustment; but the emergency at home cannot wait on that accomplishment.

Excerpt from Franklin D. Roosevelt's First Inaugural Address, 1933

1. Based on context, the word "evanescent" in this passage means:
    a. Rising
    b. Soon to disappear
    c. Falling
    d. Nonexistent

2. Based on the passage, which of these is NOT a way to solve unemployment concerns:
    a. Treat unemployment like a war
    b. Involve the federal government
    c. Increase privately owned businesses
    d. Use available people power to tackle existing projects

3. What are the two primary problems Roosevelt outlines regarding unemployment?
   a. People don't want to work and there aren't enough jobs.
   b. The age of population doesn't match the jobs available and people don't trust businesses.
   c. Citizens are more focused on high-paying jobs and material wealth.
   d. Industrial centers are overpopulated and natural resources are underutilized.

4. Based on the passage, what does Roosevelt think are the MOST important elements to restoring the U.S. economy after the depression?
   a. Creating jobs in industrial centers, restoring international trade relationships, reevaluating the focus on high paying jobs and material wealth
   b. Focusing on the national economy, restoring trust in banks and investments, reassessing employment structure and programs, investing in natural resources
   c. Learning to focus on work over profits, valuing goodwill over selfishness in business, putting an end to foreclosures, investing in farming
   d. Focusing on ethics, restructuring financial support networks, establishing public utilities for the public good

5. Roosevelt mentions we need safeguards "against a return of the evils of the old order." This suggests:
   a. Without regulation, eventually the corrupt systems will return to bad practices.
   b. The United States must start from scratch and create a new system.
   c. One of the goals should be to teach systems to police themselves.
   d. Safeguards will improve ethics, thereby solving the corruption problems.

6. Which of the following does Roosevelt suggest is one of the problems with our relief activities?
   a. They are insufficient.
   b. They are redundant.
   c. They are unequal.
   d. They are unrealistic.

7. Based on the information in the passage, one can infer that it was important for Roosevelt to act quickly because:
   a. He had limited time in office.
   b. Failure to act quickly would give corrupt systems time to defend themselves.
   c. A lot of citizens were unemployed, losing their homes, and struggling to survive.
   d. He favored action over discussion and debate.

8. In the passage, what does the phrase "making income balance outgo" mean?
   a. The United States must spend money to make money.
   b. The United States must make more money than it spends.
   c. The United States must not spend any money until the country makes some.
   d. The United States must match spending with income.

9. Roosevelt suggests that, at that moment in time, the problem with public office was:
   a. It was viewed mostly as prestigious and profitable.
   b. Too many people sought public office.
   c. It was too corrupt.
   d. It wasn't solving the problems of the populace.

The next question is based on the following directions.

Follow these instructions to transform the word into something new.

1. Start with the word AUDITORIUM.
2. Eliminate all vowels except for the letter O.
3. Eliminate the "T."

10. Which of the following is the new word?
    a. Door
    b. Torn
    c. Dorm
    d. Tram

Author User Sav127 at English Wikipedia, Wikimedia Commons,
https://commons.wikimedia.org/wiki/File:Jeep_Odometer.jpg

11. According to the image, what's the highest speed you can measure on this odometer?
    a. 120mph
    b. 125mph
    c. 128mph
    d. 130mph

12. According to the image above, how many miles have been driven on this car?
    a. 125 miles
    b. 1,308 miles
    c. 91,308 miles
    d. 3,800 miles

Questions 13–16 are based on the following passage:

Meanwhile the fog and darkness thickened so, that people ran about with flaring links, proffering their services to go before horses in carriages, and conduct them on their way. The ancient tower of a church, whose gruff old bell was always peeping slyly down at Scrooge out of a Gothic window in the wall, became invisible, and struck the hours and quarters in the clouds, with tremulous vibrations afterwards as if its teeth were chattering in its frozen head up there. The cold became intense. In the main street, at the corner of the court, some labourers were repairing the gas-pipes, and had lighted a great fire in a brazier, round which a party of ragged men and boys were gathered: warming their hands and winking their eyes before the blaze in rapture. The

349

water-plug being left in solitude, its overflowings sullenly congealed, and turned to misanthropic ice. The brightness of the shops where holly sprigs and berries crackled in the lamp heat of the windows, made pale faces ruddy as they passed. Poulterers' and grocers' trades became a splendid joke; a glorious pageant, with which it was next to impossible to believe that such dull principles as bargain and sale had anything to do. The Lord Mayor, in the stronghold of the mighty Mansion House, gave orders to his fifty cooks and butlers to keep Christmas as a Lord Mayor's household should; and even the little tailor, whom he had fined five shillings on the previous Monday for being drunk and bloodthirsty in the streets, stirred up to-morrow's pudding in his garret, while his lean wife and the baby sallied out to buy the beef.

Foggier yet, and colder. Piercing, searching, biting cold. If the good Saint Dunstan had but nipped the Evil Spirit's nose with a touch of such weather as that, instead of using his familiar weapons, then indeed he would have roared to lusty purpose. The owner of one scant young nose, gnawed and mumbled by the hungry cold as bones are gnawed by dogs, stopped down at Scrooge's keyhole to regale him with a Christmas carol: but at the first sound of

"God bless you, merry gentleman! May nothing you dismay"

Scrooge seized the ruler with such energy of action, that the singer fled in terror, leaving the keyhole to the fog and even more congenial frost.

Excerpt from <u>A Christmas Carol</u> by Charles Dickens

13. In the context in which it appears, *congealed* most nearly means which of the following?
    a. Burst
    b. Loosened
    c. Shrank
    d. Thickened

14. Which of the following can NOT be inferred from the passage?
    a. The season of this narrative is in the winter time.
    b. The majority of the narrative is located in a bustling city street.
    c. This passage takes place during the night time.
    d. The Lord Mayor is a wealthy person within the narrative.

15. According to the passage, which of the following regarding the poulterers and grocers is true?
    a. They were so poor in the quality of their products that customers saw them as a joke.
    b. They put on a pageant in the streets every year for Christmas to entice their customers.
    c. They did not believe in Christmas, so they refused to participate in the town parade.
    d. They set their shops up to be entertaining public spectacles rather than a dull trade exchange.

16. The author's depiction of the scene in the last few paragraphs does all EXCEPT which of the following?
    a. Offer an allusion to religious affiliation in England.
    b. Attempt to evoke empathy for the character of Scrooge.
    c. Provide a palpable experience through the use of imagery and diction.
    d. Depict Scrooge as an uncaring, terrifying character to his fellows.

For more than five years I maintained myself thus solely by the labor of my hands, and I found, that by working about six weeks in a year, I could meet all the expenses of living. The whole of my winters, as well as most of my summers, I had free and clear for study. I have thoroughly tried school-keeping, and found that my expenses were in proportion, or rather out of proportion, to my income, for I was obliged to dress and train, not to say think and believe, accordingly, and I lost my time into the bargain. As I did not teach for the good of my fellow-men, but simply for a livelihood, this was a failure. I have tried trade; but I found that it would take ten years to get under way in that, and that then I should probably be on my way to the devil. I was actually afraid that I might by that time be doing what is called a good business. When formerly I was looking about to see what I could do for a living, some sad experience in conforming to the wishes of friends being fresh in my mind to tax my ingenuity, I thought often and seriously of picking huckleberries; that surely I could do, and its small profits might suffice,—for my greatest skill has been to want but little,—so little capital it required, so little distraction from my wonted moods, I foolishly thought. While my acquaintances went unhesitatingly into trade or the professions, I contemplated this occupation as most like theirs; ranging the hills all summer to pick the berries which came in my way, and thereafter carelessly dispose of them; so, to keep the flocks of Admetus. I also dreamed that I might gather the wild herbs, or carry evergreens to such villagers as loved to be reminded of the woods, even to the city, by hay-cart loads. But I have since learned that trade curses everything it handles; and though you trade in messages from heaven, the whole curse of trade attaches to the business.

As I preferred some things to others, and especially valued my freedom, as I could fare hard and yet succeed well, I did not wish to spend my time in earning rich carpets or other fine furniture, or delicate cookery, or a house in the Grecian or the Gothic style just yet. If there are any to whom it is no interruption to acquire these things, and who know how to use them when acquired, I relinquish to them the pursuit. Some are "industrious," and appear to love labor for its own sake, or perhaps because it keeps them out of worse mischief; to such I have at present nothing to say. Those who would not know what to do with more leisure than they now enjoy, I might advise to work twice as hard as they do,—work till they pay for themselves, and get their free papers. For myself I found that the occupation of a day-laborer was the most independent of any, especially as it required only thirty or forty days in a year to support one. The laborer's day ends with the going down of the sun, and he is then free to devote himself to his chosen pursuit, independent of his labor; but his employer, who speculates from month to month, has no respite from one end of the year to the other.

In short, I am convinced, both by faith and experience, that to maintain one's self on this earth is not a hardship but a pastime, if we will live simply and wisely; as the pursuits of the simpler nations are still the sports of the more artificial. It is not necessary that a man should earn his living by the sweat of his brow, unless he sweats easier than I do.

One young man of my acquaintance, who has inherited some acres, told me that he thought he should live as I did, if he had the means. I would not have any one adopt my mode of living on any account; for, beside that before he has fairly learned it I may have found out another for myself, I desire that there may be as many different persons in the world as possible; but I would have each one be very careful to find out and pursue his own way, and not his father's or his mother's or his neighbor's instead. The youth may build or plant or sail, only let him not be hindered from doing that which he tells me he would like to do. It is by a mathematical point only that we are wise, as the sailor or the fugitive slave keeps the polestar in his eye; but that is sufficient guidance for all our life. We may not arrive at our port within a calculable period, but we would preserve the true course.

Excerpt from Henry David Thoreau's <u>Walden</u>, 1854

17. Based on this passage, what does Thoreau value above "rich carpets and fine furniture"?
    a. A house in Grecian or Gothic Style
    b. A teaching job
    c. His freedom
    d. Everyone to want the same things

18. In the phrase "if he had the means," how would the speaker define the word "means"?
    a. Land
    b. Skills
    c. Help
    d. Time

19. Thoreau writes, "As I did not teach for the good of my fellow-men, but simply for a livelihood, this was a failure." Based on the rest of the passage, why was teaching a failure?
    a. Because he'd rather pick berries
    b. Because he's more suited to other trades
    c. Because his parents were teachers
    d. Because it is not his true course

20. The main idea of this passage is:
    a. We should live our lives the way we want but try to live simply.
    b. It's a waste of time to learn a trade.
    c. We should follow our friends' career advice.
    d. Materialism is bad.

21. Based on the passage, which profession does Thoreau think is BEST?
    a. Teacher
    b. Day laborer
    c. Farmer
    d. Sailor

22. What does Thoreau think free time is for?
    a. Individual pursuits
    b. Study
    c. More work
    d. Sleep

23. Thoreau says his expenses were out of proportion to his income. Why does he say this?
    a. He spent a lot on furniture and cookery.
    b. He needed to live in a big house.
    c. He didn't make a lot of money.
    d. Teaching required him to spend on clothing and training.

24. What does Thoreau say of the "industrious?"
    a. They make no time for fun or leisure activities.
    b. He aspires to be like them in his own life and pursuits.
    c. They seem to love working for the sake of the work itself.
    d. They are industrious simply to get more leisure time.

Questions 25–30 are based on the following two passages:

Passage A

The exercise of the benevolent affections may be briefly treated of, under nearly the same heads as those referred to when considering the principle of Justice;—keeping in mind that they lead to greater exertion for the benefit of others, and thus often demand a greater sacrifice of self-love, than is included under the mere requirements of justice. On the other hand, benevolence is not to be exercised at the expense of Justice; as would be the case, if a man were found relieving distress by such expedients as involve the necessity of withholding the payment of just debts, or imply the neglect or infringement of some duty which he owes to another.

(1.) Compassion and benevolent exertion are due toward alleviating the distresses of others. This exercise of them, in many instances, calls for a decided sacrifice of personal interest, and, in others, for considerable personal exertion. We feel our way to the proper measure of these sacrifices, by the high principle of moral duty, along with that mental exercise which places us in the situation of others, and, by a kind of reflected self-love, judges of the conduct due by us to them in our respective circumstances.—The details of this subject would lead us into a field too extensive for our present purpose. Pecuniary aid, by those who have the means, is the most easy form in which benevolence can be gratified, and that which often requires the least, if any, sacrifice of personal comfort or self-love. The same affection maybe exercised in a degree much higher in itself, and often much more useful to others, by personal exertion and personal kindness. The former, compared with the means of the individual, may present a mere mockery of mercy; while the latter, even in the lowest walks of life, often exhibit the brightest displays of active usefulness that can adorn the human character. This high and pure benevolence not only is dispensed with willingness, when occasions present themselves; but seeks out opportunities for itself, and feels in want of its natural and healthy exercise when deprived of an object on which it may be bestowed.

(2.) Benevolence is to be exercised toward the reputation of others. This consists not only in avoiding any injury to their characters, but in exertions to protect them against the injustice of others,—to correct misrepresentations,—to check the course of slander, and to obviate the efforts of those who would poison the confidence of friends, or disturb the harmony of society.

Excerpt from Compassion and Benevolence under The Affections in Section II Of <u>The Philosophy of the Moral Feelings</u> by John Abercrombie, M.D. OXON. & EDIN

Passage B

Giving alms to common beggars is naturally praised; because it seems to carry relief to the distressed and indigent: but when we observe the encouragement thence arising to idleness and debauchery, we regard that species of charity rather as a weakness than a virtue.

Tyrannicide, or the assassination of usurpers and oppressive princes, was highly extolled in ancient times; because it both freed mankind from many of these monsters, and seemed to keep the others in awe, whom the sword or poignard could not reach. But history and experience having since convinced us, that this practice increases the jealousy and cruelty of princes, a Timoleon and a Brutus, though treated with indulgence on account of the prejudices of their times, are now considered as very improper models for imitation.

Liberality in princes is regarded as a mark of beneficence, but when it occurs, that the homely bread of the honest and industrious is often thereby converted into delicious cates for the idle and the prodigal, we soon retract our heedless praises. The regrets of a prince, for having lost a day, were noble and generous: but had he intended to have spent it in acts of generosity to his greedy courtiers, it was better lost than misemployed after that manner.

Luxury, or a refinement on the pleasures and conveniences of life, had not long been supposed the source of every corruption in government, and the immediate cause of faction, sedition, civil wars, and the total loss of liberty. It was, therefore, universally regarded as a vice, and was an object of declamation to all satirists, and severe moralists. Those, who prove, or attempt to prove, that such refinements rather tend to the increase of industry, civility, and arts regulate anew our moral as well as political sentiments, and represent, as laudable or innocent, what had formerly been regarded as pernicious and blamable.

Upon the whole, then, it seems undeniable, that nothing can bestow more merit on any human creature than the sentiment of benevolence in an eminent degree; and that a part, at least, of its merit arises from its tendency to promote the interests of our species, and bestow happiness on human society. We carry our view into the salutary consequences of such a character and disposition; and whatever has so benign an influence, and forwards so desirable an end, is beheld with complacency and pleasure. The social virtues are never regarded without their beneficial tendencies, nor viewed as barren and unfruitful. The happiness of mankind, the order of society, the harmony of families, the mutual support of friends, are always considered as the result of their gentle dominion over the breasts of men.

Excerpt from Part II in Section II Of Benevolence in <u>An Enquiry Concerning the Principles of Morals</u> by David Hume

25. According to the author of passage A, which example of benevolence is the simplest to execute?
    a. Providing money to enable a student the opportunity to attend an educational workshop
    b. Holding a lemonade stand to raise funds and awareness for pediatric cancer
    c. Shopping for new toys to donate to a fundraiser that collects gifts for needy kids
    d. Volunteering to cook and serve Thanksgiving dinner for homeless people

26. In paragraph two of passage A, what kind of "mental exercise" does the author discuss?
    a. Practicing to improve an individual's recall from memory
    b. Trying to see things from another person's point of view
    c. Strengthening a person's ability to concentrate and focus
    d. Taking only a singular perspective into consideration

27. In paragraph one of passage A, what does the term "heads" mean?
    a. Discord
    b. Unlikeness
    c. Manner
    d. Opposition

28. According to passage A, which statement accurately reflects the relationship between benevolence and justice?
    a. Benevolence can be exercised at the expense of justice
    b. Acts of justice require selflessness
    c. Benevolence can be offered in lieu of payment of debts
    d. Justice can be exercised at the expense of benevolence

29. Why does the author of passage B say that giving food and money to beggars is seen as a weakness?
    a. That type of charity encourages laziness and corruption
    b. Some individuals are not truly deserving of the charity
    c. The grants cannot reach all who are affected by poverty
    d. Individuals refuse to accept the handouts due to their pride

30. A document explaining how to use an LCD flat screen TV is known as what kind of document?
    a. Technical
    b. Persuasive
    c. Narrative
    d. Cause and effect

The next two questions are based off the following passage:

> Rehabilitation rather than punitive justice is becoming much more popular in prisons around the world. Prisons in America, especially, where the recidivism rate is 67 percent, would benefit from mimicking prison tactics in Norway, which has a recidivism rate of only 20 percent. In Norway, the idea is that a rehabilitated prisoner is much less likely to offend than one harshly punished. Rehabilitation includes proper treatment for substance abuse, psychotherapy, healthcare and dental care, and education programs.

31. Which of the following best captures the author's purpose?
    a. To show the audience one of the effects of criminal rehabilitation by comparison.
    b. To persuade the audience to donate to American prisons for education programs.
    c. To convince the audience of the harsh conditions of American prisons.
    d. To inform the audience of the incredibly lax system of Norway prisons.

32. Which of the following describes the word *recidivism* as it is used in the passage?
    a. The lack of violence in the prison system.
    b. The opportunity of inmates to receive therapy in prison.
    c. The event of a prisoner escaping the compound.
    d. The likelihood of a convicted criminal to reoffend.

The next three questions are based off the following passage:

> What a lark! What a plunge! For so it had always seemed to her, when, with a little squeak of the hinges, which she could hear now, she had burst open the French windows and plunged at Bourton into the open air. How fresh, how calm, stiller than this of course, the air was in the early morning; like the flap of a wave; the kiss of a wave; chill and sharp and yet (for a girl of eighteen as she then was) solemn, feeling as she did, standing there at the open window, that something awful was about to happen; looking at the flowers, at the trees with the smoke winding off them and the rooks rising, falling; standing and looking until Peter Walsh said, "Musing among the vegetables?"— was that it? —"I prefer men to cauliflowers"— was that it? He must have said it at

breakfast one morning when she had gone out on to the terrace — Peter Walsh. He would be back from India one of these days, June or July, she forgot which, for his letters were awfully dull; it was his sayings one remembered; his eyes, his pocket-knife, his smile, his grumpiness and, when millions of things had utterly vanished — how strange it was! — a few sayings like this about cabbages.

Excerpt from <u>Mrs. Dalloway</u> by Virginia Woolf

33. In the box below, write down what type of writing (persuasive, expository, narrative, or technical) the passage is reflective of and why:

```



```

34. The narrator was feeling _____ right before Peter Walsh's voice distracted her.
   a. A spark of excitement for the morning.
   b. Anger at the larks.
   c. A sense of foreboding.
   d. Confusion at the weather.

35. What is the main point of the passage?
   a. To present the events leading up to a party.
   b. To show the audience that the narrator is resentful towards Peter.
   c. To introduce Peter Walsh back into the narrator's memory.
   d. To reveal what mornings are like in the narrator's life.

36. Which of the following is an acceptable heading to insert into the blank space?

Chapter 5: Literature and Language Usage
    I. Figurative Speech
        1. Alliteration
        2. Metaphors
        3. Onomatopoeia
        4. _____
    a. Nouns
    b. Agreement
    c. Personification
    d. Syntax

37. Felicia knew she had to be <u>prudent</u> if she was going to cross the bridge over the choppy water; one wrong move and she would be falling toward the rocky rapids.

Which of the following is the definition of the underlined word based on the context of the sentence above?

    a. Patient
    b. Afraid
    c. Dangerous
    d. Careful

The next three questions are based on the following passage:

> There was a man named Webster lived in a town of twenty-five thousand people in the state of Wisconsin. He had a wife named Mary and a daughter named Jane and he was himself a fairly prosperous manufacturer of washing machines. When the thing happened of which I am about to write he was thirty-seven or thirty-eight years old and his one child, the daughter, was seventeen. Of the details of his life up to the time a certain revolution happened within him it will be unnecessary to speak. He was however a rather quiet man inclined to have dreams which he tried to crush out of himself in order that he function as a washing machine manufacturer; and no doubt, at odd moments, when he was on a train going some place or perhaps on Sunday afternoons in the summer when he went alone to the deserted office of the factory and sat several hours looking out at a window and along a railroad track, he gave way to dreams.

Excerpt from <u>Many Marriages</u> by Sherwood Anderson

38. What does the author mean by the following sentence?

"Of the details of his life up to the time a certain revolution happened within him it will be unnecessary to speak."

    a. The details of his external life don't matter; only the details of his internal life matter.
    b. Whatever happened in his life before he had a certain internal change is irrelevant.
    c. He had a traumatic experience earlier in his life which rendered it impossible for him to speak.
    d. Before the revolution, he was a lighthearted man who always wished to speak to others no matter who they were.

39. What Point Of View is this narrative told in?
    a. First person limited
    b. First person omniscient
    c. Second person
    d. Third person

40. In the space below, detail what it is Webster does for a living:

```



```

41. After Sheila recently had a coronary artery bypass, her doctor encouraged her to switch to a plant-based diet to avoid foods loaded with cholesterol and saturated fats. Sheila's doctor has given her a list of foods she can purchase in order to begin making healthy dinners, which excludes dairy (cheese, yogurt, cream) eggs, and meat. The doctor's list includes the following: pasta, marinara sauce, tofu, rice, black beans, tortilla chips, guacamole, corn, salsa, rice noodles, stir-fry vegetables, teriyaki sauce, quinoa, potatoes, yams, bananas, eggplant, pizza crust, cashew cheese, almond milk, bell pepper, and tempeh.

Which of the following dishes can Sheila make that will be okay for her to eat?

    a. Eggplant parmesan with a salad
    b. Veggie pasta with marinara sauce
    c. Egg omelet with no cheese and bell peppers
    d. Quinoa burger with cheese and French fries.

42. Ecologist: If we do not act now, more than one hundred animal species will be extinct by the end of the decade. The best way to save them is to sell hunting licenses for endangered species. Hunters can pay for the right to kill old and lame animals. Otherwise, there's no way to fund our conservation efforts.

Which one of the following assumptions does the ecologist's argument rely upon?
    a. Hunting licenses for non-endangered species aren't profitable.
    b. All one hundred animal species must be saved.
    c. The new hunting license revenue will fund conservation efforts.
    d. Conservation efforts should've began last decade.

The following three questions are based on the following passage:

> For any journey, by rail or by boat, one has a general idea of the direction to be taken, the character of the land or water to be crossed, and of what one will find at the end. So it should be in striking the trail. Learn all you can about the path you are to follow. Whether it is plain or obscure, wet or dry; where it leads; and its length, measured more by time than by actual miles. A smooth, even trail of five miles will not consume the time and strength that must be expended upon a trail of half that length which leads over uneven ground, varied by bogs and obstructed by rocks and fallen trees, or a trail that is all up-hill climbing. If you are a novice and accustomed to walking only over smooth and level ground, you must allow more time for covering the distance than an experienced person would require and must count upon the expenditure of more strength, because your feet are not trained to the wilderness paths with their pitfalls and traps for the unwary, and every nerve and muscle will be strained to secure a safe foothold amid the tangled roots, on the slippery, moss-covered logs, over precipitous rocks that lie in your path. It will take time to pick your way over boggy places where the water oozes up through the thin, loamy soil as through a sponge; and experience alone will teach you which hummock of grass or moss will make a safe stepping-place and will not sink beneath your weight and soak your feet with hidden water. Do not scorn to learn all you can about the trail you are to take . . . It is not that you hesitate to encounter difficulties, but that you may prepare for them. In unknown regions

take a responsible guide with you, unless the trail is short, easily followed, and a frequented one. Do not go alone through lonely places; and, being on the trail, keep it and try no explorations of your own, at least not until you are quite familiar with the country and the ways of the wild.

Excerpt from the book <u>On the Trail</u> by Lina Beard and Adelia Belle Beard

43. What does the author say about unknown regions?
   a. You should try and explore unknown regions in order to learn the land better.
   b. Unless the trail is short or frequented, you should take a responsible guide with you.
   c. All unknown regions will contain pitfalls, traps, and boggy places.
   d. It's better to travel unknown regions by rail rather than by foot.

44. Which statement is NOT a detail from the passage?
   a. Learning about the trail beforehand is imperative.
   b. Time will differ depending on the land.
   c. Once you are familiar with the outdoors you can go places on your own.
   d. Be careful for wild animals on the trail you are on.

45. Write down what type of passage this is (descriptive, persuasive, narrative, or informative) and why in the blank below.

Extended Response

For the extended response question, you will be given forty-five minutes to write one analysis essay of 500 words over two opposing passages. The ideal passage is four to seven paragraphs long, with each paragraph consisting of three to seven sentences.

Topic: Should people have smart assistants in their home?

Passage 1

Smart assistants like "Google Home" and "Amazon Echo" are great additional features to have in the home. You can buy these assistants online for a couple hundred dollars from major retailers throughout the country. These smart assistants connect wirelessly to other electronic devices in your home and act on command to make your life more efficient. For example, you can say "Alexa, turn the music up forty percent," and the music goes up. Or you can say, "OK Google, dim the lights," and your lights will dim without you having to step a foot off the couch! Devices hooked up to your TV also follow commands, such as searching for certain TV shows or playing your favorite selections.

There is really no need to worry about safety with smart assistants; we are told by people like Janie Adams and Ray Gold, both trustworthy CEO's of big tech companies, that a small percentage of our information goes to advertisers, but there is a limit to what and how much is sent to them. What's the big deal, anyway, if the shoes you have been dreaming about pop up on your google search every so often? There is no harm here because we don't have any proof that says the government has our information and is using it to monitor us.

Passage 2

Smart assistants are very interesting devices that turn your home into an AI heaven. All of your devices can be connected to one central assistant that answers your every command. These assistants make everything in the home much more efficient to manage. However, having graduated with a degree in Information Technology Management, I can confidently say that Smart Assistants like Amazon's "Alexa" and "OK Google" are dangerous to have in the home.

The danger comes from an invasion of privacy. In order for smart assistants to deliver convenience, they must be able to listen to everything you say, even if this means a personal conversation between you and your friend. Georgia Tech tells us in their recent 2016 study of smart assistants called "Smart Enough?," that the assistants work by sending whatever question you have to a server farm that stores all your questions indefinitely, for now. It also doesn't put our minds at ease to know that tech giants such as Robert Moley and Jennifer Lackey refuse to keep smart assistants in their home or office. Additionally, the article by the renowned Tech Giant Magazine called "Too Cool for Assistants?" tells us that Amazon and Google send approximately twenty percent of your information to companies who want to advertise to you in subtle ways, like on your Facebook feed or through ads on other social media services. Compromised privacy is the main concern with smart assistants, so if you don't want everything you say out in cyberspace, I would suggest buying a vacation instead.

Extended Response Prompt:

Analyze the argument in the two speeches. Develop an argument where you explain which argument is better supported and why. Be sure to present specific evidence from both passages to support the argument. Note that you shouldn't talk about the argument you most *agree* with; rather, you should talk about which argument has *better support* and why.

Science

For questions 1–2:

The annual wildland fire statistics are compiled and reported by the National Interagency Coordination Center at NIFC for federal and state agencies. Information is gathered through situation reports from many decades.

The table below lists the number of fires during the years since 2007 and the acres those fires covered:

Total Wildland Fires and Acres (1926 - 2017)

The National Interagency Coordination Center at NIFC compiles annual wildland fire statistics for federal and state agencies. This information is provided through Situation Reports, which have been in use for several decades. Prior to 1983, sources of these figures are not known, or cannot be confirmed, and were not derived from the current situation reporting process. As a result, the figures prior to 1983 should not be compared to later data.

Source: National Interagency Coordination Center

Year	Fires	Acres
2017	71,499	10,026,086
2016	67,743	5,509,995
2015	68,151	10,125,149
2014	63,312	3,595,613
2013	47,579	4,319,546
2012	67,774	9,326,238
2011	74,126	8,711,367
2010	71,971	3,422,724
2009	78,792	5,921,786
2008	78,979	5,292,468
2007	85,795	9,328,045

1. What was the average number of fires from 2007 to 2012?
    a. 67,774
    b. 76,240
    c. 70,512
    d. 78,792

2. What could be inferred about the years with the highest number of acres covered by its fires?
   a. There was a misreporting of the number of fires.
   b. The most acres reflected the greatest number of fires.
   c. The most acres reflected larger-sized fires.
   d. No conclusion can be drawn.

For questions 3–4:

Kepler's 3 Laws of Planetary Motion

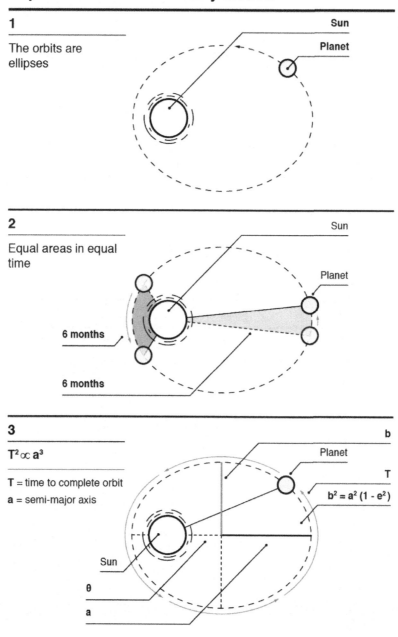

1

The orbits are ellipses

Sun

Planet

2

Equal areas in equal time

Sun

Planet

6 months

6 months

3

$T^2 \propto a^3$

T = time to complete orbit
a = semi-major axis

b

Planet

T

$b^2 = a^2 (1 - e^2)$

Sun

θ

a

3. According to the diagrams, which of the following is TRUE?

    a. When a Planet X is closest to the Sun, it travels faster.

    b. When Planet X is closest to the Sun, it travels slowest.

    c. Planet X is always the same distance away from the Sun.

    d. The distance Planet X is from the Sun is directly proportional to the square root of the time it takes to compete one orbit.

4. **Short answer.**

Would Kepler's Laws work for all of the planets in our solar system? Defend your answer.

For question 5–6:

The table below displays the decay products and associated energy million electron volts (MeV) and the type of particle emitted with the decay:

Isotope Particle	Decay Product	Energy (MeV)	Decay Type
Radon 222	Polonium 218	6.190	Alpha
Lead 210	Mercury 206	2.992	Alpha
Mercury 206	Thallium 206	0.912	Beta
Thallium 206	Lead 206	0.813	Beta

5. What can be inferred about the relationship between certain decay particles and the amount of energy emitted during radioactive decay?

    a. There is no relationship.

    b. Alpha particles are emitted with higher levels of energy.

    c. Beta particles are emitted with higher levels of energy.

    d. Alpha particles are emitted with all levels of energy.

6. The decay of Lead 210 would stop with the transition into which of the follow elements?

    a. Radon 222

    b. Mercury 206

    c. Thallium 206

    d. Lead 206

For questions 7–8:

Cell Phases

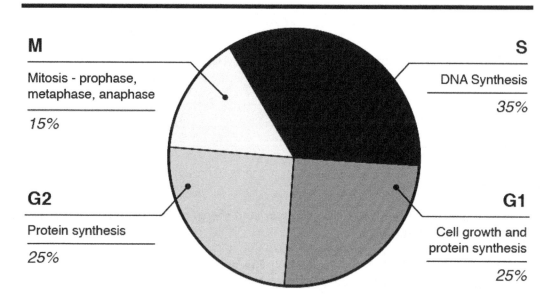

M

Mitosis - prophase, metaphase, anaphase

15%

S

DNA Synthesis

35%

G2

Protein synthesis

25%

G1

Cell growth and protein synthesis

25%

7. The chart shows the amount of time a cell spends in each part of the cyclin process. Which phase takes up the majority of the process?
   a. G1
   b. G2
   c. M
   d. S

8. If the entire cyclin time frame is completed in 120 minutes, how many minutes would the cell be in mitosis?
   a. 8 minutes
   b. 15 minutes
   c. 18 minutes
   d. 80 minutes

9. What type of volcano occurs when a crack opens in the land over a lava flow?
   a. Fissure
   b. Shield
   c. Dome
   d. Caldera

For questions 10–11:

The chart below lists temperature anomalies (both land and ocean) beginning in 1880 in degrees Celsius and degrees Fahrenheit:

Global Land and Ocean Temperature Anomalies, March

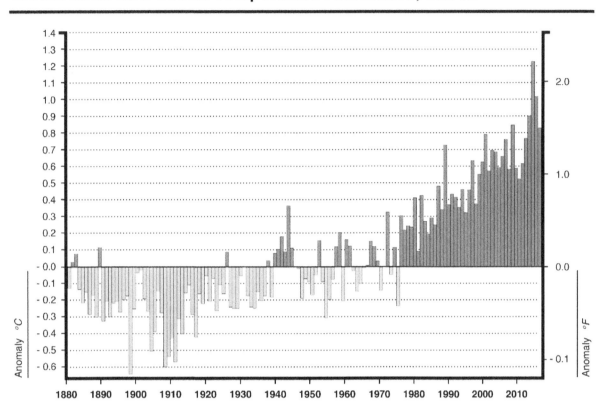

10. Approximately which year shows the highest temperature anomaly?
    a. 1900
    b. 1944
    c. 1990
    d. 2016

11. What could be inferred about the trend in temperature anomalies?
    a. Temperatures are increasing.
    b. Temperatures are decreasing.
    c. Temperatures have remained steady.
    d. Nothing can be inferred about the temperature anomalies.

For questions 12–13:

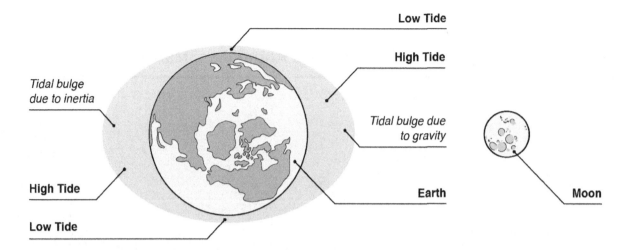

12. According to the diagram, which side of the Earth would experience a high tide?
    a. The side facing the moon
    b. The side not facing or in line with the moon
    c. The moon does not create tides on the Earth
    d. The side with the largest body of water

13. When the Sun is lined up with the moon on the same side of the Earth, what effect would this produce on the Earth's tides?
    a. No effect
    b. Higher high tides
    c. No tides due to the opposite pulls of the Sun and moon on the Earth
    d. Longer lasting high tides

For questions 14–15:

Soil samples were collected from various locations and analyzed for their composition. These minerals were sand, silt, and clay. Three types of minerals were identified and measured by percent in each sample.

Soil Sample	Sand (%)	Clay (%)	Silt (%)
1	75	5	20
2	5	80	15
3	20	35	45
4	70	15	15
5	55	25	20

14. What is the average percent of sand in all of the samples?
    a. 25
    b. 35
    c. 45
    d. 55

15. If Sample 3 weighed 5 kg, how much of that was made up of silt?
    a. 2.25 kg
    b. 3.50 kg
    c. 4.25 kg
    d. 5.50 kg

For questions 16–20:

Dendrochronology is a method of dating using tree growth through the counting of concentric bands in the trunk. This is not a completely accurate method of calculation; therefore, it is often paired with other methods of cross dating, involving referencing characteristics of rings with other samples in an area of similar conditions. Variations in bands can be caused by environmental conditions such as annual rainfall, as the rings were formed. When there is less rainfall, fewer rings are formed, and those that are formed during this time are often narrower than in times of heavier rainfall, as heavier rain can result in faster growth.

Scientists studied oak trees at three separate sites and compiled the following data. The number of trees sampled was greater than 50.

Site	Average number of growth bands per year	Average size of growth bands (mm)
1	11	2
2	14	3
3	19	11

16. Which site received the most rainfall?
    a. Site 1
    b. Site 2
    c. Site 3
    d. Cannot be determined

17. From the graph below, which would best represent the relationship between the average number of growth bands verses the average size of growth bands?

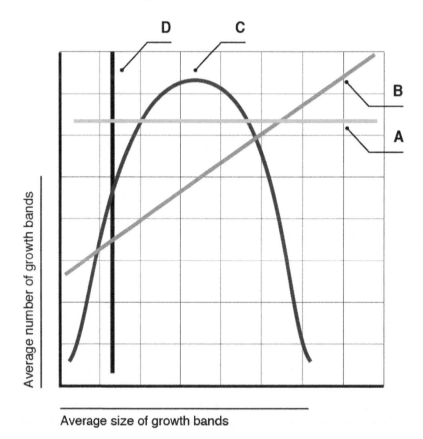

    a. A
    b. B
    c. C
    d. D

18. Which of the following would describe the trees at Site 1 when compared to those at Site 2?
    a. Trees were not homogeneous
    b. Trees experienced the same growth rate
    c. Trees experienced a slower growth rate
    d. Trees experienced a faster growth rate

19. Why is cross dating done?
    a. To decrease the number of trees necessary for study
    b. To predict the amount of rainfall for a specific area
    c. To improve the accuracy when predicting growth rate for trees
    d. To decrease the number of bands used for calculating rainfall for a specific area

20. Trees from another site were measured and found to have an average of 17 growth bands per year. When compared to the data in Table 1, what would likely be the average size of these growth bands?
    a. Less than 2 mm
    b. Between 2 mm and 4 mm
    c. Between 4 mm and 12 mm
    d. Greater than 12 mm

For questions 21–23:

Object Moving with Constant Speed		Object Moving with Changing Speed	
Time (s)	Position (m)	Time (s)	Position (m)
0	0	0	0
1	8	1	2
2	16	2	6
3	24	3	12
4	32	4	20

21. If the velocity (speed in a direction) for an object is the measurement of the change in position divided by the change in time, modeled by the equation $v = \frac{\Delta x}{\Delta t}$, what is the velocity between 3s and 4s for the object moving with a constant speed?
    a. 0 m/s
    b. 4 m/s
    c. 6 m/s
    d. 8 m/s

22. What is the average velocity for the chart of the object moving with changing speed between 1 and 3 seconds?
    a. 1 m/s
    b. 3 m/s
    c. 5 m/s
    d. 7 m/s

23. What would be the rate at which the velocity is changing in the left-hand chart display?
    a. Average velocity
    b. Average speed
    c. Force
    d. Acceleration

The following table maps the method of transportation for a new species discovered in South America.

	Male	Female	Total
Fly	34	46	80
Walk/Jump	28	17	45
Swim	15	12	27
Slither/Crawl	52	17	69
Total	129	92	221

24. What is the probability of randomly selecting a creature that swims?
    a. 0
    b. 12/221
    c. 15/221
    d. 27/221

25. What is the probability of randomly selecting a female creature or a creature that walks?
    a. 0
    b. 17/221
    c. 137/221
    d. 120/221

For questions 26–27:

The structure of the Earth has many layers. Starting with the center, or the core, the Earth comprises two separate sections: the inner core and the outer core. The innermost portion of the core is a solid center consisting of approximately 760 miles of iron. The outer core is slightly less than 1400 miles in thickness and consists of a liquid nickel-iron alloy. The next section out from the core also has two layers. This section is the mantle, and it is split into the lower mantle and the upper mantle. Both layers of the mantle consist of magnesium and iron, and they are extremely high in temperature. This hot temperature causes the metal contained in the lower mantle to rise and then cool slightly as it reaches the upper mantle. Once the metal begins to cool, it falls back down toward the lower mantle, restarting the whole process again. The motion of rising and falling with in the layers of the mantle is the cause of plate tectonics and movement of the outermost layer of the Earth. The outermost layer of the Earth is called the crust. Movements between the mantle and the crust create effects such as earthquakes and volcanoes.

26. What is the primary difference between the Earth's inner core and outer core?
    a. The inner core is made of nickel, while the outer core is made of iron
    b. The inner core is solid, while the outer core is molten
    c. The inner core is inaccessible to people, while the outer core is accessible to people
    d. The inner core is hot, while the outer core is cool

27. Which of the following generally has the greatest temperature?
   a. The oceans
   b. The lower mantle
   c. The upper mantle
   d. The crust

For questions 28–31:

When the materials of surfaces move across one another they have a specific coefficient of friction that impedes their movement. In order to begin the sliding motion of these surfaces, a coefficient of static friction must be overcome. Once the materials begin moving across each other, the coefficient of kinetic friction is less than that of what it took to begin the sliding motion. The coefficient of friction does not have units, as it represents a proportion.

The table below lists the coefficients of static and kinetic friction of rubber on various materials:

Material 1	Material 2	Kinetic friction	Static friction
Rubber	Concrete	0.67	0.89
Rubber	Asphalt	0.66	0.84
Rubber	Cardboard	0.54	0.79
Rubber	Ice	0.14	

The graph below shows the relationship between the coefficient of friction and the distance moved for rubber on three of the materials:

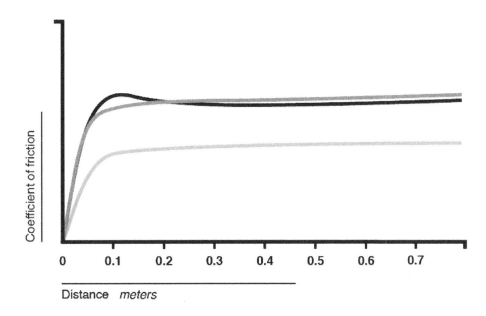

28. If material X is slightly rougher than cardboard, but smoother than asphalt, which of the following could be an approximation for the coefficient of kinetic friction of rubber on material X?
   a. 0.54
   b. 0.60
   c. 0.66
   d. Cannot be determined

371

29. Which of the following would describe the graphs of the coefficient of friction versus distance for the given materials?
    a. Linear
    b. Quadratic
    c. Exponential decay
    d. Logarithmic

30. Which color line in the graph would most likely match up to the information for concrete?
    a. Black
    b. Dark gray (the line that is higher up on the *y*-axis)
    c. Light gray (the line that is lower down on the *y*-axis)
    d. None would match concrete

31. Why do the lines in the graph all seem to level off just before a certain reading on the coefficient of friction?
    a. There are miscalculations that have altered the data
    b. They are all nearly the same material
    c. The sliding friction is overcome at that point
    d. A lubricant is added at that point

For questions 32–33:

The following diagram shows the populations of 90 different birds that were recorded across three continents. Overlap shows that a type of bird was found on all three continents.

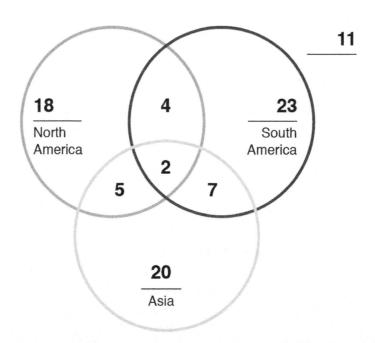

32. What is the probability that a specific bird is not found in North America, South America, or Asia?
    a. 11/90
    b. 18/90
    c. 20/90
    d. 23/90

33. What is the probability that a specific bird that has been found in North America is found when only looking in the population of birds in South America?

    a. 6/36

    b. 23/36

    c. 6/90

    d. 23/90

34. Which of the following scenarios describes mistaking correlation for causation?

    a. In the same year, the drinking water was contaminated by a nearby chicken farm, and stomach cancer in the area increased 3 percent. The contaminated water obviously caused stomach cancer to increase.

    b. Victoria notices that every time she turns on the stove burner underneath a pot with water in it, the water boils. Victoria realizes that putting heat under water causes it to boil.

    c. We saw that the thunderstorm that day caused all the neighborhoods in the East Village to flood. We could hardly get to the movie that night because of all the rain!

    d. Hector learned in school that when tectonic plates shift, it causes earthquakes to happen. There had been many earthquakes when Hector lived in California, and now he knew that the shifting of plates is what caused them.

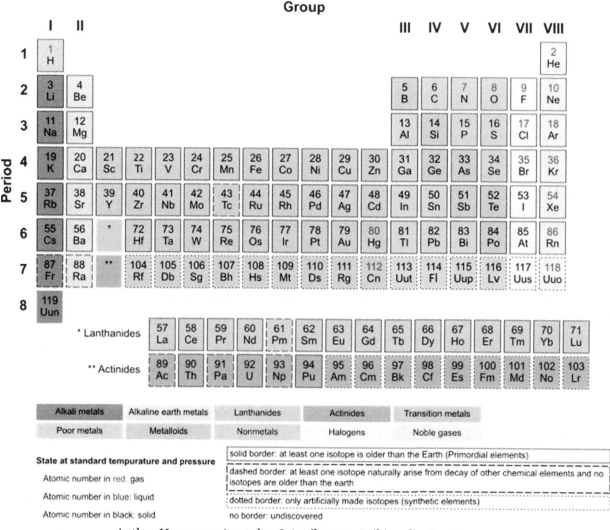

Group

Author Username Armtuk, 28 April 2007, Wikimedia Commons,
https://commons.wikimedia.org/wiki/File:Periodic_Table_Armtuk3.svg

35. Which of the following is an example of a physical property of substances?
   a. Heat of combustion
   b. Reactivity
   c. Odor
   d. Toxicity

36. Brazil nuts are a food source high in selenium, which is known for its anti-inflammatory properties. Inflammation is known to be at the source of diseases such as cancer and high blood pressure, so consuming anti-inflammatory foods would help to avoid cancer and disease. Therefore, everyone who eats brazil nuts will avoid cancer.

The pattern of flawed reasoning in which one of the following arguments is most parallel to that in the argument above?

    a. Microwaves contain high levels of RF radiation in order to warm up foods. Exposure to radiation can lead to radiation poisoning, which can cause symptoms such as vomiting, headaches, and fatigue. Therefore, some people who own microwaves may experience flu-like symptoms if exposed to too much radiation.

    b. Brown rice is a food that is known to contain arsenic, which can harm the human body. Arsenic is known to play a role in the development of cancer and diabetes, so consuming arsenic would potentially cause someone to become sick. Therefore, everyone who eats brown rice will become sick.

    c. Oranges do not contain Vitamin D. Those looking to add Vitamin D to their diets should incorporate tofu, shiitake mushrooms, or spinach. Vitamin D is important for calcium absorption and bone growth. Vitamin D deficiency is often linked to certain types of cancer and weight gain.

    d. Whales have lungs instead of gills. Whales are mammals, and therefore use lungs to breathe when they come up to the surface of the water. When whales go beneath the water, they simply hold their breath until it is time to come up again. Everyone should know about whales because they are interesting creatures.

37. A study done in 1992 delved into what would be negative effects of the drug PMR, which had just come onto the market that same year, and was used for maintaining blood pressure. Among the effects were swelling of the abdomen, skin disintegration, and complications during childbirth. Because of these disastrous findings, the drug was discontinued in 1993. Another study done in 1994 was concerned to see that complications during childbirth were up 30 percent in the year 1992; 1991 and 1993 had similar results pertaining to childbirth complications. From these findings, we can assume that the first study was correct; the negative effects of the drug PMR included complications during childbirth as one of its most catastrophic effects.

The reasoning in the argument is most vulnerable to criticism on the grounds that the argument

    a. makes circular reasoning based on the results.

    b. mistakes correlation for causation.

    c. erroneously provides the opposite of the wrong answer to be true.

    d. uses a red herring to distract from the argument's purpose.

Social Studies

The next three questions are based on the following passage:

> The old castle soon proved to be too small for the family, and in September 1853 the foundation-stone of a new house was laid. After the ceremony the workmen were entertained at dinner, which was followed by Highland games and dancing in the ballroom.

> Two years later they entered the new castle, which the Queen described as "charming; the rooms delightful; the furniture, papers, everything perfection."

> The Prince was untiring in planning improvements, and in 1856 the Queen wrote: "Every year my heart becomes more fixed in this dear Paradise, and so much more so now, that *all* has become

my dearest Albert's *own* creation, own work, own building, own laying out as at Osborne; and his great taste, and the impress of his dear hand, have been stamped everywhere. He was very busy today, settling and arranging many things for next year."

Excerpt from the biography <u>Queen Victoria</u> by E. Gordon Browne, M.A.

1. Which of the following is this excerpt considered?
    a. Primary source
    b. Secondary source
    c. Tertiary source
    d. None of these

2. It took _____ years for the new castle to be built.

3. What does the word *impress* mean in the third paragraph?
    a. To affect strongly in feeling
    b. To urge something to be done
    c. To impose a certain quality upon
    d. To press a thing onto something else

The next question is based on the following passage:

According to the plan of the convention, all judges who may be appointed by the United States are to hold their offices *during good behavior*, which is conformable to the most approved of the State constitutions and among the rest, to that of this State. Its propriety having been drawn into question by the adversaries of that plan, is no light symptom of the rage for objection, which disorders their imaginations and judgments. The standard of good behavior for the continuance in office of the judicial magistracy, is certainly one of the most valuable of the modern improvements in the practice of government. In a monarchy it is an excellent barrier to the despotism of the prince; in a republic it is a no less excellent barrier to the encroachments and oppressions of the representative body. And it is the best expedient which can be devised in any government, to secure a steady, upright, and impartial administration of the laws.

Excerpt from "The Federalist No. 78" by Alexander Hamilton

4. What is Hamilton's point in this excerpt?
    a. To show the audience that despotism within a monarchy is no longer the standard practice in the states.
    b. To convince the audience that judges holding their positions based on good behavior is a practical way to avoid corruption.
    c. To persuade the audience that having good behavior should be the primary characteristic of a person in a government body and their voting habits should reflect this.
    d. To convey the position that judges who serve for a lifetime will not be perfect and therefore we must forgive them for their bad behavior when it arises.

The next six are based on the following passage:

The box in which the President sat consisted of two boxes turned into one, the middle partition being removed, as on all occasions when a state party visited the theater. The box was on a level with the dress circle; about twelve feet above the stage. There were two entrances—the door nearest to the wall having been closed and locked; the door nearest the balustrades of the dress circle, and at right angles with it, being open and left open, after the visitors had entered. The interior was carpeted, lined with crimson paper, and furnished with a sofa covered with crimson velvet, three arm chairs similarly covered, and six cane-bottomed chairs. Festoons of flags hung before the front of the box against a background of lace.

President Lincoln took one of the arm-chairs and seated himself in the front of the box, in the angle nearest the audience, where, partially screened from observation, he had the best view of what was transpiring on the stage. Mrs. Lincoln sat next to him, and Miss Harris in the opposite angle nearest the stage. Major Rathbone sat just behind Mrs. Lincoln and Miss Harris. These four were the only persons in the box.

The play proceeded, although "Our American Cousin," without Mr. Sothern, has, since that gentleman's departure from this country, been justly esteemed a very dull affair. The audience at Ford's, including Mrs. Lincoln, seemed to enjoy it very much. The worthy wife of the President leaned forward, her hand upon her husband's knee, watching every scene in the drama with amused attention. Even across the President's face at intervals swept a smile, robbing it of its habitual sadness.

About the beginning of the second act, the mare, standing in the stable in the rear of the theater, was disturbed in the midst of her meal by the entrance of the young man who had quitted her in the afternoon. It is presumed that she was saddled and bridled with exquisite care.

Having completed these preparations, Mr. Booth entered the theater by the stage door; summoned one of the scene shifters, Mr. John Spangler, emerged through the same door with that individual, leaving the door open, and left the mare in his hands to be held until he (Booth) should return. Booth who was even more fashionably and richly dressed than usual, walked thence around to the front of the theater, and went in. Ascending to the dress circle, he stood for a little time gazing around upon the audience and occasionally upon the stage in his usual graceful manner. He was subsequently observed by Mr. Ford, the proprietor of the theater, to be slowly elbowing his way through the crowd that packed the rear of the dress circle toward the right side, at the extremity of which was the box where Mr. and Mrs. Lincoln and their companions were seated. Mr. Ford casually noticed this as a slightly extraordinary symptom of interest on the part of an actor so familiar with the routine of the theater and the play.

Excerpt from <u>The Life, Crime, and Capture of John Wilkes Booth</u> by George Alfred Townsend

5. Which of the following best describes the author's attitude toward the events leading up to the assassination of President Lincoln?
   a. Excitement due to the setting and its people.
   b. Sadness due to the death of a beloved president.
   c. Anger because of the impending violence.
   d. Neutrality due to the style of the report.

6. In the blank provided, write down what the author means by the last sentence in the passage?

7. Given the author's description of the play "Our American Cousin," which of the following is most analogous to Mr. Sothern's departure from the theater?

    a. A ballet dancer who leaves the New York City Ballet just before they go on to their final performance.

    b. A basketball player leaves an NBA team and the next year they make it to the championship but lose.

    c. A lead singer leaves their band to begin a solo career, and the band drops in sales by 50 percent on their next album.

    d. A movie actor who dies in the middle of making a movie and the movie is made anyway with an actor who resembles the deceased.

8. Based on the organizational structure of the passage, which of the following texts most closely relates?

    a. A chronological account in a fiction novel of a woman and a man meeting for the first time.

    b. A cause-and-effect text ruminating on the causes of global warming.

    c. An autobiography that begins with the subject's death and culminates in his birth.

    d. A text focusing on finding a solution to the problem of the Higgs boson particle.

9. Which of the following words, if substituted for the word *festoons* in the first paragraph, would LEAST change the meaning of the sentence?

    a. Feathers

    b. Armies

    c. Adornments

    d. Buckets

10. What is the primary purpose of the passage?

    a. To persuade the audience that John Wilkes Booth killed Abraham Lincoln.

    b. To inform the audience of the setting wherein Lincoln was shot.

    c. To narrate the bravery of Lincoln and his last days as President.

    d. To recount in detail the events that led up to Abraham Lincoln's death.

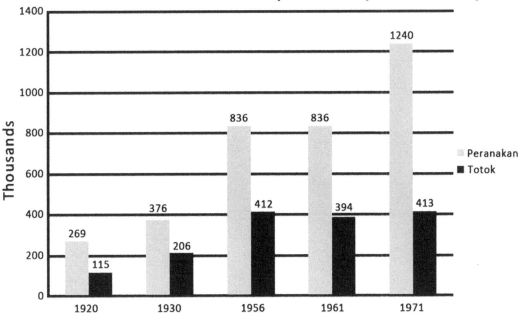

Totok and Peranakan Population (1920-1971)

11. According to the graph above, what is the population of Totok in 1961?
    a. 115,000
    b. 206,000
    c. 394,000
    d. 836,000

12. According to the graph, what is the difference between the population in Peranakan in 1971 and Peranakan in 1961?
    a. 394,000
    b. 403,000
    c. 404,000
    d. 412,000

Use this map for Question 13.

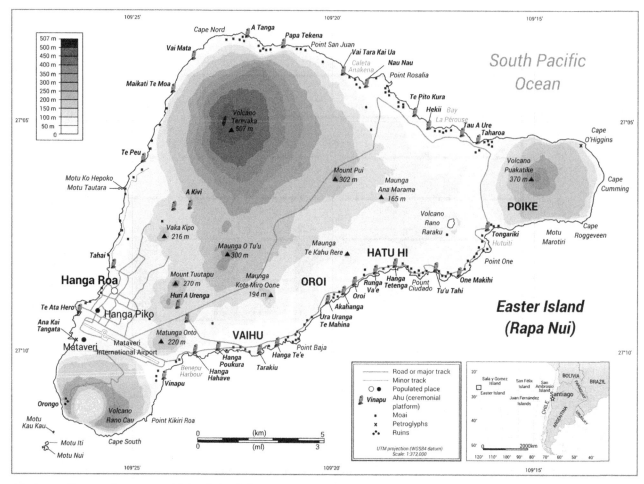

13. According to the map, which of the following is the highest points on the island?
   a. Volcano Terevaka
   b. Maunga Ana Marama
   c. Puakatike Volcano
   d. Vaka Kipo

The next seven questions are based on the following two passages, labeled "Passage A" and "Passage B":

Passage A

(from "Free Speech in War Time" by James Parker Hall, written in 1921, published in Columbia Law Review, Vol. 21 No. 6)

In approaching this problem of interpretation, we may first put out of consideration certain obvious limitations upon the generality of all guaranties of free speech. An occasional unthinking malcontent may urge that the only meaning not fraught with danger to liberty is the literal one that no utterance may be forbidden, no matter what its intent or result; but in fact it is nowhere seriously argued by anyone whose opinion is entitled to respect that direct and intentional incitations to crime may not be forbidden by the state. If a state may properly forbid murder or robbery or treason, it may also punish those who induce or counsel the commission of such crimes. Any other view makes a mockery of the state's power to declare and punish offences. And what the state may do to prevent the incitement of serious crimes which are universally condemned, it may also do to prevent the incitement of lesser crimes, or of those in regard to the bad tendency of which public opinion is divided. That is, if the state may punish John for burning straw in an alley, it may also constitutionally punish Frank for inciting John to do it, though Frank did so by speech or writing. And if, in 1857, the United States could punish John for helping a fugitive slave to escape, it could also punish Frank for inducing John to do this, even though a large section of public opinion might applaud John and condemn the Fugitive Slave Law.

Passage B

(from "Freedom of Speech in War Time" by Zechariah Chafee, Jr. written in 1919, published in Harvard Law Review Vol. 32 No. 8)

The true boundary line of the First Amendment can be fixed only when Congress and the courts realize that the principle on which speech is classified as lawful or unlawful involves the balancing against each other of two very important social interests, in public safety and in the search for truth. Every reasonable attempt should be made to maintain both interests unimpaired, and the great interest in free speech should be sacrificed only when the interest in public safety is really imperiled, and not, as most men believe, when it is barely conceivable that it may be slightly affected. In war time, therefore, speech should be unrestricted by the censorship or by punishment, unless it is clearly liable to cause direct and dangerous interference with the conduct of the war.

Thus our problem of locating the boundary line of free speech is solved. It is fixed close to the point where words will give rise to unlawful acts. We cannot define the right of free speech with the precision of the Rule against Perpetuities or the Rule in Shelley's Case, because it involves national policies which are much more flexible than private property, but we can establish a workable principle of classification in this method of balancing and this broad test of certain danger. There is a similar balancing in the determination of what is "due process of law." And we can with certitude declare that the First Amendment forbids the punishment of words merely for their injurious tendencies. The history of the Amendment and the political function of free speech corroborate each other and make this conclusion plain.

14. Which one of the following questions is central to both passages?
    a. What is the interpretation of the first amendment and its limitations?
    b. Do people want absolute liberty, or do they only want liberty for a certain purpose?
    c. What is the true definition of freedom of speech in a democracy?
    d. How can we find an appropriate boundary of freedom of speech during wartime?

15. The authors of the two passages would be most likely to DISAGREE over which of the following?
    a. A man is thrown in jail due to his provocation of violence in Washington, D.C. during a riot.
    b. A man is thrown in jail for stealing bread for his starving family, and the judge has mercy for him and lets him go.
    c. A man is thrown in jail for encouraging a riot against the U.S. government for the wartime tactics, although no violence ensues.
    d. A man is thrown in jail because he has been caught as a German spy working within the U.S. army.

16. The relationship between Passage *A* and Passage *B* is most analogous to the relationship between the documents described in which of the following?

   a. A journal article in the Netherlands about the law of euthanasia that cites evidence to support only the act of passive euthanasia as an appropriate way to die; a journal article in the Netherlands about the law of euthanasia that cites evidence to support voluntary euthanasia in any aspect.

   b. An article detailing the effects of radiation in Fukushima; a research report describing the deaths and birth defects as a result of the hazardous waste dumped on the Somali Coast.

   c. An article that suggests that labor laws during times of war should be left up to the states; an article that showcases labor laws during the past that have been altered due to the current crisis of war.

   d. A research report arguing that the leading cause of methane emissions in the world is from agriculture practices; an article citing that the leading cause of methane emissions in the world is from the transportation of coal, oil, and natural gas.

17. The author uses the examples in the last lines of Passage *A* in order to do what?

   a. To demonstrate different types of crimes for the purpose of comparing them to see by which one the principle of freedom of speech would become objectionable.

   b. To demonstrate that anyone who incites a crime, despite the severity or magnitude of the crime, should be held accountable for that crime in some degree.

   c. To prove that the definition of "freedom of speech" is altered depending on what kind of crime is being committed.

   d. To show that some crimes are in the best interest of a nation and should not be punishable if they are proven to prevent harm to others.

18. Which of the following, if true, would most seriously undermine the claim proposed by the author in Passage *A* that if the state can punish a crime, then it can punish the incitement of that crime?

   a. The idea that human beings are able and likely to change their minds between the utterance and execution of an event that may harm others.

   b. The idea that human beings will always choose what they think is right based on their cultural upbringing.

   c. The idea that the limitation of free speech by the government during wartime will protect the country from any group that causes a threat to that country's freedom.

   d. The idea that those who support freedom of speech probably have intentions of subverting the government.

19. What is the primary purpose of the second passage?

   a. To analyze the First Amendment in historical situations in order to make an analogy to the current war at hand in the nation.

   b. To demonstrate that the boundaries set during wartime are different from that when the country is at peace, and that we should change our laws accordingly.

   c. To offer the idea that during wartime, the principle of freedom of speech should be limited to that of even minor utterances in relation to a crime.

   d. To call upon the interpretation of freedom of speech to be already evident in the First Amendment and to offer a clear perimeter of the principle during war time.

20. Which of the following words, if substituted for the word *malcontent* in Passage *A*, would LEAST change the meaning of the sentence?
    a. Grievance
    b. Cacophony
    c. Anecdote
    d. Residua

The next four questions are based on the following passage and graph.

Portland is a very beautiful city of 60,000 inhabitants, and situated on the Willamette river twelve miles from its junction with the Columbia. It is perhaps true of many of the growing cities of the West, that they do not offer the same social advantages as the older cities of the East. But this is principally the case as to what may be called boom cities, where the larger part of the population is of that floating class which follows in the line of temporary growth for the purposes of speculation, and in no sense applies to those centers of trade whose prosperity is based on the solid foundation of legitimate business. As the metropolis of a vast section of country, having broad agricultural valleys filled with improved farms, surrounded by mountains rich in mineral wealth, and boundless forests of as fine timber as the world produces, the cause of Portland's growth and prosperity is the trade which it has as the center of collection and distribution of this great wealth of natural resources, and it has attracted, not the boomer and speculator, who find their profits in the wild excitement of the boom, but the merchant, manufacturer, and investor, who seek the surer if slower channels of legitimate business and investment. These have come from the East, most of them within the last few years. They came as seeking a better and wider field to engage in the same occupations they had followed in their Eastern homes, and bringing with them all the love of polite life which they had acquired there, have established here a new society, equaling in all respects that which they left behind. Here are as fine churches, as complete a system of schools, as fine residences, as great a love of music and art, as can be found at any city of the East of equal size.

But while Portland may justly claim to be the peer of any city of its size in the United States in all that pertains to social life, in the attractions of beauty of location and surroundings it stands without its peer. The work of art is but the copy of nature. What the residents of other cities see but in the copy, or must travel half the world over to see in the original, the resident of Portland has at its very door.

The city is situate on a gently-sloping ground, with, on the one side, the river, and on the other a range of hills, which, within easy walking distance, rise to an elevation of a thousand feet above the river, affording a most picturesque building site. From the very streets of the thickly settled portion of the city, the Cascade Mountains, with the snow-capped peaks of Hood, Adams, St. Helens, and Rainier, are in plain view.

Excerpt from <u>Oregon, Washington, and Alaska. Sights and Scenes for the Tourist</u>, written by E.L. Lomax in 1890

Population of Cities in 1890

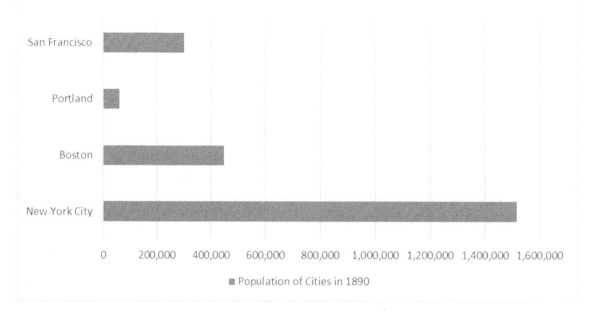

Population of Cities in 1890

21. What is a characteristic of a "boom city," as indicated by the passage?
    a. A city that is built on a solid business foundation of mineral wealth and farming.
    b. An area of land on the west coast that quickly becomes populated by residents from the east coast.
    c. A city that, due to the hot weather and dry climate, catches fire frequently, resulting in a devastating population drop.
    d. A city whose population is made up of people who seek quick fortunes rather than building a solid business foundation.

22. By stating that "they do not offer the same social advantages as the older cities of the East" in the first paragraph, the author most likely intends to suggest which of the following?
    a. Inhabitants who reside in older cities in the East are much more social that inhabitants who reside in newer cities in the West because of background and experience.
    b. Cities in the West have no culture compared to the East because the culture in the East comes from European influence.
    c. Cities in the East are older than cities in the West, and older cities always have better culture than newer cities.
    d. Since cities in the West are newly established, it takes them a longer time to develop cultural roots and societal functions than those cities that are already established in the East.

23. Based on the information at the end of the first paragraph, what would the author say of Portland?
    a. It has twice as much culture as the cities in the East.
    b. It has as much culture as the cities in the East.
    c. It doesn't have as much culture as cities in the East.
    d. It doesn't have as much culture as cities in the West.

24. How many more citizens did San Francisco have than Portland in 1890?
    a. Approximately 240,000
    b. Approximately 500,000
    c. Approximately 1,000,000
    d. Approximately 1,500,000

The next question is based on the following document, which is a section of the Constitution. This section of the Constitution focuses on the Senate. It includes directions about the methods and parameters for selecting a senator. Additionally, it provides an analysis of some of the leadership roles and duties of the US Senate.

Section. 3.

The Senate of the United States shall be composed of two Senators from each State, chosen by the Legislature thereof, for six Years; and each Senator shall have one Vote.

Immediately after they shall be assembled in Consequence of the first Election, they shall be divided as equally as may be into three Classes. The Seats of the Senators of the first Class shall be vacated at the Expiration of the second Year, of the second Class at the Expiration of the fourth Year, and of the third Class at the Expiration of the sixth Year, so that one third may be chosen every second Year; and if Vacancies happen by Resignation, or otherwise, during the Recess of the Legislature of any State, the Executive thereof may make temporary Appointments until the next Meeting of the Legislature, which shall then fill such Vacancies.

No Person shall be a Senator who shall not have attained to the Age of thirty Years, and been nine Years a Citizen of the United States, and who shall not, when elected, be an Inhabitant of that State for which he shall be chosen.

The Vice President of the United States shall be President of the Senate, but shall have no Vote, unless they be equally divided.

The Senate shall [choose] their other Officers, and also a President pro tempore, in the Absence of the Vice President, or when he shall exercise the Office of President of the United States.

The Senate shall have the sole Power to try all Impeachments. When sitting for that Purpose, they shall be on Oath or Affirmation. When the President of the United States is tried, the Chief Justice shall preside: And no Person shall be convicted without the Concurrence of two thirds of the Members present.

Judgment in Cases of Impeachment shall not extend further than to removal from Office, and disqualification to hold and enjoy any Office of honor, Trust or Profit under the United States: but the Party convicted shall nevertheless be liable and subject to Indictment, Trial, Judgment and Punishment, according to Law.

25. According to Section 3 of the U.S. Constitution, how long is the term for U.S. Senators?
    a. Two years
    b. Six years
    c. Lifelong
    d. Four years

The next question is based on the following document.

> A house divided against itself cannot stand. I believe this government cannot endure, permanently, half slave and half free. I do not expect the Union to be dissolved—I do not expect the house to fall—but I do expect it will cease to be divided. It will become all one thing or all the other. Either the opponents of slavery will arrest the further spread of it, and place it where the public mind shall rest in the belief that it is in the course of ultimate extinction; or its advocates will push it forward, till it shall become lawful in all the States, old as well as new—North as well as South.

Excerpt from Abraham Lincoln's "House Divided" speech, which was delivered to the Republican state convention on June 16, 1858

26. When President Lincoln uses the phrase a "house divided," what is he most likely referring to?
    a. Political conflicts in the White House
    b. Tensions between North and South over slavery
    c. Dissolution of the Union
    d. Ideological differences in his family

The following graphic depicts a political cartoon poster used as propaganda during World War II.

27. What is the main message of the poster?
    a. American citizens should stop driving cars.
    b. People who drive alone are Nazi sympathizers.
    c. Joining a Car-Sharing Club is a sign of American patriotism.
    d. Carpooling helps save fuel for the war effort.

28. Which term best describes the relationship between the Japanese attack on Pearl Harbor and the United States' entry into World War II?
    a. Correlation
    b. Causation
    c. Opposition
    d. Indirect connection

The next question is based on the following bar graph, which depicts the violent crime rate in the United States between 1999 and 2005:

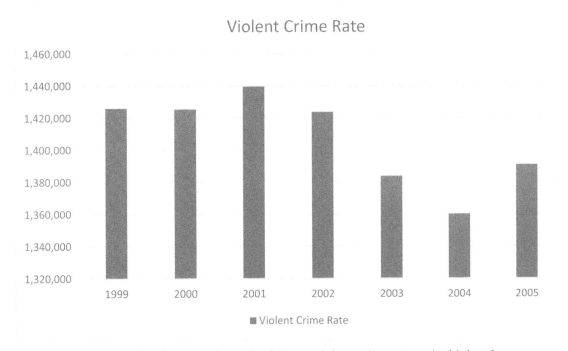

29. According to the graph, when was the United States violent crime rate at its highest?
    a. 1999
    b. 2001
    c. 2002
    d. 2005

The following graphic depicts a political cartoon that was possibly published in the Washington Evening Star, and used as propaganda during World War I.

30. Which historical event is most likely depicted by the political cartoon above?
    a. Mexican Cession
    b. Mexican-American War
    c. Zimmerman Telegram
    d. Immigration Act of 1965

The next question is based on the document below, which is a table that analyzes murder and nonnegligent manslaughter in 9 major cities over the course of a 25-year period:

Murder and Nonnegligent Manslaughter								
	1990	1995	2000	2005	2010	2012	2015	Peak Year
Chicago	850	824	631	448	432	500	473	939 in 1992
New York	2245	1177	673	539	536	419	352	2245 in 1990
Dallas	447	276	231	202	148	154	136	500 in 1991
Los Angeles	983	849	550	489	293	299	283	1094 in 1992
Milwaukee	155	138	122	121	95	91	145	163 in 1991
St. Louis	177	204	124	131	144	113	188	267 in 1993
Baltimore	305	322	261	269	223	218	344	353 in 1993
Washington D.C.	472	361	239	195	132	88	162	482 in 1991
New Orleans	304	363	204		175	193	165	424 in 1994

Source: Information comes from the Federal Bureau of Investigation

31. According to the table, how many murders and nonnegligent manslaughters occurred in Dallas during its peak year?
    a. 202
    b. 231
    c. 447
    d. 500

Use this graph for Question 32:

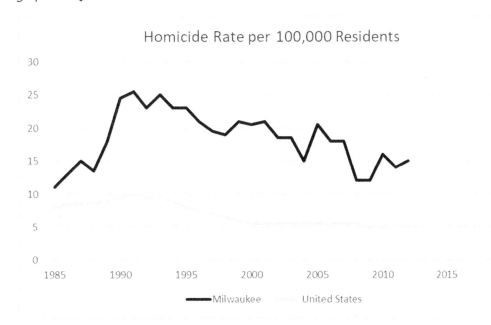

Homicide Rate per 100,000 Residents

— Milwaukee ⋯⋯ United States

32. According to the graph above, at its peak, Milwaukee had how many homicides per 100,000 residents?
    a. 10.5
    b. 20
    c. 25.5
    d. 30

389

The next question is based on the following passage:

> Those who came before us made certain that this country rode the first waves of the industrial revolutions, the first waves of modern invention, and the first wave of nuclear power, and this generation does not intend to founder in the backwash of the coming age of space. We mean to be a part of it—we mean to lead it. For the eyes of the world now look into space, to the moon and to the planets beyond, and we have vowed that we shall not see it governed by a hostile flag of conquest, but by a banner of freedom and peace. We have vowed that we shall not see space filled with weapons of mass destruction, but with instruments of knowledge and understanding.

Excerpt from John F. Kennedy's "We Choose to Go to the Moon" speech in 1962

33. Considering that this speech was delivered during the Cold War, which nation is John F. Kennedy alluding to with the phrase "hostile flag of conquest"?
   a. Great Britain
   b. Nazi Germany
   c. Soviet Union
   d. Iraq

34. In which era in American history would a sign protesting desegregation in the 1950s or 1960s likely be found?
   a. Civil Rights Movement
   b. Progressivism
   c. Emancipation
   d. Reconstruction

35. Which of the following data visualization tools would best be used for illustrating the percentage of the Black American population in the U.S. affected by unconstitutional searches during the civil rights era of the 1950s and 1960s compared to entire population of Black Americans in the U.S. (part or percentage of a whole)?
   a. Pie Chart
   b. Line Graph
   c. Table
   d. Bar Graph

The next question is based on the following speech:

> It will be recorded that the distance of Hawaii from Japan makes it obvious that the attack was deliberately planned many days or even weeks ago. During the intervening time the Japanese Government has deliberately sought to deceive the United States by false statements and expressions of hope for continued peace.

> The attack yesterday on the Hawaiian Islands has caused severe damage to American naval and military forces. I regret to tell you that very many American lives have been lost. In addition, American ships have been reported torpedoed on the high seas between San Francisco and Honolulu.

> Yesterday the Japanese Government also launched an attack against Malaya.
> Last night Japanese forces attacked Hong Kong.
> Last night Japanese forces attacked Guam.
> Last night Japanese forces attacked the Philippine Islands.

Last night the Japanese attacked Wake Island.
And this morning the Japanese attacked Midway Island.

Japan has therefore undertaken a surprise offensive extending throughout the Pacific area. The facts of yesterday and today speak for themselves. The people of the United States have already formed their opinions and well understand the implications to the very life and safety of our nation.

As Commander-in-Chief of the Army and Navy I have directed that all measures be taken for our defense, that always will our whole nation remember the character of the onslaught against us.

No matter how long it may take us to overcome this premeditated invasion, the American people, in their righteous might, will win through to absolute victory.

I believe that I interpret the will of the Congress and of the people when I assert that we will not only defend ourselves to the uttermost but will make it very certain that this form of treachery shall never again endanger us.

Hostilities exist. There is no blinking at the fact that our people, our territory and our interests are in grave danger.

With confidence in our armed forces, with the unbounding determination of our people, we will gain the inevitable triumph. So help us God.

I ask that the Congress declare that since the unprovoked and dastardly attack by Japan on Sunday, December 7th, 1941, a state of war has existed between the United States and the Japanese Empire.

Speech given by Franklin D. Roosevelt on December 8, 1941. It is about the Japanese bombing of Pearl Harbor.

36. Which causal relationship in American history does this document capture?
    a. The Japanese attack on Pearl Harbor and the United States' entry into World War II
    b. The Japanese attack on Pearl Harbor and the United States' entry into the League of Nations
    c. The Japanese attack on Pearl Harbor and the end of World War I
    d. The Japanese attack on Pearl Harbor and the failure of Congress to declare war

Answer Explanations #3

Mathematical Reasoning

1. B: The fraction $\frac{12}{60}$ can be reduced to $\frac{1}{5}$, which puts the fraction in lowest terms. First, it must be converted to a decimal. Dividing 1 by 5 results in 0.2. Then, to convert to a percentage, move the decimal point two units to the right and add the percentage symbol. The result is 20%.

2. B: Common denominators must be used. The LCD is 15, and $\frac{2}{5} = \frac{6}{15}$. Therefore, $\frac{14}{15} + \frac{6}{15} = \frac{20}{15}$, and in lowest terms, the answer is $\frac{4}{3}$. A common factor of 5 was divided out of both the numerator and denominator.

3. A: A product is found by multiplying. Multiplying two fractions together is easier when common factors are cancelled first to avoid working with larger numbers.

$$\frac{5}{14} \times \frac{7}{20} = \frac{5}{2 \times 7} \times \frac{7}{5 \times 4} = \frac{1}{2} \times \frac{1}{4} = \frac{1}{8}$$

4. D: Division is completed by multiplying by the reciprocal. Therefore,

$$24 \div \frac{8}{5} = \frac{24}{1} \times \frac{5}{8} = \frac{3 \times 8}{1} \times \frac{5}{8} = \frac{15}{1} = 15$$

5. C: Common denominators must be used. The LCD is 168, so each fraction must be converted to have 168 as the denominator.

$$\frac{5}{24} - \frac{5}{14} = \frac{5}{24} \times \frac{7}{7} - \frac{5}{14} \times \frac{12}{12}$$

$$\frac{35}{168} - \frac{60}{168} = -\frac{25}{168}$$

6. C: The correct mathematical statement is the one in which the smaller of the two numbers is on the "less than" side of the inequality symbol. It is written in answer *C* that $\frac{1}{3} > -\frac{4}{3}$, which is the same as $-\frac{4}{3} < \frac{1}{3}$, a correct statement.

7. C: $-\frac{1}{5} > \frac{4}{5}$ is incorrect. The expression on the left is negative, which means that it is smaller than the expression on the right. As it is written, the inequality states that the expression on the left is greater than the expression on the right, which is not true.

8. D: This is a one-step real-world application problem. The unknown quantity is the number of cases of cola to be purchased. Let x be equal to this amount. Because each case costs $3.50, the total number of cases multiplied by $3.50 must equal $40. This translates to the mathematical equation $3.5x = 40$. Divide both sides by 3.5 to obtain $x = 11.4286$, which has been rounded to four decimal places. Because cases are sold whole (the store does not sell portions of cases), and there is not enough money to purchase 12 cases. Therefore, there is only enough money to purchase 11 cases.

9. A: Rounding can be used to find the best approximation. All of the values can be rounded to the nearest thousand. 15,412 SUVs can be rounded to 15,000. 25,815 station wagons can be rounded to 26,000. 50,412 sedans can be rounded to 50,000. 8,123 trucks can be rounded to 8,000. Finally, 18,312 hybrids can be rounded to 18,000. The sum of the rounded values is 117,000, which is closest to 120,000.

10. D: There are 52 weeks in a year, and if the family spends $105 each week, that amount is close to $100. A good approximation is $100 a week for 50 weeks, which is found through the product $50 \times 100 = \$5,000$.

11. D: There were 48 total bags of apples sold. If 9 bags were Granny Smith and the rest were Red Delicious, then $48 - 9 = 39$ bags were Red Delicious. Therefore, the ratio of Granny Smith to Red Delicious is 9:39.

12. B: The average rate of change is found by calculating the difference in dollars over the elapsed time. Therefore, the rate of change is equal to $\$4,900 - \$4,000 \div 3$ months, which is equal to $\$900 \div 3$ or $300 a month.

13. A: Let x be the unknown, the number of hours Erin can work. We know Katie works $2x$, and the sum of all hours is less than 21. Therefore, $x + 2x < 21$, which simplifies into $3x < 21$. Solving this results in the inequality $x < 7$ after dividing both sides by 3. Therefore, Erin worked less than 7 hours.

14. 0.42: If a calculator were to be used, 33 would be divided into 14. Since a calculator is not permitted, multiply both the numerator and denominator by 3. This results in the fraction $\frac{42}{99}$, and hence a decimal of 0.42.

15. 21 questions: Gina answered 60% of 35 questions correctly; 60% can be expressed as the decimal 0.60. Therefore, she answered $0.60 \times 35 = 21$ questions correctly.

16. 16: The unknown quantity is the number of total questions on the test. Let x be equal to this unknown quantity. Therefore, $0.75x = 12$. Divide both sides by 0.75 to obtain $x = 16$.

17. A: First, the distributive property must be used on the left side. This results in $3x + 6 = 14x - 5$. The addition property is then used to add 5 to both sides, and then to subtract $3x$ from both sides, resulting in $11 = 11x$. Finally, the multiplication property is used to divide each side by 11. Therefore, $x = 1$ is the solution.

18. C: First, like terms are collected to obtain $12 - 5x = -5x + 12$. Then, if the addition principle is used to move the terms with the variable, $5x$ is added to both sides and the mathematical statement $12 = 12$ is obtained. This is always true; therefore, all real numbers satisfy the original equation.

19. D: The distributive property is used on both sides to obtain $4x + 20 + 6 = 4x + 6$. Then, like terms are collected on the left, resulting in $4x + 26 = 4x + 6$. Next, the addition principle is used to subtract $4x$ from both sides, and this results in the false statement $26 = 6$. Therefore, there is no solution.

20. A: First, the variables have to be defined. Let x be the first integer; therefore, $x + 1$ is the second integer. This is a two-step problem. The sum of three times the first and two less than the second is translated into the following expression: $3x + (x + 1 - 2)$. This expression is set equal to 411 to obtain:

$$3x + (x + 1 - 2) = 412$$

The left-hand side is simplified to obtain $4x - 1 = 411$. The addition and multiplication properties are used to solve for x. First, add 1 to both sides and then divide both sides by 4 to obtain $x = 103$. The next consecutive integer is 104.

21. B: First, the information is translated into the ratio $\frac{15}{80}$. To find the percentage, translate this fraction into a decimal by dividing 15 by 80. The corresponding decimal is 0.1875. Move the decimal point two places to the right to obtain the percentage 18.75%.

22. B: If sales tax is 7.25%, the price of the car must be multiplied by 1.0725 to account for the additional sales tax. Therefore:

$$15,395 \times 1.0725 = 16,511.1375$$

This amount is rounded to the nearest cent, which is $16,511.14.

23. A: The chart is a bar chart showing how many men and women prefer each genre of movies. The dark gray bars represent the number of women, while the light gray bars represent the number of men. The light gray bars are higher and represent more men than women for the genres of Comedy and Action.

24. B: A line graph represents continuous change over time. The line on the graph is continuous and not broken, as on a scatter plot. A bar graph may show change but isn't necessarily continuous over time. A pie graph is better for representing percentages of a whole. Histograms are best used in grouping sets of data in bins to show the frequency of a certain variable.

25. C: The mean for the number of visitors during the first 4 hours is 14. The mean is found by calculating the average for the four hours. Adding up the total number of visitors during those hours gives $12 + 10 + 18 + 16 = 56$. Then $56 \div 4 = 14$.

26. C: The mode for a set of data is the value that occurs the most. The grade that appears the most is 95. It's the only value that repeats in the set.

27. B: The relationship between age and time for attention span is a positive correlation because the general trend for the data is up and to the right. As age increases, so does attention span.

28. B: The only relation in which every x-value corresponds to exactly one y-value is the relation given in Choice *B*, making it a function. The other relations have the same first component paired up to different second components, which goes against the definition of a function.

29. A: To find a function value, plug in the number given for the variable and evaluate the expression, using the order of operations (parentheses, exponents, multiplication, division, addition, subtraction). The function given is a polynomial function and:

$$f(5) = 5^2 - 2 \times 5 + 1 = 25 - 10 + 1 = 16$$

30. C: First set the functions equal to one another, resulting in:

$$x^2 + 3x + 4 = 4x + 2$$

This is a quadratic equation, so the equivalent equation in standard form is $x^2 - x + 2 = 0$. This equation can be solved by factoring into $(x - 2)(x + 1) = 0$. Setting both factors equal to zero results in $x = 2$ and $x = -1$.

31. D: There will be no more coyotes when the population is 0, so set y equal to 0 and solve the quadratic equation:

$$0 = -(x - 2)^2 + 1600$$

Subtract 1600 from both sides, and divide through by -1. This results in $1600 = (x - 2)^2$. Then, take the square root of both sides. This process results in the following equation: $\pm 40 = x - 2$. Adding 2 to both sides results in two solutions: $x = 42$ and $x = -38$. Because the problem involves years after 2000, the only solution that makes sense is 42. Add 42 to 2000; therefore, in 2042 there will be no more coyotes.

32. B: The ball is back at the starting point when the function is equal to 800 feet. Therefore, this results in solving the equation:

$$800 = -32t^2 + 90t + 800$$

Subtract 800 off of both sides and factor the remaining terms to obtain:

$$0 = 2t(-16 + 45t)$$

Setting both factors equal to 0 results in $t = 0$, which is when the ball was thrown up initially, and:

$$t = \frac{45}{16} = 2.8 \text{ seconds}$$

Therefore, it will take the ball 2.8 seconds to come back down to its staring point.

33. A: The probability of .9 is closer to 1 than any of the other answers. The closer a probability is to 1, the greater the likelihood that the event will occur. The probability of 0.05 shows that it is very unlikely that an adult driver will wear their seatbelt because it is close to zero. A zero probability means that it will not occur. The probability of 0.25 is closer to zero than to one, so it shows that it is unlikely an adult will wear their seatbelt.

34. D: This system can be solved using the method of substitution. Solving the first equation for y results in $y = 14 - 2x$. Plugging this into the second equation gives $4x + 2(14 - 2x) = -28$, which simplifies to $28 = -28$, an untrue statement. Therefore, this system has no solution because no x value will satisfy the system.

35. B: The slopes of perpendicular lines are negative reciprocals, meaning their product is equal to -1. The slope of the line given needs to be found. Its equivalent form in slope-intercept form is $y = -\frac{4}{7}x + 23$, so its slope is $-\frac{4}{7}$. The negative reciprocal of this number is $\frac{7}{4}$. The only line in the options given with this same slope is:

$$y = \frac{7}{4}x - 12$$

36. C: This system can be solved using substitution. Plug the second equation in for y in the first equation to obtain $2x - 8x = 6$, which simplifies to $-6x = 6$. Divide both sides by 6 to get $x = -1$, which is then back-substituted into either original equation to obtain $y = -8$.

37. A: The formula for the rate of change is the same as slope: change in y over change in x. The y-value in this case is percentage of smokers and the x-value is year. The change in percentage of smokers from 2000 to 2015 was 8.1%. The change in x was $2000 - 2015 = -15$. Therefore:

$$8.1\% \div -15 = -0.54\%$$

The percentage of smokers decreased 0.54% each year.

38. A: A proportion should be used to solve this problem. The ratio of tagged to total deer in each instance is set equal, and the unknown quantity is a variable x. The proportion is $\frac{300}{x} = \frac{5}{400}$. Cross-multiplying gives $120{,}000 = 5x$, and dividing through by 5 results in 24,000.

39. B: First, the common factor 2 can be factored out of both terms, resulting in $2(y^3 - 64)$. The resulting binomial is a difference of cubes that can be factored using the rule $a^3 - b^3 = (a - b)(a^2 + ab + b^2)$ with a = y and b = 4. Therefore, the result is:

$$2(y - 4)(y^2 + 4y + 16)$$

40. D: The exponential rules $(ab)^m = a^m b^m$ and $(a^m)^n = a^{mn}$ can be used to rewrite the expression as:

$$4^4 y^{12} \times 3^2 y^{14}$$

The coefficients are multiplied together and the exponential rule $a^m a^n = a^{m+n}$ is then used to obtain the simplified form $2304y^{26}$.

41. B: The number of representatives varies directly with the population, so the equation necessary is $N = k \times P$, where N is number of representatives, k is the variation constant, and P is total population in millions. Plugging in the information for New York allows k to be solved for. This process gives $27 = k \times 20$, so $k = 1.35$. Therefore, the formula for number of representatives given total population in millions is $N = 1.35 \times P$. Plugging in $P = 11.6$ for Ohio results in $N = 15.66$, which rounds up to 16 total Representatives.

42. B: The outlier is 35. When a small outlier is removed from a data set, the mean and the median increase. The first step in this process is to identify the outlier, which is the number that lies away from the given set. Once the outlier is identified, the mean and median can be recalculated. The mean will be affected because it averages all of the numbers. The median will be affected because it finds the middle number, which is subject to change because a number is lost. The mode will most likely not change because it is the number that occurs the most, which will not be the outlier if there is only one outlier.

43. C: The mean is found by adding all the times together and dividing by the number of times recorded. $25 + 18 + 23 + 28 + 30 + 22.5 + 23 + 33 + 20 = 222.5$, divided by $9 = 24.7$. Rounding to the nearest minute, the mean is 25 minutes.

44. C: The mode is the time from the data set that occurs most often. The number 23 occurs twice in the data set, while all others occur only once, so the mode is 23 minutes.

45. A: To find the median of a data set, you must first list the numbers from smallest to largest, and then find the number in the middle. If there are two numbers in the middle, as in this data set, add the two numbers in the middle together and divide by 2. Putting this list in order from smallest to greatest yields 18, 20, 22.5, 23, 23, 25, 28, 30, and 33, where 23 is the middle number.

46. B: Look on the horizontal axis to find 3:00 p.m. Move up from 3:00 p.m. to reach the dot on the graph. Move horizontally to the left to the horizontal axis to between 20 and 25; the best answer choice is 22. The answer of 25 is too high above the projected time on the graph, and the answers of 20 and 16 degrees are too low.

Reasoning Through Language Arts

1. B: Part of the argument here is that the profits are fleeting, whereas the other values of work Roosevelt discusses are longer lasting and more important. As such, the correct choice is Choice *B*.

2. C: At no point does Roosevelt suggest involving private businesses in the solution (Choice *C*). He does, however, suggest Choices *A*, *B*, and *D* as parts of the solution.

3. D: Midway through the passage, Roosevelt discusses that the two problems that go "hand in hand" are too many people in industrial centers (more people than jobs) and too few people in rural areas where there is more potential for job creation.

4. B: In Choice A, we see the first item on the list runs contrary to what Roosevelt states in the passage. In Choice C, Roosevelt discusses foreclosures as part of the larger problem but falls short of recommending they be stopped. In Choice D, Roosevelt acknowledges that ethics are an issue, but they are not to be the focal point as the problems are in the system as much as our values. In Choice B, we see four items all discussed as parts of Roosevelt's plan.

5. A: Roosevelt does not recommend either Choice *B* or Choice *C*. Though he discusses a need to evaluate ethics, he does not propose that safeguards will force this process. However, he does suggest (Choice A) that failure to institute safeguards will result in a slide back to the corrupt policies and procedures that resulted in the economic crash.

6. C: Roosevelt lists three criticisms of the relief efforts: it is unequal (Choice *C*), scattered, and uneconomical. While they might also be insufficient (Choice *A*), he doesn't discuss this. Further, he doesn't mention that they might be redundant (Choice *B*) or unrealistic (Choice *D*).

7. C: Throughout the passage, Roosevelt refers to the struggles Americans were facing (Choice *C*). While it is true that presidential terms are limited in the United States (Choice *A*), he does not use this to justify the need for quick action. While it is possible that corrupt systems might look to entrench themselves to fight against regulation (Choice *B*), this isn't stated in the passage. Finally, while Roosevelt may have been a man to prefer action, that's not offered as information in this passage (Choice *D*).

8. D: In this phrase, Roosevelt is looking to encourage the United States to balance its budget, which means its spending matches its income (Choice *D*) or that the country should not be spending more than it makes. Choice *A* suggests that the United States should spend money and wait for expenditures to make money. Choice *B* suggests that the country should have more income than it is spending, and while that may be ideal, it's not what's suggested in the passage. Finally, Choice *C* suggests ceasing all spending until income is generated and, again, while that might be good in theory, it is not what Roosevelt says.

9. A: Roosevelt makes no mention of who or how many citizens are running for office (Choice *B*). While both Choice *C* and Choice *D* might have been true, they are not explicitly discussed in this passage. However, Roosevelt does state "public office and high political position are to be valued only by the standards of pride of place and personal profit" (Choice *A*).

10. C: Dorm. Eliminating all vowels but the "O" makes Choice *D* incorrect. The "T" is also eliminated, which makes Choice *B* incorrect. Choice *A* is incorrect because there is only one "O" in "auditorium."

11. B: 125mph. We see the number on the odometer reach 120mph, and then there are five marks after the 120, making the top speed 125 mph.

12. C: 91,308 miles. We can see the number of miles driven by the digital numbers underneath the odometer.

13. D: *Congealed* in this context most nearly means *thickened*, because we see liquid turning into ice. Choice *B*, *loosened*, is the opposite of the correct answer. Choices *A* and *C*, *burst* and *shrank*, are also incorrect.

14. C: Choice *A* is incorrect. We cannot infer that the passage takes place during the night time. While we do have a statement that says that the darkness thickened, this is the only evidence we have. The darkness could be thickening because it is foggy outside. We don't have enough proof to infer this otherwise. We *can* infer that the season of this narrative is in the winter time. Some of the evidence here is that "the cold became intense," and people were decorating their shops with "holly sprigs,"—a Christmas tradition. It also mentions that it's Christmastime at the end of the passage. Choice *B* is incorrect; we *can* infer that the narrative is located in a bustling city street by the actions in the story. People are running around trying to sell things, the atmosphere is busy, there is a church tolling the hours, etc. The scene switches to the Mayor's house at the end of the passage, but the answer says "majority," so this is still incorrect. Choice *D* is incorrect; we *can* infer that the Lord Mayor is wealthy—he lives in the "Mansion House" and has fifty cooks.

15. D: The passage tells us that the poulterers' and grocers' trades were "a glorious pageant, with which it was next to impossible to believe that such dull principles as bargain and sale had anything to do," which means they set up their shops to be entertaining public spectacles in order to increase sales. Choice *A* is incorrect; although the word "joke" is used, it is meant to be used as a source of amusement rather than something made in poor quality. Choice *B* is incorrect; that they put on a "pageant" is figurative for the public spectacle they made with their shops, not a literal play. Choice *C* is incorrect, as this is not mentioned anywhere in the passage.

16. B: The author, at least in the last few paragraphs, does not attempt to evoke empathy for the character of Scrooge. We see Scrooge lashing out at an innocent, cold boy, with no sign of affection or feeling for his harsh conditions. We see Choice *A* when the author talks about Saint Dunstan. We see Choice *C* providing a palpable experience and imaginable setting and character, especially with the "piercing, searching, biting cold," among other statements. Finally, we see Choice *D* when Scrooge chases the young boy away.

17. C: Thoreau writes in the second paragraph that he values freedom above material items and specifically mentions Choice *A*, *B*, and *D* as items he has little interest in.

18. B: We learn early in the sentence that the speaker has the land (Choice *A*). As Thoreau discusses his work alone, we can infer that help is not what the speaker means (Choice *C*). Thoreau writes earlier in the passage that living without many material goods, he has a lot of time (Choice *D*). However, in the sentence that follows, Thoreau mentions specifically "before he has fairly learned it" which implies that there are skills and knowledge (Choice *B*) needed to live life as he does.

19. D: Teaching was a failure for Thoreau because it was not his true course. Early in the passage, Thoreau writes that he entertained the idea of berry picking (Choice *A*) and learning a trade (Choice *B*), but neither of these suited him, though a connection between that and his teaching is never established in the passage. We also never learn anything about his parents in the passage (Choice *C*).

20. A: While Choices *B* and *D* are discussed in the passage, they are not the main point. Thoreau actually writes that we should not follow our friends into careers (Choice *C*). He does, however, stress that we should pursue the lives that will make us happy and find the simplest way to do that, and that brings us the most joy, whatever it may be. Therefore, Choice *A* is correct.

21. B: Thoreau writes, "For myself I found that the occupation of a day-laborer was the most independent of any, especially as it required only thirty or forty days in a year to support one. The laborer's day ends with the going down of the sun, and he is then free to devote himself to his chosen pursuit, independent of his labor; but his employer, who speculates from month to month, has no respite from one end of the year to the other."

22. A: Though Thoreau mentions he used his free time for study (Choice *B*), he suggests later that free time left in the day for workers should be used to pursue individual interests (Choice *A*).

23. D: Thoreau writes, specifically, that he was "obliged to dress and train" (Choice *D*), whereas Choices *A* and *B* are things he specifically avoided. He does not mention how much money he made (Choice *C*).

24. C: Choices *A*, *B*, and *D* are not stated in the passage. However, Thoreau does note Choice *C*, that they "love labor for its own sake."

25. A: Providing money to allow a student the opportunity to attend an educational workshop is the simplest way to do something benevolent, according to the author of passage A. In the second paragraph of the passage, the author states that pecuniary (or monetary) aid is the easiest form of benevolence and that it requires the least amount of sacrifice. The remaining answer choices would all involve much greater amounts of sacrifice.

26. B: The phrase "mental exercise" is used in paragraph two of passage A to represent trying to see things from another person's point of view. In paragraph two, the author is discussing practicing benevolence by placing oneself in the situations of others. The author does not make mention of a person only taking their own perspective into consideration, improving an individual's memory recall, or strengthening a person's concentration.

27. C: In passage A, the term "heads" means manner. The author is discussing treating the practice of benevolent affections in almost the same way or manner as when considering the principles of justice. The remaining answer choices (discord, unlikeness, and opposition) do not make sense in this context.

28. D: The author of passage A states that justice can be exercised at the expense of benevolence, but this premise is not true the other way around ("benevolence is not to be exercised at the expense of justice"). Additionally, acts of benevolence require selflessness (not acts of justice). Finally, benevolence cannot take the place of the payment of debts that are owed per acts of justice.

29. A: The author of passage B says that giving food and money to beggars is seen as a weakness because that type of charity encourages laziness and corruption ("giving alms to common beggars" leads to an increase in "idleness and debauchery"). Although the remaining answer choices may indeed be true, they are not mentioned by the author of passage B.

30. A: Technical document. Technical documents are documents that describe the functionality of a technical product, so Choice *A* is the best answer. Persuasive texts, Choice *B*, try to persuade an audience to follow the author's line of thinking or to act on something. Choice *C*, narrative texts, seek to tell a story. Choice *D*, cause and effect, try to show why something happened, or the causes or effects of a particular thing.

31. A: To show the audience one of the effects of criminal rehabilitation by comparison. Choice *B* is incorrect because although it is obvious the author favors rehabilitation, the author never asks for donations from the audience. Choices *C* and *D* are also incorrect. We can infer from the passage that American prisons are probably harsher than Norway prisons. However, the best answer that captures the author's purpose is Choice *A*, because we see an effect by the author (recidivism rate of each country) comparing Norwegian and American prisons.

32. D: The likelihood of a convicted criminal to reoffend. The passage explains how a Norwegian prison, due to rehabilitation, has a smaller rate of recidivism. Thus, we can infer that recidivism is probably not a positive attribute. Choices *A* and *B* are both positive attributes, the lack of violence and the opportunity of inmates to receive therapy, so Norway would probably not have a lower rate of these two things. Choice *C* is possible, but it does not make sense in context, because the author does not talk about tactics in which to keep prisoners inside the compound, but ways in which to rehabilitate criminals so that they can live as citizens when they get out of prison.

33. The passage is reflective of a narrative. A narrative is used to tell a story, as we see the narrator trying to do in this passage by using memory and dialogue. Persuasive writing uses rhetorical devices to try and convince the audience of something, and there is no persuasion or argument within this passage. Expository writing is a type of writing used to inform the reader. Technical writing is usually used within business communications and uses technical language to explain procedures or concepts to someone within the same technical field.

34. C: A sense of foreboding. The narrator, after feeling excitement for the morning, feels "that something awful was about to happen," which is considered foreboding. The narrator mentions larks and weather in the passage, but there is no proof of anger or confusion at either of them.

35. C: To introduce Peter Walsh back into the narrator's memory. Choice *A* is incorrect because, although the novel *Mrs. Dalloway* is about events leading up to a party, the passage does not mention anything about a party. Choice *B* is incorrect; the narrator calls Peter *dull* at one point, but the rest of her memories of him are more positive. Choice *D* is incorrect; although morning is described within the first few sentences of the passage, the passage quickly switches to a description of Peter Walsh and the narrator's memories of him.

36. C: Personification. Figurative language uses words or phrases that are different from their literal interpretation. Personification is included in figurative speech, which means giving inanimate objects human characteristics. Nouns, agreement, and syntax all have to do with grammar and usage, and are not considered figurative language.

37. D: Felicia had to be prudent, or careful, if she was going to cross the bridge over the choppy water. Choice *A*, patient, is close to the word careful. However, careful makes more sense here. Choices *B* and *C* don't make sense within the context—Felicia wasn't hoping to be *afraid* or *dangerous* while crossing over the bridge, but was hoping to be careful to avoid falling.

38. B: Whatever happened in his life before he had a certain internal change is irrelevant. Choices *A, C,* and *D* use some of the same language as the original passage, like "revolution," "speak," and "details," but they do not capture the meaning of the statement. The statement is saying the details of his previous life are not going to be talked about—that he had some kind of epiphany, and moving forward in his life is what the narrator cares about.

39. B: First-person omniscient. This is the best guess with the information we have. In the world of the passage, the narrator is first-person, because we see them use the "I," but they also know the actions and thoughts of the protagonist, a character named "Webster." First-person limited tells their own story, making Choice *A* incorrect. Choice *C* is incorrect; second person uses "you" to tell the story. Third person uses "them," "they," etc., and would not fall into use of the "I" in the narrative, making Choice *D* incorrect.

40. Webster is a washing machine manufacturer. This question depends on reading comprehension. We see in the second sentence that Webster "was a fairly prosperous manufacturer of washing machines."

41. B: Veggie pasta with marinara sauce. Choices *A* and *D* are incorrect because they both contain cheese, and the doctor gave Sheila a list *without* dairy products. Choice *C* is incorrect because the doctor is also having Sheila stay away from eggs, and the omelet has eggs in it. Choice *B* is the best answer because it contains no meat, dairy, or eggs.

42. C: Choice *C* correctly identifies a necessary assumption in the argument. The argument is that hunting licenses for endangered species should be sold to support conservation efforts. If the revenue from those licenses isn't funding conservation efforts, then the ecologist's entire argument falls apart.

43. B: Choice *B* is the best answer here; the sentence states "In unknown regions take a responsible guide with you, unless the trail is short, easily followed, and a frequented one." Choice *A* is incorrect; the passage does not state that you should try and explore unknown regions. Choice *C* is incorrect; the passage talks about trails that contain pitfalls, traps, and boggy places, but it does not say that *all* unknown regions contain these things. Choice *D* is incorrect; the passage mentions "rail" and "boat" as means of transport at the beginning, but it does not suggest it is better to travel unknown regions by rail.

44. D: Choice *D* is correct; it may be real advice an experienced hiker would give to an inexperienced hiker. However, the question asks about details in the passage, and this is not in the passage. Choice *A* is incorrect; we do see the author encouraging the reader to learn about the trail beforehand . . . "wet or dry; where it leads; and its length." Choice *B* is also incorrect, because we do see the author telling us the time will lengthen with boggy or rugged places opposed to smooth places. Choice *C* is incorrect; at the end of the passage, the author tells us "do not go alone through lonely places . . . unless you are quite familiar with the country and the ways of the wild."

45. This is an informative passage. Informative passages explain to the readers how to do something; in this case, the author is attempting to explain the fundamentals of camping and hiking. Descriptive is a type of passage describing a character, event, or place in great detail and imagery, so this is incorrect. A persuasive passage is an argument that tries to get readers to agree with something. A narrative is a passage that tells a story, so this is also incorrect.

Science

1. B: Looking at Table 1, the number listed for 2012 is Choice *A,* 67,774. The number listed for 2008 is Choice *D,* 78,792. Neither of these represents the average. In order to calculate the average, the following formula should be used:

$$\frac{\Sigma \, fires}{Number \, of \, values} = \frac{457,437}{6} = 76,240$$

Choice *B* lists the correct value for the average number of fires and Choice *C* is the average for 2007 to 2017, which is erroneous because the question only requests the average of number of fires from 2007 to 2012.

2. C: Because the highest number of acres that are listed are not paired with the largest numbers of fires, another factor would have to contribute to the size of the fires. Many of the larger numbers of acres are after a year with a higher temperature anomaly. This potentially indicates possible drought conditions, which would make it more difficult to contain or extinguish even a small number of fires.

3. A: According to Kepler's three laws of planetary motion, a planet's orbit is elliptical in shape around the Sun, making Choice *C* incorrect. The laws also state that the square of the period of a planet's orbit is proportional to the cube of the distance the planet is from the Sun, making Choice *D* incorrect. If a planet is to cover the same distance in equal lengths of time, then it would move faster when closer to the Sun and slower when farther away from the Sun, making Choice *B* incorrect. Choice A is the only correct option.

4. Yes. Kepler studied the motion of heavenly bodies in order to devise laws that would work for all celestial planets. These laws work for every planet and were expanded upon by Newton and Einstein for further study when they discovered the laws may have been broken by Mercury, before the scientists were able to observe unseen portions of the sky to confirm their application.

5. B: According to the chart, alpha particles are emitted with higher levels of energy during radioactive decay. Beta particles are emitted with the second highest levels of energy. While not on the chart, gamma rays are emitted with the lowest levels of energy.

6. D: According to the chart, Lead 210 does transition through both Mercury 206 (Choice *B*) and Thallium 206 (Choice *C*), but only Choice *D* gives the correct stable ending element for the radioactive decay of Lead 210.

7. D: Choice *A*—G1—only takes up 25 percent of the time; Choice *B*—G2—also takes up 25 percent of the time; and Choice *C*—M—only takes up 15 percent of the time. The correct answer, S, which is Choice *D,* takes up the majority of time at 35 percent.

8. C: The mitosis phase only takes up 15 percent of the entire time. In order to calculate 15 percent of 120 minutes, use the following equation: $120 \times 0.15 = 18$ minutes.

9. A: A fissure volcano occurs when the ground above an active volcano flow cracks, thus exposing the lava to the air. Choice *B,* a shield volcano, has a vent that gradually slopes down to the land surrounding it, and will mostly produce a slow oozing of lava to the area around it. Choices *C* and *D* both have a more mountainous exterior and typical "Hollywood" style explosions.

10. D: For the years listed on the chart, 2016 (Choice *D)* shows an anomaly of approximately 2.22 degrees Fahrenheit, Choice *A* shows an anomaly of approximately -1.22 degrees Fahrenheit, 1944 shows an anomaly of approximately 0.7 degrees Fahrenheit, and 1990 shows an anomaly of approximately 1.46 degrees Fahrenheit. Overall, Choice *D* is the highest.

11. A: The chart shows an increase in temperatures, thus causing an increase in anomalies. There have not been any decreases in temperatures significant enough to cause an anomaly since the 1970s; thus, Choice *B* is not correct. The chart displays multiple anomalies, so Choices *C* and *D* are also not correct.

12. A: The diagram shows that the moon creates a bulge on the surface of the Earth it faces. This bulge, or pull in the waters on that side, also creates high tides on the side opposite of the moon, or in line with the moon. The low tides are created at the ends not facing, or in line with, the moon.

13. B: When the Sun and the moon line up on the same side of the Earth, the combination of their gravitational pulls creates higher high tides on the side facing the Sun and the moon, and on the opposite side of the Earth.

14. C: To find the average, add all of the values in the sand column and then divide by the number of items in the column.

$$\frac{\Sigma \text{ sand values}}{Number\ of\ values} = \frac{225}{5} = 45$$

15. A: Sample 3 is 45% silt. In order to calculate 45% of 5 kg, use the following equation: $5 \times 0.40 = 2.25$ kg.

16. C: According to the information provided, more rainfall produced more growth bands per year and an increased size of the growth bands. Site 3 had both the greatest number of growth bands and the largest average size of growth bands; therefore, Choice *C* is the best possible answer.

17. B: The graph for Choice *B* is the only one that displays a linear relationship, which shows that as the number of growth bands increases, so does the size of the growth bands. Graphs for Choices *A* and *D* show that one of the factors is held constant (which is not the case for any of the sites) and the graph for Choice *C* shows an increase in both factors and then a decrease in the number of growth bands.

18. C: Site 1 had smaller growth bands in both number and size than those of Site 2. This leads to the conclusion that the trees at Site 1 did not grow as fast as the trees at Site 2.

19. C: The information provided relays that counting the number of growth bands is not completely accurate on its own. Therefore, the cross dating can help to increase the accuracy for predicting the growth rate of trees. Decreasing the number of trees studied or the number of bands studied could decrease the accuracy of a study. There is no way to predict the amount of rainfall associated with the band number or size according to this study; therefore, Choice *C* is the best answer.

20. C: According to the data in Table 1, the number of growth bands at the new site falls between the number found at Site 2 and the number found at Site 3. Therefore, the average size of these growth bands should be somewhere between the sizes for Site 1 and Site 3, which is between 4 mm and 12 mm. The other choices are too small or too big for a comparison between Site 1 and Site 2.

21. D: In order to calculate the velocity, the following equation should be utilized:

$$v = \frac{\Delta x}{\Delta t} = \frac{x_2 - x_1}{t_2 - t_1} = \frac{32m - 24m}{4s - 3s} = \frac{8m}{1s} = 8\ m/s$$

22. C: In order to calculate this, the same equation in question 21 can be used, but the values used will be pulled from the right half of the chart:

$$v = \frac{\Delta x}{\Delta t} = \frac{x_2 - x_1}{t_2 - t_1} = \frac{12m - 2m}{3s - 1s} = \frac{10m}{2s} = 5\ m/s$$

23. D: The change in velocity over the change in time would display the acceleration, which is modeled by the equation $a = \frac{\Delta v}{\Delta t}$. Choice *A* is calculated through finding the average of all of the velocity values, Choice *B* involves calculating the magnitude of the average without the direction, and Choice *C* cannot be calculated with the information provided.

24. D: The total number of creatures that can swim is 27 and the total number of creatures tracked is 221. So, the chance of selecting any one creature that could swim would be 27/221. Choice *A* is not correct, because there are creatures that can swim. Choice *B* and Choice *C* are not correct because they display only how many males and females, respectively, out of all of the creatures that can swim relative to the total.

25. D: Choice *D* shows that selecting from the whole group of 221, there are 92 total females and an additional 28 creatures that can walk, for a total of 120 creatures out of a possible 221. Choice *A* is not correct because there are female creatures and creatures that can walk. Choice *B* shows the probability of selecting females from just the group that walks, and Choice *C* shows the total number of creatures that walk and the females (double counting the females that walk) from the whole group.

26. B: According to the passage, the inner and outer core are both made up of the same elements (iron and nickel), are extremely hot, and both are inaccessible to humans. However, the inner core is solid while the outer core is molten.

27. B: The passage states that both layers of the mantle are extremely high in temperature. However, it says that this hot temperature causes the metal contained in the lower mantle to rise and then cool slightly as it reaches the upper mantle, which means that it's hotter in the lower mantle. Once the metal begins to cool as it reaches the upper mantle, it falls back down toward the lower mantle, restarting the whole process again.

28. B: Choice *B* gives a value that is partway between the coefficient of kinetic friction of cardboard and the coefficient of kinetic friction of asphalt, which would be a reasonable approximation for material X. Choice *A* would be too low, representing the coefficient of kinetic friction of cardboard, and Choice *C* is too high for the coefficient of kinetic friction of asphalt. Choice *D* is not correct, because a relationship can be established between the materials.

29. D: The shape of the graph is most like that of a logarithmic function; therefore, Choice *D* is the only possible correct answer. Choice *A*, linear, would look like:

Linear function

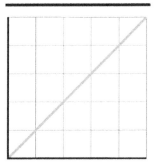

Choice *B*, quadratic, would look like:

Quadratic function

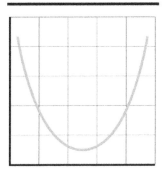

Choice *C*, exponential decay, would look like:

Exponential Decay

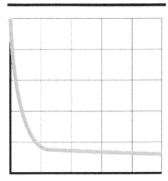

30. A: Looking at the table, rubber and concrete have the highest coefficient of sliding friction, and therefore would match the graph that matches the highest value on the *y*-axis (vertical). None of the other choices models this extreme.

31. C: The value of static friction for surfaces in contact must be overcome in order to start motion. After this coefficient of static friction is overcome, the materials begin moving across each other, and the coefficient of kinetic friction is less than that of what it took to begin the sliding motion. As two out of the three graphed lines are similar in value as to where the coefficient of static friction is overcome, it shows that the surfaces move more easily across each other with the lower coefficient of kinetic friction.

32. A: Choice *A* shows the number of birds not in any of the circles (are any of the listed continents), which is the number in the upper right corner, 11. Choice *B* is the number of birds found only in North America. Choice *C* is the number of birds found only in Asia. Choice *D* is the number of birds found only in South America.

33. A: Choice *A* shows the correct number of birds that overlap between North America and South America, and also contains the correct sample set, which would be the total number of birds found in South America, 36. Choice *C* shows the correct overlapping number, but mistakenly uses the entire bird population as the sample set to choose from. Choice *B* has the correct sample set, but gives the number of birds found only in South America. Choice *D* does not give either correct value.

34. A: The drinking water being contaminated and the stomach cancer is not necessarily determined by cause and effect. The two events could have just happened at the same time, which is correlation, not causation. The other three examples are very clear examples of cause and effect; we know for a fact that when water becomes hot enough, it boils, or that earthquakes are caused by tectonic plates shifting, and that thunderstorms bring rain..

35. C: Physical properties of substances are those that can be observed without changing that substance's chemical composition, such as odor, color, density, and hardness. Heat of combustion, reactivity, flammability, and toxicity are all chemical properties of substances. They describe the way in which a substance may change into a different substance. They cannot be observed without changing the identity of the substance.

36. B: The best answer choice is *B*. Let's look at it in simpler terms: Brown rice contains arsenic. Arsenic can harm the human body by causing cancer and diabetes. Therefore, everyone who eats brown rice will become sick. Or, B contains A. A causes C. Therefore, everyone who eats B will get C. Let's look at the original argument in simpler terms: BN contain S. S causes AI. Therefore, everyone who eats BN will get AI (anti-inflammatory effects). These two arguments are the closest in reasoning as any of the others. Let's look at the rest:

Choice *A* is close to the correct answer. However, we have a lack of absolutes at the conclusion of the argument: Microwaves contain R. Exposure can cause V, H, and F. Therefore, some who own M may experience V, H, and F. The wording of "some" and "may" provides a different argument structure than Choice *B* and the original passage.

Choice *C* is not the correct answer choice because it does not parallel the reasoning in the original passage. Choice *C* says: O does not contain D. Those looking to have D should eat T, M, and S. D is important for C and BG. Not enough D could cause C and WG. This reasoning is much different than what we have in the original paragraph. This is more of an informative passage on Vitamin D rather than a persuasive argument.

Choice *D* also does not parallel the reasoning found in the original passage. Choice *D* says that Whales have lungs instead of gills. Therefore, they use lungs to breathe when they come up from the water. Already our argument looks different than the original "BN contains S. S Causes AI. Therefore, everyone who eats BN will get AI." The passage ends with what everyone should do, mirroring the absolute language, but on the whole the arguments are structured differently.

37. B: Choice *B* is the best answer here; we see the argument mistaking correlation for causation. The arguer knows that the drug PMR had what "would be" negative effects that included complications in childbirth. Then, the arguer ran across an article that showed a rise in complications in childbirth in the same year the drug was being used by the population. The arguer then assumes that the two *must* be related; that the drug PMR was *the cause* of complications during childbirth. However, there is a possibility that this is just an incidence of correlation that the argument failed to account for. We have no way of knowing whether complications during childbirth were a direct effect of the drug or an effect of something else going on in that same year.

Social Studies

1. B: Secondary source. This excerpt is considered a secondary source because it actively interprets primary sources. We see direct quotes from the queen, which would be considered a primary source. But since we see those quotes being interpreted and analyzed, the excerpt becomes a secondary source. Choice *C*, tertiary source, is an index of secondary and primary sources, like an encyclopedia or Wikipedia.

2. It took **two years** for the new castle to be built. It states this in the first sentence of the second paragraph. In the third year, we see the Prince planning improvements, and arranging things for the fourth year.

3. C: To impose a certain quality upon. The sentence states that "the impress of his dear hand [has] been stamped everywhere," regarding the quality of his tastes and creations on the house. Choice *A* is one definition of *impress*, but this definition is used more as a verb than a noun: "She impressed us as a songwriter." Choice *B* is incorrect because it is also used as a verb: "He impressed the need for something to be done." Choice *D* is incorrect because it is part of a physical act: "the businessman impressed his mark upon the envelope." The phrase in the passage is meant as figurative, since the workmen did most of the physical labor, not the Prince.

4. B: To convince the audience that judges holding their positions based on good behavior is a practical way to avoid corruption.

5. D: Neutrality due to the style of the report. The report is mostly objective; we see very little language that entails any strong emotion whatsoever. The story is told almost as an objective documentation of a sequence of actions—we see the president sitting in his box with his wife, their enjoyment of the show, Booth's walk through the crowd to the box, and Ford's consideration of Booth's movements. There is perhaps a small amount of bias when the author mentions the president's "worthy wife." However, the word choice and style show no signs of excitement, sadness, anger, or apprehension from the author's perspective, so the best answer is Choice *D*.

6. Mr. Ford assumed Booth's movement throughout the theater was due to being familiar with the theater.

7. C: A lead singer leaves their band to begin a solo career, and the band drops in sales by 50 percent on their next album. The original source of the analogy displays someone significant to an event who leaves, and then the event becomes worse for it. We see Mr. Sothern leaving the theater company, and then the play becoming a "very dull affair." Choice *A* depicts a dancer who backs out of an event before the final performance, so this is incorrect. Choice *B* shows a basketball player leaving an event, and then the team makes it to the championship but then loses. This choice could be a contestant for the right answer; however, we don't know if the team has become worse for his departure or better for it. We simply do not have enough information here. Choice *D* is incorrect. The actor departs an event, but there is no assessment of the quality of the movie. It simply states that another actor filled in instead.

8. A: A chronological account in a fiction novel of a woman and a man meeting for the first time. It's tempting to mark Choice A wrong because the genres are different. Choice *A* is a fiction text, and the original passage is not a fictional account. However, the question's stem asks specifically for organizational structure. Choice *A* is a chronological structure just like the passage, so this is the correct answer. The passage does not have a cause and effect or problem/solution structure, making Choices *B* and *D* incorrect. Choice *C* is tempting because it mentions an autobiography; however, the structure of this text starts at the end and works its way toward the beginning, which is the opposite structure of the original passage.

9. C: The word *adornments* would LEAST change the meaning of the sentence because it's the most closely related word to *festoons*. The other choices don't make sense in the context of the sentence. *Feathers* of flags, *armies* of flags, and *buckets* of flags are not as accurate as the word *adornments* of flags. The passage also talks about other décor in the setting, so the word adornments fits right in with the context of the paragraph.

10. D: To recount in detail the events that led up to Abraham Lincoln's death. Choice *A* is incorrect; the author makes no claims and uses no rhetoric of persuasion towards the audience. Choice *B* is incorrect, though it's a tempting choice; the passage depicts the setting in exorbitant detail, but the setting itself is not the primary purpose of the passage. Choice *C* is incorrect; one could argue this is a narrative, and the passage is about Lincoln's last few hours, but this isn't the *best* choice. The best choice recounts the details that leads up to Lincoln's death.

11. C: 394,000. Choice *A* is the population of Totok in 1920. Choice *B* is the population of Totok in 1930. Choice *D* is the population of Peranakan in 1956 and 1961.

12. C: The answer is 404,000. This question requires some simple math by subtracting the total population of Peranakan in 1971 (1,240,000) from the total population of Peranakan in 1961 (836,000) which comes to 404,000.

13. A: According to the map, Volcano Terevaka is the highest point on the island, reaching 507m. Volcano Puakatike is the next highest, reaching 307m. Vaka Kipo is the next highest reaching 216m. Finally, Maunga Ana Marama is the next highest, reaching 165m.

14. A: A central question to both passages is: What is the interpretation of the first amendment and its limitations? Choice *B* is incorrect; a quote mentions this at the end of the first passage, but this question is not found in the second passage. Choice *C* is incorrect, as the passages are not concerned with the definition of freedom of speech, but how to interpret it. Choice *D* is incorrect; this is a question for the second passage, but is not found in the first passage.

15. C: The authors would most likely disagree over the situation where the man is thrown in jail for encouraging a riot against the U.S. government for the wartime tactics although no violence ensues. The author of Passage A says that "If a state may properly forbid murder or robbery or treason, it may also punish those who induce or counsel the commission of such crimes." This statement tells us that the author of Passage A would support throwing the man in jail for encouraging a riot, although no violence ensues. The author of Passage B states that "And we can with certitude declare that the First Amendment forbids the punishment of words merely for their injurious tendencies." This is the best answer choice because we are clear on each author's stance in this situation.

Choice A is tricky; the author of Passage A would definitely agree with this, but it's questionable whether the author of Passage B would agree with this. Violence does ensue at the capitol as a result of this man's provocation, and the author of Passage B states "speech should be unrestricted by censorship . . . unless it is clearly liable to cause direct . . . interference with the conduct of war." This answer is close, but it is not the *best* choice. Choice B is incorrect because we have no way of knowing what the authors' philosophies are in this situation. Choice D is incorrect because, again, we have no way of knowing what the authors would do in this situation, although it's assumed they would probably both agree with this.

16. A: Choice A is the best answer. To figure out the correct answer choice we must find out the relationship between Passage A and Passage B. Between the two passages, we have a general principle (freedom of speech) that is questioned on the basis of interpretation. In Choice A, we see that we have a general principle (right to die, or euthanasia) that is questioned on the basis of interpretation as well. Should euthanasia only include passive euthanasia, or euthanasia in any aspect? Choice B is incorrect because it does not question the interpretation of a principle, but rather describes the effects of two events that happened in the past involving contamination of radioactive substances. Choice C begins with a principle—that of labor laws during wartime—but in the second option, the interpretation isn't questioned. The second option looks at the historical precedent of labor laws in the past during wartime. Choice D is incorrect because the two texts disagree over the cause of something rather than the interpretation of it.

17. B: This is the best answer choice because the author is trying to demonstrate by the examples that anyone who incites a crime, despite the severity or magnitude of the crime, should be held accountable for that crime in some degree. Choice A is incorrect because the crimes mentioned are not being compared to each other, but they are being used to demonstrate a point. Choice C is incorrect because the author makes the same point using both of the examples and does not question the definition of freedom of speech but its ability to be limited. Choice D is incorrect because this sentiment goes against what the author has been arguing throughout the passage.

18. A: The idea that human beings are able and likely to change their minds between the utterance and execution of an event that may harm others. This idea most seriously undermines the claim because it brings into question the bad tendency of a crime and points out the difference between utterance and action in moral situations. Choice B is incorrect; this idea does not undermine the claim at hand, but introduces an observation irrelevant to the claim. Choices C and D would most likely strengthen the argument's claim; or, they are at least supported by the author in Passage A.

19. D: To call upon the interpretation of freedom of speech to be already evident in the First Amendment and to offer a clear perimeter of the principle during war time. Choice A is incorrect; the passage calls upon no historical situations as precedent in this passage. Choice B is incorrect; we can infer that the author would not agree with this, because they state that "In war time, therefore, speech should be unrestricted . . . by punishment." Choice *C* is incorrect; this is more consistent with the main idea of the first passage.

20. A: The word that would least change the meaning of the sentence is *A*, grievance. *Malcontent* is a complaint or grievance, and in this context would be uttered in advocation of absolute freedom of speech. Choice *B*, *cacophony*, means a harsh or discordant noise; someone may express or "urge" a cacophony, but it would be an awkward word in this context. Choice *C*, *anecdote*, is a short account of an amusing story. Since the word is a noun, it fits grammatically in the sentence, but anecdotes are usually thought out, and this word is considered "unthinking." Choice *D*, *residua*, means an outcome, and also does not make sense within this context.

21. D: A city whose population is made up of people who seek quick fortunes rather than building a solid business foundation. Choice *A* is a characteristic of Portland, but not that of a boom city. Choice *B* is close—a boom city is one that becomes quickly populated, but it is not necessarily *always* populated by residents from the east coast. Choice *C* is incorrect because a boom city is not one that catches fire frequently, but one made up of people who are looking to make quick fortunes from the resources provided on the land.

22. D: Choice *D* is the best answer because of the surrounding context. We can see that the fact that Portland is a "boom city" means that the "floating class"— a group of people who only have temporary roots put down—go through. This would cause the main focus of the city to be on employment and industry, rather than society and culture. Choice *A* is incorrect, as we are not told about the inhabitants being social or antisocial. Choice *B* is incorrect because the text does not talk about the culture in the East regarding European influence. Finally, Choice *C* is incorrect; this is an assumption that has no evidence in the text to back it up.

23. B: The author would say that it has as much culture as the cities in the East. The author says that Portland has "as fine churches, as complete a system of schools, as fine residences, as great a love of music and art, as can be found at any city of the East of equal size," which proves that the culture is similar in this particular city to the cities in the East.

24. A: Approximately 240,000. We know from the image that San Francisco has around 300,000 inhabitants at this time. From the text (and from the graph) we can see that Portland has 60,000 inhabitants. Subtract these two numbers to come up with 240,000

25. B: Senators serve six-year terms. Choice *A* is an incorrect answer; members of the House of Representative serve two-year terms. Choice *C* is an incorrect answer; Supreme Court Justices hold lifelong terms, but Senators only serve six-year terms. Choice *D* is an incorrect answer; the President of the United States has a four-year term length.

26. B: The speech was made before the Civil War, which began in 1861; it captures the rising tension over slavery between the North and the South. Choices *A* and *D* are incorrect because Lincoln is not talking about an actual house. Instead, he is talking about a nation divided over slavery by using the metaphor of a house. Choice *C* is incorrect because the speech was given in 1858, before the secession of South Carolina and the dissolution of the Union in 1861.

27. D: The political propaganda poster is trying to get American citizens to carpool with one another in order to save fuel for the war effort. Choice *A* is incorrect because cars were an essential part of American culture and lifestyle at the time. If citizens stopped driving cars altogether, it would put a strain on the economy and the work flow. Choice *B* is incorrect because the message the poster is trying to convey is not that extreme. Even though the poster says, "When you ride ALONE you ride with Hitler!", it does not mean the government is accusing its citizens of treason. It was a hyperbole used to drastically encourage its citizens to understand the importance of conserving fuel to aid the war effort. Choice *C* is incorrect because, even though American patriotism is an important message behind all propaganda posters, it is not the main message of this particular poster.

28. B: The Japanese attack on Pearl Harbor and the United States' entry into World War II have a cause-and-effect relationship. Choice *A* is incorrect because the relationship is causation, not correlation. Choice *C* is incorrect because these two events are causally related rather than in opposition to one another. Choice *D* is incorrect because this particular cause-and-effect relationship is direct.

29. B: The bar graph shows that the violent crime rate in the United States was at its peak in 2001. Choice *A* is incorrect because, despite its high numbers, the violent crime rate in 1999 was still lower than it was in 2001. Choice *C* is incorrect because it is almost equal to the violent crime rate in 1999. Choice *D* is incorrect because the graph shows a decrease in the violent crime rate in 2005.

30. C: This political cartoon symbolically depicts the Zimmerman Telegram, in which Germany tried to recruit Mexico to join the Central powers during World War I. The image shows a hand carving up a map of the southwestern United States, representing the portion of the country that Germany promised to give Mexico if they joined their cause. Choice *A* is incorrect because, although the territory depicted was included in the Mexican Cession, the incorporation of the hand wearing a gauntlet with the imperial German eagle on it clearly alludes to the Zimmerman Telegram. Choice *B* is incorrect because Germany had no involvement in the war between the U.S. and Mexico. Choice *D* is incorrect because, although the act aimed to be more permissive in terms of immigration by eliminating quotas, it had nothing to do with Germany.

31. D: According to the last column, entitled "Peak Year," Dallas had an all-time high of 500 murders in 1991. Choice *A* is incorrect because that is the number of murders in 2005 in Dallas, but it is not an all-time high. Choice *B* is incorrect because that is the number of murders in 2000 in Dallas, but it is not an all-time high. Choice *C* is incorrect because 447 is the number of murders in Dallas, but it is not an all-time high. Choice *C* is tempting because it presents the highest number of murders for the 5-year periods listed for Dallas on the table, but the "Peak Year" column on the right offers dates that exist outside these 5-year frames.

32. C: According to the graph, Milwaukee had a peak of about 25.5 homicides per 100,000 in 1991. All other answers can be eliminated by finding the highest peak (designated by Milwaukee's darker line) on the graph.

33. C: The United States and the Soviet Union were the two superpowers of the Cold War, racing to go to the moon. Choice *A* is incorrect because the United States was an ally of Great Britain at this time. Choice *B* is incorrect because Nazi Germany had already fallen by 1962. Choice *D* is incorrect because tensions between Iraq and the United States did not begin until Desert Storm of the 1990s.

34. A: The sign would likely be a representation of the vitriolic reaction against federal desegregation (officially ending segregation by race) than any other category because it takes place in the 1950s or 1960s according to the introduction. The fact that the sign would be protesting desegregation confirms the racial tensions of the time. Choice *B* is incorrect because, although Progressivism was marred by racial tensions, it was mostly a positive reform era. Choice *C* is incorrect because emancipation occurred in the 1860s and focused on ending slavery; Choice *D* is incorrect because Reconstruction happened between 1865 and 1877.

35. A: Pie charts are best for comparing parts or percentages of the whole. While all other data visualization tools could likely convey this comparison in some way, the pie chart is best because of its part-to-whole focus.

36. A: The Japanese attack on Pearl Harbor and the United States' entry into World War II are causally related, as conveyed by this document. All other answer choices are factually incorrect and not cited in the text.

Greetings!

First, we would like to give a huge "thank you" for choosing us and this study guide for your GED exam. We hope that it will lead you to success on this exam and for years to come.

Our team has tried to make your preparations as thorough as possible by covering all of the topics you should be expected to know. In addition, our writers attempted to create practice questions identical to what you will see on the day of your actual test. We have also included many test-taking strategies to help you learn the material, maintain the knowledge, and take the test with confidence.

We strive for excellence in our products, and if you have any comments or concerns over the quality of something in this study guide, please send us an email so that we can improve.

As you continue forward in life, we would like to remain alongside you with other books and study guides in our library. We are continually producing and updating study guides in several different subjects. If you are looking for something in particular, all of our products are available on Amazon. You can also send us an email!

Sincerely,
APEX Publishing
info@apexprep.com

FREE

Free Study Tips DVD

In addition to the tips and content in this guide, we have created a FREE DVD with helpful study tips to further assist your exam preparation. **This FREE Study Tips DVD provides you with top-notch tips to conquer your exam and reach your goals.**

Our simple request in exchange for the strategy-packed DVD is that you email us your feedback about our study guide. We would love to hear what you thought about the guide, and we welcome any and all feedback—positive, negative, or neutral. It is our #1 goal to provide you with top quality products and customer service.

To receive your **FREE Study Tips DVD**, email freedvd@apexprep.com. Please put "FREE DVD" in the subject line and put the following in the email:

> a. The name of the study guide you purchased.
>
> b. Your rating of the study guide on a scale of 1-5, with 5 being the highest score.
>
> c. Any thoughts or feedback about your study guide.
>
> d. Your first and last name and your mailing address, so we know where to send your free DVD!

Thank you!

Made in the USA
Monee, IL
01 September 2021